Contents

Real-time PCR

Current technology and applications

Edited by

Julie Logan

Applied and Functional Genomics
Centre for Infections
Health Protection Agency
London
UK

Kirstin Edwards

Applied and Functional Genomics
Centre for Infections
Health Protection Agency
London
UK

and

Nick Saunders

Applied and Functional Genomics
Centre for Infections
Health Protection Agency
London
UK

 Caister Academic Press

Copyright © 2009

Caister Academic Press
Norfolk, UK

www.caister.com

British Library Cataloguing-in-Publication Data
A catalogue record for this book is available from the British Library

ISBN: 978-1-904455-39-4

Printed and bound in Great Britain

Contributors

Tom Brown
Department of Chemistry
University of Southampton
Southampton
UK

tb@soton.ac.uk

Stephen A. Bustin
The Royal London Hospital
London
UK

s.a.bustin@qmul.ac.uk

Nick M. Cirino
Wadsworth Centre
New York State Department of Health
Albany, NY
USA

ncirino@wadsworth.org

Kirstin J. Edwards
Applied and Functional Genomics
Centre for Infections
Health Protection Agency
London
UK

kirstin.edwards@hpa.org.uk

Christina Egan
Wadsworth Centre
New York State Department of Health
Albany, NY
USA

eganc@wadsworth.org

Julie D. Fox
Provincial Laboratory for Public Health
(Microbiology) and University of Calgary
Calgary, AB
Canada

J.Fox@provlab.ab.ca

Mikael Kubista
TATAA Biocenter
Goteborg
Sweden

mikael.kubista@tata.com

Martin A. Lee
Enigma Diagnostics Ltd
Salisbury
UK

martinalanlee@aol.com

Dario L. Leslie
Dstl Porton Down
Salisbury
UK

centralenquiries@dstl.gov.uk
(attn DL Leslie)

Julie M. J. Logan
Applied and Functional Genomics
Centre for Infections
Health Protection Agency
London
UK

julie.logan@hpa.org.uk

Elaine Lyon
ARUP Laboratories,
Salt Lake City, UT
USA

lyone@aruplab.com

Rong Mao
ARUP Laboratories,
Salt Lake City, UT
USA

rong.mao@aruplab.com

Catherine Moore
Molecular Diagnostics Unit
Specialist Virology Center
NPHS Microbiology
Cardiff University Hospital of Wales
Cardiff
UK

catherine.moore@nphs.wales.nhs.uk

Alison Millson
ARUP Laboratories,
Salt Lake City, UT
USA

milsoa@aruplab.com

Kimberlee A. Musser
Wadsworth Centre
New York State Department of Health
Albany, NY
USA

musser@wadsworth.org

Tania Nolan
Sigma-Aldrich
Cambridge
UK

tania.nolan@sial.com

David S. Perlin
Public Health Research Institute
International Center for Public Health
Newark, NJ
USA

perlin@phri.org

Michael W. Pfaffl
Physiology Weihenstephan
Center of Life and Food Science Weihenstephan
Technical University of Munich
Weihenstephan
Germany

michael.pfaffl@wzw.tum.de

Genevieve Pont-Kingdon
ARUP Laboratories,
Salt Lake City, UT
USA

pontkig@aruplab.com

Andrew David Sails
Health Protection Agency
Newcastle Laboratory
Institute of Pathology
Newcastle General Hospital
Newcastle upon Tyne
UK

andrew.sails@hpa.org.uk

Nick A. Saunders
Applied and Functional Genomics
Centre for Infections
Health Protection Agency
London
UK

nick.saunders@hpa.org.uk

David J. Squirrell
Enigma Diagnostics Ltd
Bldg 224
Tetricus Science Park
DSTL Porton Down
Salisbury
Wiltshire
SP4 0JQ
UK

Jeffrey Swensen
Departments of Paediatrics and Pathology
University of Utah
Salt Lake City, UT
USA

jeffrey.swensen@aruplab.com

Jo Vandesompele
Center for Medical Genetics Ghent
Ghent University Hospital
Ghent
Belgium

Joke.Vandesompele@UGent.be

Diana Westmoreland
Molecular Diagnostics Unit
Specialist Virology Center
NPHS Microbiology
Cardiff University Hospital of Wales
Cardiff
UK

diana.westmoreland@nphs.wales.nhs.uk

Gordon Wiseman
Premier Foods Group Ltd
The Lord Rank Centre
High Wycombe
UK

gordon.wiseman@premierfoods.co.uk

Elisa Wurmbach
Pathology Research
Forensic Biology
Office of Chief Medical Examiner
New York, NY
USA

ewurmbach@ocme.nyc.gov

Preface

The polymerase chain reaction (PCR) is a technique so commonplace in the modern-day laboratory that it is easy to forget its revolutionary impact. Real-time PCR has removed many of the limitations of standard end-point PCR, and since its introduction in the mid-1990s there has been an explosion both in the number of publications and available instrumentation describing real-time PCR applications across many disciplines.

This book aims to provide both the novice and experienced user with an invaluable point of reference to the technology, instrumentation and its wide range of applications. The initial chapters cover the important aspects of real-time PCR, from choosing an instrument and probe system to set-up, controls, validation and data analysis. It then goes on to give a comprehensive overview of important real-time methodologies such as quantitation, expression analysis and mutation detection. This is complemented by the final chapters, which address the application of real-time PCR to diagnosis of infectious diseases, biodefence, food authenticity and molecular haplotyping. This essential manual should serve both as a basic introduction to real-time PCR and a source of current trends and applications for those already familiar with the technology. We also hope this text will stimulate readers of all levels to develop their own innovative approaches to real-time PCR.

Julie M. J. Logan
Kirstin J. Edwards
Nick A. Saunders

Current Books of Interest

Caister Academic Press   www.caister.com

An Introduction to Real-Time PCR

1

Nick A. Saunders

Abstract

The development of instruments that allowed real-time monitoring of fluorescence within PCR reaction vessels was a significant advance. The technology is flexible, and many alternative instruments and fluorescent probe systems have been developed and are currently available. Real-time PCR assays can be completed rapidly since no manipulations are required post-amplification. Identification of the amplification products by probe detection in real time is highly accurate compared with size analysis on gels. Analysis of the progress of the reaction allows accurate quantification of the target sequence over a very wide dynamic range, provided suitable standards are available. Further investigation of the real-time PCR products within the original reaction mixture using probes and melting analysis can detect sequence variants including single base mutations. Since the first practical demonstration of the concept real-time PCR has found applications in many branches of biological science. Applications include gene expression analysis, the diagnosis of infectious disease and human genetic testing. Owing to their fluorimetry capabilities, these real-time machines are also compatible with alternative amplification methods such as NASBA, provided a fluorescence end-point is available.

PCR: the early years

The theoretical concept of producing many copies of a specific DNA molecule by a cycling process using DNA polymerase and oligonucleotide primers was first expounded in a paper by Kleppe and colleagues in 1971 (Kleppe *et al.*, 1971). At that time, the practical exploitation of such a process must have seemed remote to the biologists who read the paper. This was due to the difficulty and cost of producing oligonucleotides, the non-availability of thermostable DNA polymerases and the lack of automated thermocycling instruments. By the time of the first demonstration of the PCR process by Saiki and colleagues in 1985 (Saiki *et al.*, 1985) automated oligonucleotide synthesizers were commonly available. This meant that the potential of PCR in a wide range of applications was recognized. However, it was still necessary to inject fresh thermolabile polymerase prior to each elongation step and thermal cyclers were still in development. Consequently, the key step in realizing the potential of the PCR was probably the use of a thermostable polymerase which was first described in 1988 (Saiki *et al.*, 1988). Since the first description of a practical DNA amplification process many refinements have been described and automatic thermal cyclers have become standard laboratory equipment. PCR is now an essential tool for many biologists and the standard protocols are simple and user friendly. The exponential amplification process provides nanogram quantities of essentially identical DNA molecules starting from a few copies of a target sequence. The amplified material (the PCR amplicon) is available in sufficient quantity to be identified by size analysis, sequencing, amplicon melting or by probe hybridization. It can also be cloned readily or used as a reagent.

The need for real-time PCR

Much of the technical effort involved in standard PCR is now directed towards positive recognition of the amplicons. The important methods of post-PCR analysis rely on either the size or sequence of the amplicon. Gel electrophoresis is often used to measure the amplicon size and this is both inexpensive and simple to implement. Unfortunately, size analysis has limited specificity since different molecules of approximately the same molecular weight cannot be distinguished. Consequently, gel electrophoresis alone is not a sufficient PCR end-point in many instances, including most clinical applications. Characterization of the product by its sequence is far more reliable and informative. Probe hybridization assays for this purpose are available but many are multistep procedures. Such methods are time-consuming and care must be taken to ensure that amplicons accidentally released into the laboratory environment do not contaminate the DNA preparation and clean rooms. Real-time PCR machines greatly simplify amplicon recognition by providing the means to monitor the accumulation of specific products continuously during cycling. All current instruments designed for real-time PCR, measure the progress of amplification by monitoring changes in fluorescence within the PCR reaction vessel. Changes in fluorescence can be linked to product accumulation by a variety of methods. A further advantage of the real-time format is that the analysis can be performed without opening the tube which can then be disposed of without the risk of dissemination of PCR amplicons or other target molecules into the laboratory environment. Although alternative methods for avoiding PCR contamination are available, containment within the PCR vessel is likely to be the most efficient and cost-effective. A major drawback of standard PCR formats that rely on end-point analysis is that they are not quantitative because the final yield of product is not primarily dependent upon the concentration of the target sequence in the sample. Real-time PCR overcomes this limitation.

Chemistries

There are two general approaches used to obtain a fluorescent signal from the synthesis of product in PCR. The first depends upon the property of fluorescent dyes such as SYBR Green I to bind to double-stranded DNA and undergo a conformational change that results in an increase in their fluorescence. The second approach is to use fluorescent resonance energy transfer (FRET). These methods use a variety of ways to alter the relative spatial arrangement of photon donor and acceptor molecules. These molecules are attached to probes, primers or the PCR product and are usually selected so that amplification of a specific DNA sequence brings about an increase in fluorescence at a particular wavelength. A major advantage of the real-time PCR instruments and signal transduction systems currently available is that it is possible to characterize the PCR amplicon *in situ* on the machine. This is done by analysis of the melting temperature and/or probe hybridization characteristics of the amplicon within the PCR reaction mixture. In the intercalating dye system, the melting temperature of the amplicon can be estimated by measuring the level of fluorescence emitted by the dye as the temperature is increased from below to above the expected melting temperature. The methods that rely upon probe hybridization to produce a fluorescent signal are generally less liable to produce false-positive results than alternative methods such as the use of intercalating dyes to detect net synthesis of double-stranded DNA (dsDNA) followed by melting analysis of the product. Hybridization, ResonSense™ and hydrolysis probe systems give fluorescent signals that are only produced when the target sequence is amplified and are unlikely to give false-positive results. An additional feature of the hybridization, ResonSense™ and related methods is the possibility to measure the temperature at which the probes dissociate from their complementary sequences. This measurement gives a further verification of the specificity of the amplification reaction. An important feature of many of the probe systems is their compatibility with multiplexing due to the availability of fluorophores with resolvable emission spectra. The chemistries available are discussed in detail in Chapter 3, 'Homogeneous Fluorescent Chemistries.'

Real-time PCR instrumentation

Thermal cyclers with integrated fluorimeters and some arrangement for transferring excitation light from a source into the reaction vessel and then from the sample to a detector are required for real-time PCR. The heating blocks that are the mainstay of the standard PCR instrument market present several technical challenges in conversion to application in real-time machines. The main problem being that the light must be channelled through the lid of the block and the cap of the reaction vessel across an air gap and then into the sample. Emitted light must then take the return path. Although blocks are used by several real-time machines including the first commercial instrument (Applied Biosystems 7700), the difficulties associated with them have led to the development of alternative designs. The LightCycler® (LC24) was the forerunner of machines that use air as the heating/cooling medium. Thermal transfer via air has the advantage of greater uniformity and rapidity than can be achieved on block-based cyclers, besides allowing shortening of the light path. Besides differing in the choice of heating medium real-time PCR machines also provide a range of options for the light source and detection of fluorescence. Current machines tend to allow the excitation and detection of multiple dyes so that internal standards and multiplex reactions are possible. There is also a tendency to build in a bias toward the use of either universal donor or universal recipient chemistry (see Chapter 3).

Since their introduction and continuing since the first edition of this book, the cost of real-time PCR instruments has fallen in tandem with continual improvement in their capability and accuracy. This has been the result of competition, the volume of sales and the introduction into the marketplace of improved designs dependent on new technology. These trends are unlikely to be reversed and will contribute to the growth in its popularity. Instrumentation for real-time PCR is described and discussed in more detail in Chapter 2, 'An Overview of Platforms'.

Quantification

Unlike standard PCR, real-time PCR instruments measure the kinetics of product accumulation in each PCR reaction tube. Generally, no product is detected during the first few temperature cycles as the fluorescent signal is below the detection threshold of the instrument. However, most combinations of machine and fluorescence reporter are capable of detecting the accumulation of amplicons before the end of the exponential amplification phase. During this time the efficiency of PCR is often close to 100%, giving a doubling of the quantity of product at each cycle. As product concentrations approach the ng/µl level the efficiency of amplification falls primarily because the amplicons re-associate during the annealing step. This leads to a phase during which the accumulation of product is approximately linear with a constant level of net synthesis at each cycle. Finally, a plateau is reached when net synthesis approximates zero. Quantification in real-time PCR is done by measuring the number of cycles required for the fluorescent signal to reach a threshold level or the second derivative maximum of the fluorescence versus cycle curve. This cycle number is proportional to the number of copies of template in the sample. Real-time quantification applications are discussed in detail in several chapters, see Chapter 9, 'Analysis of mRNA Expression', Chapter 10, 'Array Verification', and Chapter 16, 'Food Authenticity and Verification.

SNP detection

The methods used to verify the identity of the amplicon(s) produced in real-time PCR are also often sufficiently powerful to detect small variations between sequences. Variations in sequence, including single nucleotide polymorphisms (SNPs) have been successfully identified in real-time PCR assays. One common approach to the detection of sequence variation is to compare melting curves. In general, the effect of base substitutions on the melting kinetics of PCR products is too small to be detected reliably (if at all). However, one group (Wittwer et al., 2003) has demonstrated that heteroduplexes of relatively long amplicons differing by a SNP can be distinguished from the homoduplexes on the basis of their melting curves. This was presented as the basis of a method for mutation screening. More commonly, the melting curves of short fluorescent probes are used to distinguish between amplicons, for example (Edwards et al.,

2001; Whalley *et al.*, 2001). The latter method is sensitive to SNPs, which usually cause a shift in the melting peak of several degrees. A common alternative to the melting curve approach is to use hydrolysis (TaqMan™) probes. The efficiency of the 5′–3′ endonuclease reaction is greatly impaired when a well-designed probe mismatches its target sequence by even a single base. The detection of mutations by real-time PCR is discussed in Chapter 11, 'Mutation Detection', and Chapter 17, 'Molecular Haplotyping'. Although the melting curve and hydrolysis probe methods for mutation analysis are widely used they are only able to detect sequences that represent a large proportion of the population. Elsewhere a quantitative real-time ARMS method is described (see Chapter 11). ARMS assays are designed to detect the emergence of significant sequence mutants within a background that remains mainly of the parent type (Punia *et al.*, 2004).

Real-time PCR data analysis

The software provided with real-time PCR instruments allows four principal types of data analysis: (1) normalization of the raw data; (2) measurement of the cycle number at which any increase in the fluorescence within each reaction vessel reaches significance; (3) the data are used in conjunction with the results from internal or external standards to estimate the original number of template copies; (4) melting curves are transformed to provide plots of $-dF/dT$ against T (F = fluorescence and T = temperature) in which a peak (melting peak) occurs at the equilibrium temperature for each duplex. In general, instrument-specific software is easy to use and allows rapid and reproducible data analysis. In addition to the bundled software, a range of third-party utilities are now available to improve the flexibility of real-time PCR data analyses. Software is discussed in two chapters of this edition, Chapter 4, 'Normalization Software', and Chapter 5, 'Data Analysis Software'.

Non-PCR applications

Real-time PCR machines are also capable of use as real-time fluorimeters. For example, one simple application is estimation of the melting temperature (T_m) of an oligonucleotide. The oligonucleotide is mixed with its complementary sequence in the presence of a dye such as SYBR Green I, the temperature is increased and the level of fluorescence is measured to give a melting

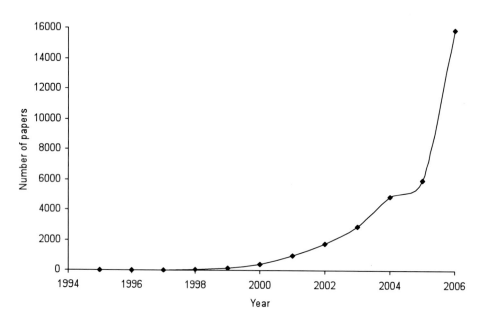

Figure 1.1 PubMed at NCBI (http://www.ncbi.nlm.nih.gov/) was searched by year for the term 'real-time PCR'. The result for 2006 was for an incomplete year.

curve from which the T_m may be deduced. This procedure is explained in more detail in Chapter 6, 'Performing Real-time PCR and Kits'. Chapter 12, 'Real-time NASBA', presents an alternative application using a real-time PCR instrument that relies on real-time fluorimetry. NASBA is a method for the isothermal amplification of RNA that produces quantities of antisense RNA copies. Molecular beacons complementary to the product are used to give a fluorescent signal.

The growth in the use of real-time PCR

In a relatively short time since their first introduction in the mid-1990s real-time PCR machines have become widely available to biologists. This has led to an explosion in the number of publications describing applications of the method. Indeed, a graph of number of papers against time resembles a real-time PCR plot (Fig. 1.1). The blip in 2005 suggests an artefact but may be associated with the increased use of real-time PCR for array validation (Chapter 10, 'Array Verification'). Initially the main applications that exploited real-time PCR previously relied on standard PCR. Thus the main fields of application were diagnostic microbiology and human genetic analysis. However, the decreased hands-on time, increased reliability and improved quantitative accuracy of real-time PCR methods are contributing to their adoption for gene expression analysis as described in Chapter 9, 'Analysis of mRNA Expression', and Chapter 10, 'Array Verification'.

References

Edwards, K.J., Metherell, L.A., Yates, M. and Saunders, N.A. 2001. Detection of *rpoB* mutations in *Mycobacterium tuberculosis* by biprobe analysis. J. Clin. Microbiol. 39, 3350–3352.

Kleppe, K., Ohtsuka, E., Kleppe, R., Molineux, I. and Khorana, H.G. 1971. Studies on polynucleotides. XCVI. Repair replications of short synthetic DNAs as catalyzed by DNA polymerases. J. Mol. Biol. 56, 341–361.

Punia, P., Cane, P., Teo, C.-G. and Saunders, N.A. 2004. Quantitation of hepatitis B lamivudine resistant mutants by real-time ARMS PCR. J. Hepatol. 40, 986–992.

Saiki, R.K., Gelfand, D.H., Stoffel, S., Scharf, S.J., Higuchi, R., Horn, G.T., Mullis, K.B. and Erlich, H.A. 1988. Primer-directed enzymatic amplification of DNA with a thermostable DNA polymerase. Science 239, 487–491.

Saiki, R.K., Scharf, S., Faloona, F., Mullis, K.B., Horn, G.T., Erlich, H.A. and Arnheim, N. 1985. Enzymatic amplification of beta-globin genomic sequences and restriction site analysis for diagnosis of sickle cell anemia. Science 230, 1350–1354.

Whalley, S.A., Brown, D., Teo, C.G., Dusheiko, G.M. and Saunders, N.A. 2001. Monitoring the emergence of hepatitis B virus polymerase gene variants during lamivudine therapy using the LightCycler. J. Clin. Microbiol. 39, 1456–1459.

Wittwer, C.T., Reed, G.H., Gundry, C.N., Vandersteen, J.G. and Pryor, R.J. 2003. High-resolution genotyping by amplicon melting analysis using LCGreen. Clin. Chem. 49, 853–860.

An Overview of PCR Platforms

Julie M. J. Logan and Kirstin J. Edwards

2

Abstract

Real-time PCR continues to have a major impact across many disciplines of the biological sciences, and this has been a driver to develop and improve existing instruments. From the first two commercial platforms introduced in the mid-1990s, there is now a wide choice of instruments, which continues to increase. Advances include faster thermocycling times, higher throughput, flexibility, expanded optical systems, increased multiplexing and more user-friendly software. The main features of each instrument are compared and factors important to weigh up when deciding on a platform are highlighted.

History of real-time PCR

Today it is clear that few techniques have had such a powerful impact on biology than the development of the polymerase chain reaction (PCR). More recently the PCR has become even more sophisticated with the introduction of real-time PCR. Initial work by Higuchi and colleagues (Higuchi *et al.*, 1992) first demonstrated the simultaneous amplification and detection of specific DNA sequences in real-time by simply adding ethidium bromide (EtBr) to the PCR reaction so that the accumulation of PCR product could be visualized at each cycle. When EtBr is bound to double-stranded DNA and excited by UV light it fluoresces, therefore an increase in fluorescence in such a PCR indicates positive amplification. Soon afterwards they introduced the idea of real-time PCR product quantitation or 'kinetic PCR', by continuously measuring the increase in EtBr intensity during amplification

with a charge-coupled device camera (Higuchi *et al.*, 1993). By creating amplification plots of fluorescence increase versus the cycle number, they demonstrated that the kinetics of EtBr fluorescence accumulation during thermocycling was directly related to the starting number of DNA copies. Fewer cycles are needed to produce a detectable signal, when a greater number of target molecules are present. Kinetic monitoring also provided a means whereby the efficiency of amplification under different conditions could be determined, providing for the first time insight into the fundamental PCR processes. Therefore, the principle underlying real-time PCR can simply be defined as the monitoring of fluorescent signal from one or more PCRs cycle-by-cycle to completion, where the amount of product produced during the exponential amplification phase can be used to determine the amount of starting material.

The approach described above was not ideal since EtBr binds non-specifically to DNA duplexes and non-specific amplification products, such as primer-dimers, can contribute to the fluorescent signal and result in quantification inaccuracies. Subsequent refinements, the most significant of which was the introduction of fluorogenic probes to monitor product accumulation, added a greater element of specificity to real-time PCR and provided greater quantitative precision and dynamic range than previous methods.

These significant advances to the basic PCR technique not surprisingly led to the development of a new generation of PCR platforms and reagents, which allowed simultaneous amplification and quantification of specific nucleic acid sequences cycle by cycle. Indeed, a few years after Higuchi coined the term 'kinetic' or 'real-time PCR' the first commercial platforms were released on the market. The first was the Applied Biosystems ABI Prism 7700 Sequence Detection System, followed by the Idaho Technology LightCycler (now manufactured and sold by Roche Diagnostics) (Wittwer et al., 1997). Both of these platforms utilized fluorogenic chemistry and like any real-time PCR platform, they basically consist of a thermal cycler with an integrated optical detection system, which can heat, cool, detect and report. New and improved models have now superseded these two instruments and several other manufacturers have introduced their own real-time PCR platforms. Choosing a suitable instrument is now a complex task. Real-time PCR offers many advantages that include:

* amplification and detection in an integrated system;
* fluorescent dyes/probes allowing constant reaction monitoring;
* rapid cycling times (20–40 min for 35 cycles);
* high sample throughput (~200–5000 samples/day);
* low contamination risk due to sealed reactions;
* increased sensitivity (~ 3 pg or 1 genome equivalent of DNA);
* detection across a broad dynamic range of $10-10^{10}$ copies;
* reproducibility with a CV < 2.0%;
* allows for quantification of results;
* software-driven operation.

Current disadvantages include:

* limited capacity for multiplexing using all chemistries;
* development of protocols can requires a high level of technical skill and/or support (R&D capacity and capital)
* medium capital equipment costs
* analysis requires skill.

This chapter is intended to provide an overview of the main features of real-time platforms, as well as highlighting aspects to consider when introducing real-time PCR to your laboratory. Information on the available fluorescent chemistries, their principles and methods is detailed in Chapter 3.

Real-time PCR platforms

A real-time PCR instrumentation platform consists of a thermal cycler, optics for both fluorescence excitation and emission collection, together with a computer and software for data acquisition and analysis. A wide range of systems are now available (see References) and these differ in their design and level of sophistication, providing users with several choices, which include: format, reaction vessels, emission and excitation wavelengths, throughput, level of control, chemistry, software, speed and applications. They all have in common the ability to measure the accumulation of PCR product during the exponential phase of the reaction using online fluorescence monitoring, whether specific or non-specific and hence provide accurate data on initial starting copy numbers. As amplification and detection are combined in a single step, the process can occur in a single closed reaction vessel, which eliminates any need for numerous post-PCR manual manipulations, as well as reducing the possibility of introducing contamination or variability. Additional technical advantages include both qualitative and quantitative PCR, mutation analysis, multiplexing and high-throughput analysis. Although the fluorescence chemistries used in different platforms are similar, their mechanics and methodologies are wide-ranging. A summary of the features of each platform is detailed in Table 2.1 and the reader is encouraged to refer to it throughout this chapter.

Real-time PCR thermocycling

The first component to consider in a PCR platform is the thermal engine. Successful thermal cycling is dependent on the accurate regulation of temperature in the sample vessels and the speed at which these target temperatures can be achieved. The majority of real-time platforms use advanced heating block technology based on the Peltier effect, to actively transfer heat in

Table 2.1 Features of real-time PCR platforms

Company	Applied Biosystems	Applied Biosystems
Model	ABI 7300	ABI 7500
Laser/lamp	Tungsten halogen lamp	Tungsten halogen lamp
Detector	CCD camera	CCD camera
Thermocycling	Peltier element block	Peltier element block
Excitation spectrum	Single excitation filter	Five excitation filters
Filters/detection channels	Four-position filter wheel: FAM/SYBR Green I, VIC/JOE, NED/TAMRA and ROX	Five-position filter wheel: FAM/SYBR Green I, VIC/JOE, NED/TAMRA/Cy3, ROX/Texas Red and Cy5 dyes
Format	96-well plates or 0.2 ml tubes	96-well plates or 0.2 ml tubes
Time (40 cycles)	1.75 hours	1.75 hours
Reaction volume	20–100 µl	20–100 µl
Fluorescence chemistry	Hydrolysis probes and SYBR Green I, other chemistries possible but not supported	Hydrolysis probes and SYBR Green I, other chemistries possible but not supported
Multiplexing supported	4-plex	5-plex
Passive reference	ROX (optional)	ROX (optional)
Dimensions (H × W × D)	34 × 49 × 45 cm	34 × 49 × 45 cm
Weight	29 kg	34 kg
Other features	Primer/probe design software included	Primer/probe design software included
	PCR mastermix kits, parameter-specific kits	PCR mastermix kits and parameter-specific kits
	Assay by Design and Assay on Demand Services	Assay by Design and Assay on Demand Services
	Software: gene expression analysis, absolute quantification, allelic discrimination and plus/minus assays	Software: absolute quantification, relative quantification, allelic discrimination, isothermal and plus/minus assays
		SDS v1.4 21CFRp11 Module assists with 21CFRp11 compliance

Abbreviations: LED, light-emitting diode; CCD, charge-coupled device camera; PMT, photomultiplier tube; UPL, Universal Probe Library.

Table 2.1 *continued*

	Applied Biosystems	Applied Biosystems
Company		
Model	ABI 7500 Fast	ABI 7900 Fast HT with automation accessory
Laser/lamp	Tungsten halogen lamp	Argon ion laser
Detector	CCD camera	CCD camera
Thermocycling	Peltier element block	Peltier element 9700 block
Excitation spectrum	Five excitation filters	488 nm
Filters/detection channels	5-position filter wheel: FAM/SYBR Green I, VIC/JOE, NED/TAMRA/Cy3, ROX/Texas Red and Cy5 dyes	500–660 nm continuous wavelength detection, SYBR Green I, FAM, VIC, JOE, NED, ROX, TAMRA, TET
Format	Fast 96-well plates optimized for 10 µl reactions	96- and 384-well plates (interchangeable blocks)
Time (40 cycles)	< 30 min	33 min (96-well plate) or 52 min (384-well plate)
Reaction volume	10–30 µl	5–20 µl
Fluorescence chemistry	Hydrolysis probes and SYBR Green I, other chemistries possible but not supported	Hydrolysis probes and SYBR Green I, other chemistries possible but not supported
Multiplexing supported	5-plex	6-plex
Passive reference	ROX (optional)	ROX (optional)
Dimensions (H × W × D)	34 × 49 × 45 cm	64 × 125 × 84 cm
Weight	34 kg	114 kg
Other features	Primer/probe design software included	Primer/probe design software included
	PCR mastermix kits, parameter-specific kits	PCR mastermix kits, parameter-specific kits
	Assay by Design and Assay on Demand Services	Assay by Design and Assay on Demand Services
	Software: absolute quantitation, relative quantification, allelic discrimination, isothermal, melt curve analysis and plus/minus assays	Software: absolute quantitation, relative quantitation, allelic discrimination, isothermal, plus/minus assays, high-throughput SNP genotyping and gene expression
	SDS v1.4 21CFRp11 Module assists with 21CFRp11 compliance	SDS v1.4 21CFRp11 Module assists with 21CFRp11 compliance
		Robotic plate un/loading for up to 84 plates, high-throughput: real-time PCR >5000 wells per day and end-point 30,000 wells every 3 hours

	Applied Biosystems	Roche
Company		
Model	ABI StepOne	LightCycler 480
Laser/Lamp	LED	High-intensity xenon lamp
Detector	Three emission filters and photodiode	CCD camera
Thermocycling	Peltier element block	Peltier element block incorporating Therma-Base technology
Excitation spectrum	480 nm	430–630 nm, 5-position filter wheel: 450, 483, 523, 558 and 615 nm
Filters/detection channels	Three filters: SYBR Green I/FAM, VIC/JOE and ROX	Six-position filter wheel: 500, 533, 568, 610, 640, 670 nm
Format	48-well plates, 0.1 ml individual or 8-strip tubes	96- and 384-well plates (interchangeable blocks)
Time (40 cycles)	< 40 min	< 1 hour (96-well plate) or < 40 min (384-well) plate
Reaction volume	10–30 μl	10–100 μl (96-well plate), 3–20μl (384-well plate)
Fluorescence chemistry	Hydrolysis probes, SYBR Green I, other chemistries possible but not supported	Hydrolysis probes, hybridization probes, Simple Probes, SYBR Green I, UPL
Multiplexing supported	3-plex	3-plex hybridization and 6-plex hydrolysis probes
Passive reference	ROX (optional)	Not required
Dimensions (H × W × D)	24.6 × 51.2 × 42.7 cm	54.5 × 60 × 60 cm
Weight	23.6 kg	55 kg
Other features	Primer/probe design software included	Primer/probe design software available
	Touchscreen, wizard-driven software with networking capabilities	System compatible with 21 CFR part 11 requirements
	PCR mastermix kits	PCR mastermix kits, parameter-specific kits
	Software: standard curve, relative standard curve, comparative C_t (DDCt), allelic discrimination, melt curve analysis and plus/minus assays	Software: absolute quantification, T_m calling, colour compensation, qualitative detection, statistical analysis, relative quantitation, high-resolution melting, genotyping, LIMS/bar-code module

Abbreviations: LED, light-emitting diode; CCD, charge-coupled device camera; PMT, photomultiplier tube; UPL, Universal Probe Library.

Table 2.1 *continued*

	Roche	Roche
Company		
Model	LightCycler 1.5	LightCycler 2.0
Laser/lamp	LED	LED
Detector	Three photodetection diodes	Six photodetection diodes
Thermocycling	Heated air	Heated air
Excitation spectrum	470 ± 10 nm	470 ± 10 nm
Filters/detection channels	Three channels: 530, 640, 710 nm	Six channels: 530, 560, 610, 640, 670, 710 nm
Format	32 glass capillaries	32 glass capillaries
Time (40 cycles)	30 min	30 min
Reaction volume	20 µl	20 µl or 100 µl
Fluorescence Chemistry	Hybridization probes, hydrolysis probes, molecular beacons, SYBR Green I	Hybridization probes, hydrolysis probes, molecular beacons, SYBR Green I, Simple Probes, UPL
Multiplexing supported	2-plex hybridization probes	4-plex hybridization and 2-plex hydrolysis probes
Passive reference	Not required	Not required
Dimensions (H × W × D)	45 × 30 × 40 cm	45 × 30 × 40 cm
Weight	20 kg	20 kg
Other features	Primer/probe design software available	Primer/probe design software included
	PCR mastermix kits, parameter-specific kits	PCR mastermix kits, parameter-specific kits
	Software: absolute quantitation, genotyping analysis, melting curve analysis, multiplex analysis	Software: absolute quantification, relative quantification, genotyping analysis, T_m calling analysis, multiplex analysis, plus/minus assays
	Relative quantification software available	Nucleic acid quantitation
		Conforms to European directive for *in vitro* diagnostic medical devices 98/79/EC

Company	Stratagene	Stratagene
Model	Mx4000	Mx3000P
Laser/lamp	Quartz tungsten halogen lamp	Quartz tungsten halogen lamp
Detector	Four PMTs	One scanning PMT
Thermocycling	Solid-state resistive/convective Peltier hybrid block	Solid-state based-based block
Excitation spectrum	Four interchangeable filter wheels in the range 350–750 nm	Four user defined matched filter wheels in the range 350–750 nm
Filters/detection channels	Four interchangeable filter wheels in the range 350–830 nm	Four user defined matched filter wheels in range 350–700 nm (default: FAM, HEX, ROX, Cy 3)
Format	96-well plates, 0.2 ml tubes, 8 × 0.2ml strips	96-well plates, 0.2 ml tubes, 8 × 0.2ml strips
Time (40 cycles)	< 1.5 hours	< 1.5 hours
Reaction Volume	10–50 µl	10–50 µl
Fluorescence Chemistry	Hydrolysis probes, FRET probes, molecular beacons, Scorpions, Amplifluor, SYBR Green I	Hydrolysis probes, FRET probes, molecular beacons, Scorpions, Amplifluor, SYBR Green I
Multiplexing Supported	4-plex	4-plex
Passive Reference	Optional ROX	Optional ROX
Dimensions (H x W x D)	76 × 46 × 51cm	33 × 46 × 43 cm
Weight	50 kg	20 kg
Other features	PCR mastermix kits	PCR mastermix kits
	System compatible with 21 CFR part 11 requirements	System compatible with 21 CFR part 11 requirements
	Software: gene expression analysis, allelic discrimination, melting curve analysis	Software: gene expression analysis, allelic discrimination, melting curve analysis
	Integrated computer	Integrated computer

Abbreviations: LED, light-emitting diode; CCD, charge-coupled device camera; PMT, photomultiplier tube; UPL, Universal Probe Library.

Table 2.1 *continued*

	Stratagene	Cepheid
Company		
Model	Mx3005P	SmartCycler
Laser/Lamp	Quartz tungsten halogen lamp	Four high-intensity LEDs
Detector	One scanning PMT	Silicon photodetectors
Thermocycling	Solid-state Peltier-based block	Resistive heating of ceramic plates with forced-air cooling
Excitation spectrum	Five user-defined independent filter wheels in the range 350–750 nm	Four channels (450–495, 500–550, 565–590, 630–650)
Filters/detection channels	5 user defined independent filter wheels in the range 350–700 nm (default: FAM, HEX, ROX, Cy5, Cy 3)	4 channels (510–527, 565–590, 606–650, 670–750) (default: FAM, Cy3, TET, Alexa 532, Texas Red, Cy5, Alexa 647 and Eva Green)
Format	96-well plates, 0.2 ml tubes, 8 × 0.2 ml strips	16 proprietary tubes
Time (40 cycles)	< 1.5 hours	20–40 min
Reaction volume	10–50 μl	25 or 100 μl
Fluorescence chemistry	Hydrolysis probes, FRET probes, molecular beacons, Scorpions, Amplifluor, SYBR Green I	Hydrolysis probes, molecular beacons, Eclipse probes, Amplifluor and Scorpion primers, SYBR Green I
Multiplexing supported	5-plex	4-plex
Passive reference	Optional ROX	Not required
Dimensions (H × W × D)	33 × 46 × 43 cm	30 × 30 × 25 cm
Weight	20 kg	10 kg
Other features	PCR mastermix kits	PCR mastermix kits and parameter-specific kits
	System compatible with 21 CFR part 11 requirements	Random access (16 independently programmable reactions sites)
	Software: gene expression analysis, allelic discrimination, melting curve analysis	Software: gene expression analysis, allelic discrimination, melting curve analysis and end-point
	Integrated computer	Compact and portable for field use
		Modular system – flexibility to expand in multiples of 16 module units up to 96 reaction sites

	Corbett	Eppendorf
Company / Model	Rotor-Gene 6000	Mastercycler ep *realplex*[4]
Laser/Lamp	Six high-power LEDs	96 LEDs
Detector	PMT	Two-channel PMT
Thermocycling	Resistive heater with air cooling and centrifugation	Gradient Peltier element block
Excitation spectrum	365, 470, 530, 585, 625, 680 nm,	470 nm
Filters/detection channels	460, 510, 555, 610, 660, 712 nm	520/550/580/605 nm
Format	36 0.2 ml or 72, 0.1 ml plastic tubes; 72 or 100 well Gene-Disc plastic rings	96-well plates, 0.2 ml tubes
Time (40 cycles)	40 min	30 min
Reaction volume	10–100 µl, 10–50 µl, 15–50 µl, 10–25 µl	5–100 µl
Fluorescence chemistry	Hydrolysis probes, molecular beacons, hybridization probes, SYBR Green I	Open system all formats supported, except hybridization probes
Multiplexing supported	6-plex	4-plex
Passive reference	Not required	Not required
Dimensions (H × W × D)	27.5 × 37 × 42 cm	39.6 × 26 × 41 cm
Weight	14 kg	24 kg
Other features	Centrifugal rotary design Software: HRM (high-resolution melt), end-point, association/disassociation (i.e. melting/annealing) kinetics, relative expression and nucleic acid concentration measurement	PCR mastermix kits Software: raw data, melting curve analysis, quantification/relative quantification, allele discrimination, endpoint, ± assays. Gradient range of 24°C

Abbreviations: LED, light-emitting diode; CCD, charge-coupled device camera; PMT, photomultiplier tube; UPL, Universal Probe Library.

Table 2.1 *continued*

Company	BioRad	BioRad
Model	MiniOpticon	MyiQ
Laser/Lamp	48 LEDs	Tungsten-halogen lamp
Detector	Two photodiodes	12-bit CCD camera
Thermocycling	MJ Mini Peltier element block	iCycler Peltier element block
Excitation spectrum	470–500 nm	475–495 nm
Filters/detection channels	523–543 nm 540–700 nm	515–545 nm
Format	48 well PCR plate, 6 × 8 tube strips	96 well plates, 0.2 ml tubes
Time (40 cycles)	< 40 min	< 40 minutes
Reaction volume	10–100 µl (20 µl recommended)	10–100 µl
Fluorescence chemistry	Hydrolysis probes, molecular beacons, SYBR Green I	Hydrolysis probes, molecular beacons, SYBR Green I
Multiplexing supported	2-plex	Single-plex
Passive reference	Not required	Fluorescein
Dimensions (H × W × D)	33 × 18 × 32 cm	39 × 29 × 58 cm
Weight	6.8 kg	17.6 kg
Other features	PCR mastermix kits	PCR mastermix kits
	Software: allelic discrimination, gene expression analysis, melting curve analysis	Software: allelic discrimination, gene expression analysis, melting curve analysis, end-point detection, absolute quantification
	Connect four instruments to one PC	Optional software for 21 CFR part 11 compliance
	Gradient range of 16°C	Gradient range of 25°C

	BioRad	BioRad
Company		
Model	Opticon2	Chromo4
Laser/lamp	96 LEDs	Four LEDs in photonics shuttle
Detector	Dual PMTs	Four photodiodes
Thermocycling	DNA Engine Peltier element block	DNA Engine Dyad Disciple Peltier element block
Excitation spectrum	470–505 nm	450–650 nm
Filters/detection channels	523–543, 540–700 nm	515–730 nm
Format	96-well plate, 8 × 0.2ml strip tubes	96-well PCR plate or 96 × 0.2 ml tubes
Time (40 cycles)	< 40 min	< 40 min
Reaction volume	10–100 µl (20 µl recommended)	10–100 µl (20 µl recommended)
Fluorescence chemistry	Hydrolysis probes, molecular beacons, SYBR Green I	Hydrolysis probes, molecular beacons, SYBR Green I
Multiplexing supported	2-plex	4-plex
Passive reference	Not required	Not required
Dimensions (H × W × D)	60 × 34 × 47 cm	37 × 24 × 35 cm
Weight	29 kg	12.6 kg
Other features	PCR mastermix kits	PCR mastermix kits
	Software: allelic discrimination, gene expression analysis, melting curve analysis	Software: allelic discrimination, gene expression analysis, melting curve analysis
	Gradient range of 24°C	Customisable filter sets
		Modular design for maximal flexibility, 2 × 96 Dyad
		Gradient range of 24°C

Abbreviations: LED, light-emitting diode; CCD, charge-coupled device camera; PMT, photomultiplier tube; UPL, Universal Probe Library.

Table 2.1 *continued*

Company	BioRad
Model	iQ5
Laser/Lamp	Tungsten halogen lamp
Detector	12-bit CCD camera
Thermocycling	iCycler Peltier element block
Excitation spectrum	Five filter positions in range 475–645 nm
Filters/detection channels	Five filter positions available in range 515–700 nm, two provided
Format	96- or 384-well plate, 8 × 0.2ml strip tubes
Time (40 cycles)	< 40 minutes
Reaction volume	15–100 µl
Fluorescence chemistry	Hydrolysis probes, molecular beacons, hybridization probes, SYBR Green I
Multiplexing supported	5-plex
Passive reference	Fluorescein
Dimensions (H × W × D)	36 × 33 × 62 cm
Weight	17.6 kg
Other features	PCR mastermix kits
	Software: allelic discrimination, gene expression analysis, melting curve analysis, end-point detection, absolute quantification
	Gradient range of 25°C
	Optional software for 21 CFR part 11 compliance

Abbreviations: LED, light-emitting diode; CCD, charge-coupled device camera; PMT, photomultiplier tube; UPL, Universal Probe Library.

and out of thin-walled plastic reaction vessels (e.g. ABI 7300, ABI 7500, ABI 7900HT, ABI Step One, Opticon 2, MiniOpticon, Chromo 4, Mx4000, Mx3000P, Mx3005P, Mastercycler ep, MyiQ, iQ5, LightCycler 480). Peltier devices transfer heat from one side of a semiconductor to another. In general, blocks have significant mass and consequently a degree of thermal inertia. Furthermore, the plastic insulating layer between the reaction vessel and the heater produces an additional thermal lag. As a consequence of this, the temperature transitions are relatively slow and blocks must be very carefully designed to minimize well-to-well variation. Other advances on the Peltier-based technology include its combination with Joule, resistive or convective technology to give improved temperature control and performance across the block. More recently has been the inclusion of patented Therma-Base technology, whose working principle is based on the evaporation and condensation of a working fluid in a thin vacuum cavity to accurately control well-to-well variation. Three platforms employ alternative heat exchange technologies which permit more rapid thermal ramp rates than blocks, resulting in significantly increased thermocycling speeds. These include a stationary turbulent air-heated glass capillary format (LightCyclers 1.5 and 2.0), a centrifugal air-heated plastic tube format (Rotor-Gene) and a high-thermal-conductivity ceramic heating plate plastic tube format (SmartCycler). For example, the time taken to equilibrate at 72°C using a Rotor-Gene is 0 s compared to 15 s with a standard 96-well Peltier block, resulting in run times that are on average 50% faster. Detailed information on the temperature specifications for each platform is shown in Table 2.2, which highlights that the LightCyclers 1.5 and 2.0 have the capacity to perform the fastest PCR and the Rotor-Gene has the smallest variation in temperature uniformity.

Real-time PCR optics

An integrated fluorimeter is required to detect and monitor the levels of fluorescence during the PCR process for real-time PCR and there is a range of options available for both the excitation light source and fluorescent emission detection. The light sources that cause fluorophore excitation can be classed as narrow- or broad-spectrum.

If a broad-spectrum light source is employed (e.g. Mx4000, Mx3000P, Mx3005P, MyiQ, iQ5, ABI 7300, ABI 7500, LightCycler 480) then filters can be used to provide light tuned to the excitation spectrum of a specific individual fluorophore. Such a system provides the user with a wider choice of available fluorophores, although it is best to select those with good separation of their emission spectra. A disadvantage of this optical system is that the light intensity passing through the filters can be limited and this could in theory limit the sensitivity of detection. There are currently two narrow-spectrum light sources used in real-time platforms, these can be light-emitting diodes (LEDs) (e.g. LightCyclers 1.5 and 2.0, Opticon 2, MiniOpticon, Chromo4, SmartCycler, Rotor-Gene, StepOne, Mastercycler ep) or laser (ABI 7900HT). The SmartCycler and Rotor-Gene have multiple LEDs (4 and 6 respectively) that excite at different wavelengths, providing a greater selection of fluorophores and giving these instruments capabilities similar to those of the broad-spectrum platforms above. The LightCyclers 1.5 and 2.0, Opticon2 and ABI 7900HT have single light source excitation, which ultimately limits the choice of fluorophores.

In general, the detectors used in real-time platforms are set to measure narrow bands of the spectrum, although filter sets that can be customized by the user are available for the iQ5, Mx4000, Mx3005P and Mx3000P, Chromo4. The number of detection channels that can be effective is dependent on the available range of excitation wavelengths. For example, if a single narrow range excitation source is available, one approach is to use fluorophores that are all excited to some extent in the same range and then to rely on software correction to deconvolute the light emitted from a given area of the spectrum, as was employed successfully with the now discontinued ABI 7700. Another approach is demonstrated with the LightCycler 1.5, where a narrow-spectrum light source excites the fluorophores SYBR or fluorescein and emitted light is collected via three discrete optical detectors. Two of these detect long wavelength light emissions from fluorophores which are only minimally excited by the blue LED light source, which are instead excited using FRET technology (see Chapter 3).

Table 2.2 Temperature specifications of real-time PCR platforms

Platform	Max. heating/cooling rate (°C/s)	Temperature accuracy (°C)	Temperature uniformity (°C)
ABI 7900HT Fast	3.0/3.0	± 0.25	± 0.5
ABI 7300	1.1/1.1	± 0.25	± 0.5
ABI 7500	0.8/1.6	± 0.25	± 0.5
ABI 7500 Fast	3.5/3.5	± 0.25	± 0.5
ABI StepOne	2.2/2.2	± 0.25	± 0.5
Mastercycler ep *realplex*[4]	6.0/4.5	± 0.2	± 0.3
Roche LightCycler	20.0/20.0	± 0.4	± 0.2
Roche LightCycler 480	4.8/2.5	± 0.2	NA
BioRad MiniOpticon	2.5/2.0	± 0.2	± 0.4
BioRad MyiQ	3.3/2.0	± 0.3	± 0.4
BioRad Opticon 2	3.0/2.0	± 0.4	± 0.4
BioRad Chromo4	3.0/2.0	± 0.3	± 0.4
BioRad iQ5	3.3/2.0	± 0.3	± 0.4
Stratagene Mx4000	2.2/2.2	± 0.25	± 0.25
Stratagene Mx3000P	2.5/2.5	± 0.25	± 0.25
Stratagene Mx3005P	2.5/2.5	± 0.25	± 0.25
Corbett Rotor-Gene 6000	15.0/20.0	± 0.25	± 0.01
Cepheid SmartCycler	10.0/2.5	± 0.5	± 0.5

NA, not available

Real-time PCR chemistries

As the technology has advanced rapidly, second, third and fourth generation real-time PCR platforms have been developed with improvements seen in multiplexing and increased throughput capabilities. The optical characteristics of a given platform clearly have an impact on the ability to multiplex and also determine which probe systems are compatible. In addition, the analysis software may also predetermine the appropriate chemistries. The ability to mulitiplex the available fluorophores is fully discussed in Chapter 3. However, it is important to point out that the platform and choice of fluorescent chemistry are strongly linked. Indeed some platforms are biased towards a particular probe system and whilst the optics permit different probe chemistries to be excited and detected, often the analysis software does not and the user is required to export the data to a spreadsheet program for detailed user analysis. For example, the Applied Biosystems platforms do not officially support any chemistries other than hydrolysis probes and SYBR Green and only support duplexing. Therefore, the reporting chemistry required for an application should be strongly considered before a choice of platform is made.

Additional platform features

Several of the platforms employ a standard 96-well block format or interchangeable 384-well block and offer a medium to high throughput. An advantage of employing a block format is that standard PCR plates and tubes can be used and these tend to be cheaper than instrument-specific plastics (SmartCycler, Rotor-Gene) or glass capillaries (LightCyclers 1.5 and 2.0). Also, the LightCyclers 1.5 and 2.0 and SmartCycler require centrifugation to move sample into the reaction vessel and these alternative designs may not be suitable for all applications. For the highest throughput, the ABI 7900HT combines a 384-well plate format with an automation accessory, which allows for up to 84 plates to be loaded and unloaded, providing a throughput of >5000 wells per 8-hour day or 30,000 wells for end-point analysis only. The LightCycler 480 also offers a similar level of automation and high-throughput workflow capability. These high-throughput instruments are ideal for dedicated laboratories where large batches of samples are run, with few different cycling parameters. An additional specification of a few platforms is the ability to perform gradient thermocycling, which can be very useful at the assay optimization stage.

Some platforms offer a low to medium throughput but are more flexible. Although the format may allow only 32 or even 16 samples, thermocycling times are faster and multiple runs can be performed thereby increasing the potential throughput. Performing multiple runs rapidly may be an advantage when several applications are employed which require different cycling parameters. The SmartCycler offers a different concept to real-time PCR of random access, which means independent programming of cycling parameters for 16 different assays. Each reaction vessel has its own element so that runs in available slots may be started any time, whilst other reaction sites are already in use. It is also modular allowing up to six units to be operated by a single computer. Others, such as the Chromo4 offers a modular design for maximum flexibility.

Ideally, the analysis software supplied with the platform should be as user-friendly as possible but it is also important to check that the software can fully analyse results of the chosen probe chemistry. As already mentioned, some platforms and analysis software suites are biased towards certain chemistries. Some real-time instruments also have specific primer and probe design software that is either supplied with the hardware or available at extra cost. Such software can help simplify and speed up the assay design process and is optimized for that system and reagents. The LightCyclers also have specific relative quantification software that is designed to determine the exact relative nucleic acid concentration normalized to a calibrator sample. This software speeds up and greatly simplifies relative quantification.

The majority of instrument manufacturers supply optimized real-time PCR mastermixes, these reagents benefit from being quality controlled, are easy to use, and usually offer reproducible and reliable results. However, for certain laboratories the additional cost can be prohibitive; however, with an array of third-party companies (e.g. Qiagen, Eurogentec, Invitrogen) now supplying real-time reagents, competition in this market should lead to reduced costs. Target-specific kits for a range of applications are available from some manufacturers and other companies (e.g. Roche, Applied Biosystems, Qiagen). Additionally, Applied Biosystems offer Assays-on-Demand and Assays-by-Design services for SNP genotyping by real-time PCR. A number of companies now offer real-time PCR services from primer design right through to assay design and validation, quality assurance and other custom services.

Finally, new introductions include the Cepheid GeneXpert and Enigma Diagnostics Enigma ML instruments, which combine automated sample preparation and real-time PCR into a single integrated compact bench top instrument, which opens new avenues for real-time PCR in field-based and point-of-care environments. At the other end of the scale is the introduction of nanolitre high-throughput quantitative PCR (Morrison *et al.* 2006). The Biotrove OpenArray system allows 3072 33-nl reactions on a microscope slide array format, which equates to a single operator performing 27,000 qPCR reactions in a working day. Another interesting development on the horizon is ultra fast real-time PCR chips capable of performing 40 cycles in under 6 minutes (Neuzil *et al.* 2006).

Final considerations

In weighing up the pros and cons of the different platforms for your laboratory, factors to consider include: supported chemistries; multiplex capability for that chemistry; throughput; flexibility; format; ease-of-use and robust software package; reproducibility; speed; size; technical support; customer support and not least the cost, not only of the initial equipment outlay and servicing but also the associated cost of consumables and reagents. It is also possible to 'try before you buy', most companies will provide a loan machine. It is wise to test a few of these once you have narrowed down your choice. User experiences should not be overlooked and there are now a number of useful websites and news groups where you can address you questions and queries (see References).

Clearly real-time PCR has undergone significant development over the last ten years, which has resulted in a wide range of different platforms becoming available. The drive to improve current technology is likely to continue for the foreseeable future as real-time PCR finds it way into more laboratories and specialist niches. As competition intensifies, users should benefit from more sophisticated high-throughput platforms, with increasing levels of automation and cost-effectiveness, although making a choice of platform will continue to be difficult.

References

Papers
Higuchi, R., Fockler, C., Dollinger, G. and Watson, R. 1993. Kinetic PCR analysis: real-time monitoring of DNA amplification reactions. Biotechnology *11*, 1026–1030.
Higuchi, R., Dollinger, G., Walsh, P.S. and Griffith, R. 1992. Simultaneous amplification and detection of specific DNA sequences. Biotechnology *10*, 413–417.
Morrison, T, Hurley, J., Garcia, J., Yoder, K., Katz, A., Roberts, D., Cho, J., Kanigan, T., Iilyin, S.E., Horowitz, D., Dixon, J.M and Brenan J.H.B. 2006. Nanoliter high throughput quantitative PCR. Nucleic Acids Res. *34*, e123.
Neuzil, P., Zhang, C., Pipper, J., Oh, S. and Zhuo, L. 2006. Ultra fast miniaturised real-time PCR: 40 cycles in less than six minutes. Nucleic Acids Res. *34*, e77.

Wittwer, C.T., Ririe, K.M., Andrew, R.V., David, D.A., Gundry, R.A. and Balis, U.J. 1997. The LightCycler: a microvolume multisample fluorimeter with rapid temperature control. Biotechniques *22*, 176–181.

Manufacturers' websites
Applied Biosystems www.appliedbiosystems.com
BioGene www.biogene.com
Bio-Rad Laboratories www.bio-rad.com
BioTrove www.biotrove.com
Cepheid www.cepheid.com
Corbett Research Ltd www.corbettresearch.com
Eppendorf www.eppendorf.com
Enigma Diagnostics www.enigmadiagnostics.com
Eurogentec www.eurogentec.com
Idaho Technologies www.idahotech.com
Invitrogen www.invitrogen.com
Qiagen www.qiagen.com
Roche Diagnostics Ltd www.roche-applied-science.com
Stratagene www.stratagene.com

Web resources
www.fluoresentric.com
www.gene-quantification.info
www.lightcycler-online.com
www.meltcalc.de
www.primerdesign.co.uk
www.tataa.com

News groups
http://groups.yahoo.com/group/lightcycler/
http://groups.yahoo.com/group/qpcrlistserver/
http://groups.yahoo.com/group/realtimepcr/
http://groups.yahoo.com/group/taqman/

Homogeneous Fluorescent Chemistries for Real-Time PCR

Martin A. Lee, David J. Squirrell, Dario L. Leslie and Tom Brown

Abstract

The development of fluorescent methods for a closed tube polymerase chain reaction has greatly simplified the process of nucleic acid quantification. Current approaches use fluorescent probes that interact with the amplification products during the PCR allowing kinetic measurement of product accumulation. These probe methods include generic approaches to DNA quantification such as fluorescent DNA binding dyes. There are also a number of strand-specific probes that use the phenomenon of Fluorescent Energy Transfer. In this chapter we describe these methods in detail, outline the principles of each process, and describe published examples. This text has been written to provide an impartial overview of the utility of different assays and to show how they may be used on various commercially available thermal cyclers.

Introduction

A fluorescent real-time polymerase chain reaction (PCR) (Saiki et al., 1985; Mullis et al., 1987) can provide both qualitative and quantitative analysis in various applications. Real-time PCR differs from earlier methods of analysis in that additional components are required to carry out the process. These components include an optical system integrated into the thermal cycler, and a probe that reports amplification during the course of the PCR process. The real-time PCR thermal cycler is discussed in Chapter 2 in detail. In this chapter we discuss the probe technologies or 'reporting chemistries.' Since the first edition of this book, new methods have been reported in the literature and emerged as commercial products and this edition is expanded to include these developments.

In order to give the reader a complete understanding of current technology, the principles of real-time analysis will initially be presented outside the context of any specific commercial platform. The number of these is increasing and it is certain that the capabilities on offer will undergo continued improvement and allow new fluorescent approaches to be realized. At this point it is important to highlight to new users the relationship between the choice of probe system and the instrument. These are inextricably linked: the optical specification and the analysis tools on any given platform greatly influence the applicability and utility of different probe systems. Whilst the main factors for the choice of instrument are often driven by throughput requirements and the initial purchase cost, careful consideration of the reporting chemistry for the required application should be made before purchase since the operation of one chemistry or another may be greatly compromised on some platforms. Equally important is the technical support provided from suppliers. This may be limited for non-supported chemistries. So called 'open platforms' from manufacturers that do not support any one chemistry may not be able to provide technical advice for the chemistry of choice for the required application. In this chapter we break down probe technology into a number of components to enable the reader to understand how the different assay systems may be used on current and future fluorimetric thermal cyclers. For new users this will facilitate the implementation of various assays on commercial platforms.

We first present a background to the limitations of PCR without real-time detection and introduce the two main classes of real-time chemistries. A second section on the basics of fluorescent resonance energy transfer (FRET) will allow the reader to understand how dye technologies may be applied in strand-specific applications. An outline of the function and utility of real-time PCR probe methods will provide the reader with an understanding of how such technology may be best applied. Finally, experimental considerations and specific examples will be provided although the reader should refer to the relevant chapters in this text for detailed information. It should be noted that it can be difficult to describe some of these processes in words. Diagrams make it easier to understand how the chemistries work, however, it should be noted that the diagrams, with the DNA strands depicted as simple straight lines on the paper, do not give credit to the dynamic nature of single-stranded DNA molecules which in reality have the freedom to twist and turn and gyrate. Even double-stranded DNA, which is inherently more rod-like, can 'breathe' with the strands partially melting and re-annealing and can adopt more than one helical form. Nothing's simple! But it is hoped that this guide to the various reporting permutations available for real-time PCR will be of practical use.

Background to real-time PCR

Prior to the introduction of the first commercial real-time PCR instruments in 1996 the utility of the PCR was limited. It could only be easily applied as a qualitative method. Analysis of amplification products was carried out at the end of thermal cycling using techniques such as gel electrophoresis or PCR ELISA. The dynamic range for quantitative PCR using these approaches was limited because the accumulation of specific product plateaus when the reaction is allowed to progress to completion. At high cycle numbers the amount of product is often unrelated to the initial amount of target nucleic acid. Factors that contribute to the plateau effect include, amongst others, substrate depletion (nucleotides, amplimers, etc.), specific inhibition

(competitive binding of products-to-products rather than amplimer-to-product) non-specific product inhibition (amplimer artefact accumulation and mis-priming), and the accumulation of pyrophosphate. Quantitative analysis, with a limited dynamic range, could only be achieved by stopping thermal cycling before this plateau was achieved. Competitive PCR (utilizing molecular mimics that compete for amplimers) could be used to achieve a wider dynamic range. However, this approach is less accurate and more time consuming than the real-time methods described here.

The fluorescent real-time approach requires a thermal cycler that can interrogate the sample throughout the course of amplification. These instruments are discussed in Chapter 2 and the basic kinetic theory of real-time PCR is in Chapter 2. The signal collected during the pre-plateau stage of amplification is of most use for the determination of the number of initial target nucleic acid molecules. At this stage the reaction efficiency is so high that the number of amplicon molecules effectively doubles every cycle. Amplification is observed as a twofold increase in signal as it rises above that of the background noise. The amount of noise in a real-time assay is a function of the type of chemistry utilized as well as other experimental considerations such as volume effects and the degree of mixing of reaction cocktails. It is also affected by the optical and thermal performance of the instrument.

There are two main classes of real-time fluorescent chemistries: 'generic' and 'strand-specific' methods. Generic methods use probes that bind non-specifically to DNA and include the intercalators and other DNA binding dyes. Strand-specific methods use nucleic acid probes that target the amplicon (product) between the amplimer binding regions. It should be noted that most of the probes described utilize DNA for the probe. The use of modified bases was proposed in 1991 (Holland *et al.*, 1991). Methods for peptide nucleic acid (PNA) probes utilizing FRET have been reported (Ortiz *et al.*, 1998) and described for application in real-time PCR (Svanik *et al.*, 2000), and for those probes that use hybridization alone (rather than hydrolysis) it is possible to utilize PNA and other analogues such as locked nucleic acid (LNA) (Koshkin *et al.*, 1998). These nucleotide analogues can be used to increase the T_m of the probe to improve signalling particularly where the target is AT rich. The strand-specific methods have the advantage that amplification of reaction artefacts such as amplimer-dimers do not contribute to the observed signal and can provide higher specificity and better signal–noise ratios than the generic methods that are described next. The type and mode of action of different probe systems is illustrated in Table 3.1.

Generic detection using DNA binding probes and melting point analysis

DNA binding dyes have been used extensively in molecular biology research for the direct analysis and quantification of nucleic acids. These dyes bind to double-stranded DNA (dsDNA) with enhanced fluorescence. Ethidium bromide, which is used conventionally for staining agarose gels, has been used in real-time PCR (Higuchi *et al.*, 1992, 1993). Others, such as those from Mo-

Table 3.1 Summary of fluorogenic reporting chemistries for real-time PCR. The schematics show the probe structure and mode of action

Method	Principle	Self probing equivolent	Probe Structure	Nucleic Acid Interaction
DNA binding agents	Enhancement of Fluorescence	n/a		
5′ Nuclease Assay	Hydrolysis of Probe	Intrataq		
Dual Hybridization probes	Hybridization of linear probes	Embodiment of Scorpions		
ResonSense	Hybridization of linear probes	Angler		
Hybeacons	Hybridization of linear probes	Not reported		
Eclipse	Hybridization of linear probes with conformational change	Not reported		
Light-up	Hybridization with intercalator	Not reported		
Molecular Beacon	Hybridization and conformational change	Scorpions		
Ying-Yang Probes	Compelative binding of linear probes	Duplex Scorpions		

lecular Probes (YoPro® and YoYo®) (Ogura *et al.*, 1994), have also been reported. The fluorescent signal is usually monitored towards the end of the extension step in a three-step PCR.

Of these probes by far the most reported are the minor-groove binding dyes SYBR®Green-1 (Becker *et al.*, 1996) and SYBR®Gold. These dyes typically exhibit 20- to 100 fold fluorescence enhancement and are commonly used because their emission maxima closely match that of fluorescein and the optics in most commercial real-time instruments are set to detect in this (circa 520 nm) wavelength range. The SYBR® dyes are popular for real-time detection because they are readily available from PCR reagent suppliers and their use requires little additional experimental design.

The optimum concentration for these dyes for a number of instrument platforms is published in the open literature. For SYBR® dyes this is typically 1:30,000 to 1:100,000 dilution of the reference solution supplied by the manufacturer, although the concentration is dependent on the tube format (composite glass/native polypropyl-

ene) and the optical efficiency of the instrument. Ready-to-go cocktails are available from suppliers with the dye already in the reagent mix.

Binding dye chemistries are excellent for assay optimization. The signal is proportional to the total nucleic acid concentration and therefore directly related to the PCR process. In an optimized assay, they may thus be used to determine accurately the reaction efficiency. It should be noted that this is not always the case when using labelled nucleic acid probes, where the signal generated is related to both the assay and the 'probe efficiency'. In some assay systems the probe efficiency can have a major effect on the signal obtained and this is discussed later for each assay type. In both strand-specific and generic RT-PCR, the signal obtained is additionally dependent on the efficiency of the reverse transcriptase step.

Melting point analysis may be carried out on most commercial real-time instruments, usually at the end of the PCR amplification (Fig. 3.1). Temperature-dependent fluorescence measurements are made whilst slowly increasing

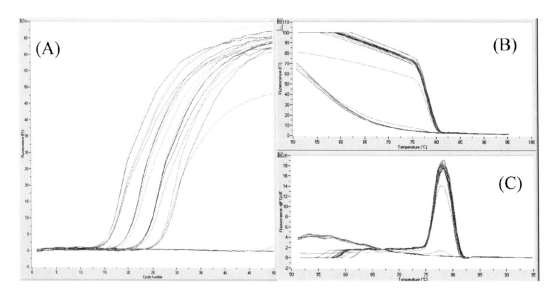

Figure 3.1 Amplification using generic binding dyes (minor-groove binding dye SYBR®Gold) on the LightCycler®. (A) The amplification plot shows 50 cycles (to completion) of amplification of four dilutions (four replicates) of a 10-fold dilution series and four no-template controls, one of which produces a low-amplification signal. The reaction utilized anti-Taq antibody hot start and UNG carry-over protection and the data utilized the background subtraction algorithm. (B) Melting point analysis of the amplified products and (C) the first negative differential of the fluorescence with respect to temperature, plotted against temperature. From this plot it can be observed that the signal in the positive no-template control has the same melting point as the specific amplification in positive samples, and therefore amplification is most likely a result of cross-contamination. A colour version of this figure is located in the plate section at the back of the book.

the temperature of the reaction products from around 50°C to around 95°C. As DNA duplexes melt apart the fluorescence decreases as the bound dye is released. Most instruments provide an analysis of this data by plotting the first negative differential of the fluorescence signal with respect to temperature, against temperature. This plot appears as one or more peaks representing the point(s) at which the maximum rate(s) of change in fluorescence occurs corresponding to a particular dsDNA product. Specific reaction products generally melt at a higher temperature than artefacts such as amplimer dimers and mis-primed products. T_m is a function of GC content and, to a lesser extent, product length, which allows one or the other to be assessed in some reactions.

The melting peak method is analogous to that of agarose gel electrophoresis in that it does not unequivocally determine the presence of the correct sequence. Products with similar molecular masses or melting peaks may not be resolved by either technique. Gel electrophoresis has a higher resolution than that of melting point analysis and the quality of T_m data collected varies greatly between hardware platforms. Those instruments where the temperature is accurately controlled and recorded, and where the correlation between temperature and fluorescence is good, will produce the highest resolution T_m data.

The major limitation with binding dye chemistries is that they bind to total nucleic acid, and in a PCR of sub-optimal efficiency a significant amount of signal will be derived from reaction artefacts. This leads to a dramatic loss in the dynamic range of quantification of samples and standards with low (<1000) initial target copy numbers. However, a melting point analysis allows non-specific artefacts to some degree to be discriminated from specific amplicons (Ririe et al., 1997). For many dyes, the concentration that can be used in reactions is constrained. At higher concentrations they may inhibit amplification. At lower concentrations, resolution can be lost in melting analysis as the dye is titrated out by competition between products. New dyes such as LC-green can be used at higher concentrations and have facilitated high resolution melting to enable, on specialist instruments, the analysis of product polymorphism without the need for probes.

Melting point analysis adds significantly to the utility of fluorescent DNA binding dyes as the data can be used to determine a number of important variables. Amplicon complements can be synthesized and used in melting point experiments to empirically determine the T_m. Knowing the T_m of the amplicon then allows the reaction to be improved in three ways. First, by sequentially lowering the denaturing step to a few degrees above the amplimer T_m, high molecular mass sample DNA is not denatured in later (e.g. >5) cycles and this can be used to reduce the risk of mis-priming events later on. Second, lowering of the denaturing temperature allows for shorter cycle times. (NB: Loss of polymerase activity through enzyme denaturation, which might otherwise affect later cycles, can be beneficially reduced by this cooler, faster cycling). Third, if the T_m of reaction artefacts is significantly lower than the T_m of the amplicon, fluorescence can be monitored at a temperature just below the T_m of the amplicon, but higher than that of artefacts (Morrison et al., 1998). This can reduce interference from artefacts and improve the dynamic range of quantification.

An important consideration when using such dyes is the incorporation of a 'hot-start' to reduce the formation of artefacts and so improve the dynamic range of quantification (Morrison et al., 1998; Lee et al., 1999). It should be noted that most of these DNA-binding reagents are, to varying degrees, both toxic and mutagenic. Storage conditions are critical since they may be photo-bleached by both ambient and artificial light. The use of light-tight containers such as darkened microfuge tubes will reduce this, but most are labile in long-term storage.

When optimized, a SYBR®Green-1 or SYBR®Gold assay can be used for most applications including expression studies. The ability to multiplex these probes is limited because the binding, which generates the fluorescence, is non-specific. However, it is possible to multiplex by melting point providing that the melting peak of each species can be resolved. Allelic variation and mutation detection can be carried out, for example, using the amplification refractory mutation system (ARMS) approach (Newton et al., 1989). However, this requires the assays for the variants to be carried out in separate tubes, which is not ideal. Because of these limitations, assays that require high-confidence results are best addressed using the strand-specific approaches that are discussed next.

Strand-specific detection using labelled nucleic acid probes

The strand-specific methods employ fluorophore-coupled nucleic acids to interact with reaction products, probing accumulating amplicons for the presence of the target sequence. Such PCRs exhibit greater specificity than the respective assay without the probe component. However, the probe is targeted at only a portion of the amplicon sequence and therefore the signal reports the presence of complementary sequences rather than the amplicon *per se*. The basis of detection of the product can be classed by the mode of action of the probe (e.g. hydrolysis or hybridization, etc.).

The 'efficiency' of the probe is important as the signal it produces needs to be closely correlated with the concentration of amplicon in the exponential stage of amplification for the number of initial target copies to be extrapolated. Inefficient probes produce an inaccurate value for the PCR amplification efficiency. The signal generated by a probe is dependent on, for example, sufficient hybridization and/or cleavage as is later described. The target sequence is the main factor affecting probe efficiency because of different sequence dependent factors for a given probe type. Therefore the benefits of selecting a probe type should be carefully considered when designing an experiment and also when purchasing an instrument since the preferred assay system may not easily be applied on every instrument.

The optimum composition of the core chemistry may vary with different platforms and this is discussed in Chapter 2. The optical filter specifications also differ such that they will be optimized for specific fluorophores. The concentration of probes may also vary between platforms. Typically probes are used at a final concentration of around 0.1–1 μM depending on the sensitivity of the real-time fluorimeter.

In this text probe types are grouped for discussion by mode of action. The evolution of real-time methods is not presented chronologically, although reference is made to how and when they were implemented. Most of these methods use

the phenomenon of FRET. This phenomenon will be discussed first and then the application and experimental considerations of the various methods will be outlined.

Fluorescent resonance energy transfer probes and the passive reference

FRET, or Förster energy transfer, involves the non-radiative (no photon emitted) transfer of energy between fluorophores and has been used for a number of biological assays. There are excellent reviews of principles and applications (Cardullo, *et al.*, 1988; Selvin, 1995; Clegg, 1992), so there is no need to include an in depth description of the process in this text. Here we present an overview and the two main approaches to detection and multiplexing using FRET as used in real-time PCR experiments. Reference will be made to specific instruments to illustrate the process.

For FRET to occur there must be at least two fluorophores in close proximity. One fluorophore, called the 'donor,' is excited by an incident light source and (in isolation) emits light at a longer wavelength that falls within the excitation spectrum of the second fluorophore, called the 'acceptor.' When the donor and acceptor are on the same or neighbouring molecule then the emission energy of the excited donor may be transferred through a number of distance and orientation dependent effects including dipole-dipole interactions between the donor and acceptor. Variations in the intensity of emissions of fluorophores on nucleic acids resulting from spatial changes during the course of amplification provide the basis for detecting amplicon accumulation (Fig. 3.2). The choice of dye pairs is important for achieving a good signal–noise ratio. The Förster distance (the distance between the dye pair at which the efficiency of energy transfer is 50%) varies for different dye combinations. Förster coupling is highly distance dependent being proportion to $1/R^6$, where R is the separation between the fluorophores.

Analysis utilizing FRET involves monitoring either a rise or drop in fluorescence of either fluorophore depending on the assay type and instrument. On some platforms it is possible to present these data as a change in the quotient of the emissions of the two dyes. For example, if the

(A)

(B)

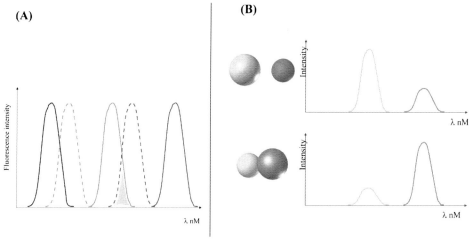

Figure 3.2 Fluorescence energy resonance transfer. (A) The emissions of a light source (blue line) falls within the excitation spectrum (green line dashed) of a fluorophore and cause this dye to fluoresce. The emissions (green solid) of this fluorophore fall within the excitation spectrum (red dashed line) of a second fluorophore that emits fluorescence at longer wavelengths (red solid). When the second fluorophore is close to the first on the same or neighbouring molecule, the energy that would be emitted by the first fluorophore ((B) top) can be transferred to the second through a number of energy transfer mechanisms such that the second fluorophore emits this energy at its own emission wavelengths ((B) bottom). A colour version of this figure is located in the plate section at the back of the book.

basis of detection is the donor fluorophore being liberated spatially from the acceptor, one can express the data as donor fluorescence/acceptor fluorescence. Where the basis of detection is the donor being brought close to the acceptor, one can express the data as acceptor fluorescence/donor fluorescence. This simple type of analysis is useful in a non-multiplexed real-time PCR since subtle changes in sample volume due to pipetting errors and/or changes in sample volume due to evaporation can be normalized out.

Of importance to the signal–noise ratio of assay systems utilizing FRET is the actual ratio of dyes in a FRET pair. In particular this is important where two dyes may be used on the same probe (dual labelled) as described later. These probes are more complicated in terms of synthesis since they require efficient incorporation of both dyes into the probe molecule. Their manufacture is more expensive as a result. They are usually made by one of two methods. The first dye, if this is at the 3'-end of the probe, is part of the support matrix on which the oligonucleotide is synthesized by sequential addition of each base. The second dye may either be directly incorporated using a fluorophore labelled monomer, or a monomer with a reactive group may be used for subsequent covalent cross-linking of the

dye after synthesis. In the latter case any free dye needs to be removed by a separation step and this is usually carried out using high-performance liquid chromatography (HPLC). In cases where neither dye is at the 3'-end of the probe (e.g. Scorpions), both dyes are incorporated as fluorophore labelled monomers. In dual labelled probe assays the donor is liberated or spatially separated from the acceptor. If the stoichiometry of donor and acceptor fluorophore incorporation is not 1:1, a low signal and/or a high background may be obtained. The quality of probes can vary greatly between suppliers and even between scales of synthesis. Quality control of such probes is important for avoiding time wasting in efforts to optimize assays where the fundamental problem is poor dual labelling of the probes. The authors recommend that probes should be checked following manufacture by any one or more of a number of techniques to ensure the integrity of each probe. These include mass spectrometry, and HPLC with analysis of the fractions at the relevant wavelengths. Degradation and subsequent analysis by spectrofluorimetry is the ideal method of determining dual labelled dye quality. Some suppliers offer one or more methods as a service or as part of their own quality control procedures.

Another experimental consideration important in improving the signal–noise ratio of assays is sequence dependent quenching of fluorescence. In particular, a fluorescein-based dye may be partially quenched by a neighbouring G in the sequence, which should be therefore avoided in the design of the probe. Whilst both dyes in a FRET pair may be fluorescent it is becoming increasingly common to find the use of non-fluorescent acceptors (quenchers). These are often termed 'dark quenchers' and provide collisional, or 'Dexter', quenching. No light is emitted from these molecules, rather the energy is dissipated to solvent. Their use is becoming common because a non-fluorescent moiety on a dual labelled probe frees the instrument's spectral bandwidth, increasing both the utility of instrument for multiplexing, and the dynamic range of a given detector where fluorescent emissions may have otherwise overlapped. Such quenching moieties are now commonly employed on dual labelled probes. One limitation of this approach we have observed is that it is difficult for the end user to determine the efficiency of the incorporation of these quenchers into dual labelled probes. In practice, this may lead to a higher noise for the

reasons described above (Zhang *et al.*, 2001). The use of gold nanoparticles as efficient quenching moieties has also been reported (Dubertret *et al.*, 2001); however, this has yet to be demonstrated in a homogeneous PCR.

Since most instruments can excite and detect the emissions of fluorescein, a combination of fluorescein and any dye that can act as an acceptor may be used to report amplification. On most instruments and for various assay types the rise or drop in signals from either fluorescein or the acceptor would be suitable to report amplification. However, when a multiplexed assay is required, the instrument specification narrows the choice of dyes and assay types that can be used. Two principal approaches are used by different manufactures to allow multiplexing. Here we describe these as the universal donor method, and the universal acceptor method (Fig. 3.3):

•   *Universal donor.* In this the system is set up to multiplex assays by monitoring the increase in acceptor emissions. Different acceptors are used to report amplification from two or more amplifications in the same tube. Typically the machine has one excita-

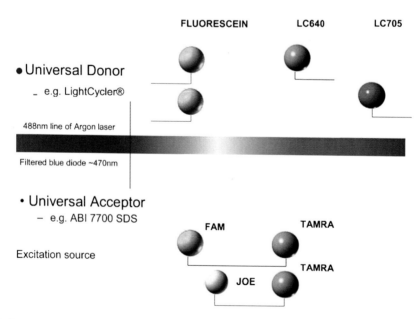

Figure 3.3 Schematic showing the two principal methods for multiplexing using FRET. Top, the Universal Donor on the LightCycler®, and bottom, the Universal Acceptor on the ABI 7700. A colour version of this figure is located in the plate section at the back of the book.

tion source with detection at separate longer wavelengths that are used when multiplexing e.g. the LightCycler®.

- *Universal acceptor.* In this arrangement the system is set up to multiplex by monitoring the increase in donor emissions that are quenched by a longer wavelength or non-fluorescent acceptor. Different donors are used to report amplification from two or more amplifications in the same tube. Typically the systems detect donors that have short wavelength emission and are all excited by the one excitation source e.g. Applied Biosystems' 7700/7000 machines. Alternatively dyes that are spectrally well separated with longer wavelengths may be used on some instruments that have the ability to both excite at more than one wavelength and measure emission at more than one wavelength e.g. BioRad iCycler® and Cepheid SmartCycler®.

In theory, most fluorimetric instruments can be used to detect either universal donor or universal donor dye combinations, but the software on most platforms is not flexible enough to allow practical application and analysis using both approaches without first exporting the data. The universal acceptor arrangement is the most common because of its effectiveness with quenched dual labelled probes (5′ nuclease probes, molecular beacons etc) that are widely used. However, instruments such as the Corbett RotorGene® are flexible enough in optical design and analysis to allow the use of most, if not all fluorescent methods. It should be noted that generally the signal obtained when measuring acceptor fluorescence in universal donor methods, is of a lower value (because of energy transfer inefficiencies) and may be inherently noisier than the universal acceptor methods, where donor fluorescence is measured. That is not to say that a comparable signal–noise cannot be achieved with universal donor methods. In particular this is possible where the probe and FRET efficiencies are good and the dyes are spectrally well separated, such that the donor moieties do not significantly contribute to the high background when detecting the acceptor emissions, as can be achieved with the LightCycler® dye combinations of fluorescein with LC640 and LC705.

One way to allow the universal acceptor method to be used on a system optimized for the universal donor method is the 'big dye' approach to transferring the energy from the excitation source to the reporting dye (Solinas *et al.*, 2001). This was applied previously in DNA sequencing chemistries and instruments (Ju *et al.*, 1995). For example, on the LightCycler® the dyes for the longer wavelength detectors are only minimally excited by the blue LED. A fluorescein molecule that is chemically coupled to such a dye may now be detected in the longer wavelength detectors since the energy will be transferred by the fluorescein, in turn this energy may be transferred to a longer wavelength or dark acceptor dye. Although less efficient the method does work and these probes are available from a few suppliers. However, the trend with new instruments is to move away from supporting Universal Donor chemistries. For example, the LightCycler® 480 instrument has multiple excitation wavelengths to facilitate multiplex Universal Acceptor methods. This makes dye deconvolution algorithms unnecessary for some applications. Table 3.2 provides a summary of commonly used dye pairs.

A passive reference is used in some systems. This dye is detected by the instrument and its signal is used principally to normalize for optical variations in the instrument (such as edge effects). It also normalizes for differences in volume that may occur between wells and non-uniformity in the optical properties of tubes and wells. The differences may be due either to volumetric errors in pipetting during preparation and/or to changes that occur during the amplification from evaporation which causes a drift in the background signal. The reference dye takes no part in the reaction itself. On Applied Biosystems and similar instruments ROX (carboxy-X-rhodamine) dye is commonly used for this purpose. On instruments where there is little surface area for evaporation to occur, such as capillary systems like the Idaho and Roche LightCycler®, and in centrifugal systems where the volume and concentration is maintained in the reaction well, such as the Corbett RotorGene®, a passive reference is less important.

A prerequisite for multiplexing using fluorescence is discriminating the emission signals from different fluorescent dyes at different wave-

Table 3.2 Some of the common fluorophore combinations that can be used effectively in energy transfer pairs

Fluorophore	Excitation Maxima (nm)	Emmission Maxima (nm)	As a DONOR when used with	As an Acceptor when used with
Cascade Blue	396	410	DABCYL	
Cy3	552	570	DABCYL, BHQ-2, ElleQuencher, Eclipse	Fluorescein, SYBR® Gold, SYBR® Green I
Cy3.5	581	596	DABCYL, ElleQuencher	Fluorescein, SYBR® Gold, SYBR® Green I
Cy5	643	667	DABCYL, BHQ-2, ElleQuencher, Eclipse	Fluorescein, SYBR® Gold, SYBR® Green I
Cy5.5	675	694	DABCYL, BHQ-2, ElleQuencher	Fluorescein, SYBR® Gold, SYBR® Green I
EDANS(5-((2-aminoethyl)amino) naphthalene-1- sulfonic acid)	335	493	DABCYL	n/a
FAM (carboxyfluorescein)	494	518	TAMRA, ROX, Cy3, Cy3.5, Cy5, Cy5.5, DABCYL, BHQ-1/2, Eclipse	n/a
Fluorescein	492	520	TAMRA, ROX, Cy3, Cy3.5, Cy5, Cy5.5, DABCYL, BHQ-1/2, Eclipse	
HEX (carboxy-2',4,4',5',7,7'- hexachlorofluorescein)	535	556	TAMRA, DABCL, BHQ-1/2, Eclipse	n/a
JOE (carboxy-4', 5'-dichloro-2',7'- dimethoxyfluorescein	520	548	TAMRA, DABCL, BHQ-1, ElleQuencher	n/a
LC640	625	640	n/a	Fluorescein
LC705	685	705	n/a	Fluorescein
ROX (carboxy-X-rhodamine)	585	605	DABCYL, BHQ-2, ElleQuencher	Fluorescein
SYBR® Gold	495	537	Cy5, Cy5.5, LC640, LC705	n/a
SYBR® Green I	494	521	Cy5, Cy5.5, LC640, LC705	n/a
TAMRA (carboxytetramethylrhodamine)	565	605	DABCYL, BHQ-2, ElleQuencher	Fluorescein
TET (carboxy-2',4,7,7'- tetrachlorofluorescein)	521	544	TAMRA, DABCL, BHQ-1/2, Eclipse	n/a
Texas Red	583	603	BHQ-2	n/a
VIC	538	554	TAMRA	n/a
Yakima Yellow	526	549	Eclipse	n/a

Note: DABCYL(((4-(dimethylamino)phenyl)azo)benzoic acid), Black Hole Quencher 1 & 2 (BHQ), ElleQuencher™ and Eclipse™ are non-fluorescent "dark" quenching moieties.

lengths. Approaches to enable this discrimination with examples include:

1 a spectrofluorimeter that can measure a suitable range of the visible spectrum with a calibration algorithm and software to deconvolute the different dye components (Applied Biosystems 7700)
2 a fluorimeter with discrete optical detectors using a calibration algorithm and software to deconvolute the different dye components (Applied Biosystems 7000 and Roche LightCycler®)
3 a fluorimeter with discrete optical detectors and a 'factory set' calibration to deconvolute the different dye components (Cepheid SmartCycler®)
4 a fluorimeter with discrete optical detectors with sufficient bandwidth between dyes and narrow cut-off filters to minimize cross-talk (Corbett RotorGene®).

All approaches are equally applicable but we would recommend for high-confidence applications the use of an algorithm to minimize the risk of a false positive due to erroneous data interpretation. A good algorithm is one that completely compensates for the contribution from spectral overlap of other dyes at the wavelength of the dye being monitored. This is usually achieved using a calibration with pure dyes. A number of factors can affect the requirement for re-calibration. These include the drift in detector sensitivities and ageing of the excitation sources. Effects of specific sequences on dye emissions may also necessitate that a re-calibration be carried out to account for any shift in emission wavelength.

The 5′ nuclease assay

The principle of a DNA synthesis dependent 5′–3′ exonuclease process using the enzyme *Taq* polymerase was first described in 1989 (Gelfand, 1989). The application of this process for the analysis of a PCR amplification using isotopic labelling and a separation analysis was later reported (Holland *et al.*, 1991). A fluorogenic approach for monitoring the hydrolysis of HIV-1 peptides in real-time (Matayoshi *et al.*, 1990) had been described and this principle was subsequently applied to the 5′ nuclease assay. The first

report describing a fluorogenic 5′ nuclease PCR assay was by Lee *et al.* in 1993.

The 5′ nuclease assay is now by far the most commonly used strand-specific probe method, in part because it was the principal detection chemistry for the first commercial real-time PCR instrument, Applied Biosystems 7700 launched in 1996 (Bassam *et al.*, 1996). Applied Biosystems market the assay for research applications under the registered TaqMan® trademark of Roche Molecular Systems Inc.

The probe is dual labelled. Typically, one dye of a FRET pair, usually the donor, is located at the 5′ end of the probe. The second, usually the acceptor, is labelled at or near the 3′ end. However, the reverse is acceptable and more efficient quenching may be obtained by locating the acceptor closer to the donor. When dyes are coupled in close proximity the signal–noise ratio of the assay improves compared to terminally labelled probes. The optimum distance of the two dyes on the probe is dependent upon the properties of the dyes used in the FRET pair. Typically a spacing of six nucleotides is optimum. If the dyes are too close then there may be inefficient transfer of energy (Heller *et al.*, 1987) and they risk ending up on the same cleavage fragment. The probe is protected with a phosphate group or another blocker at the 3′ end to prevent it being extended as an amplimer by the *Taq* polymerase.

The 5′ nuclease assay probe is designed to be cleaved by the exonuclease activity of *Taq* polymerase (Holland *et al.*, 1991) which occurs whilst the enzyme is extending the upstream amplimer in a 5′ to 3′ direction. Probe hybridization is necessary, but not sufficient, for this to occur. The products of hydrolysis have been shown to include 1–2 base products and it is proposed that that the polymerase displaces the first one or two base pairs it encounters before cleavage begins (Holland *et al.*, 1991). The 5′ nuclease assay is inherently different to other strand-specific probe methods because signal generation requires the probe to be destroyed. It is therefore not possible to carry out melting point analyses either during or post-amplification.

The first commercially available 5′ nuclease probes used FAM (carboxyfluorescein), JOE (carboxy-4′,5′-dichloro-2′,7′ dimethoxyfluorescein), TET (carboxy-

2′,4,7,7′-tetrachlorofluorescein), and HEX (carboxy-2′,4,4′,5′,7,7′-hexachlorofluorescein) as the donors, and TAMRA (carboxytetramethylrhodamine) as the universal acceptor since Applied Biosystems 7700 used dyes that had previously been optimized for their sequencing machines. The ROX sequencing dye was used as a passive reference. In practice a large number of dye combinations is possible and a multiplex of up to six reporter dyes in one reaction has been used for SNP scoring (Lee et al., 1999).

A number of software tools can be used to design 5′ nuclease probes. Similar instruments that followed the 7700 were developed with optics that supported dual-labelled probe systems using the a universal acceptor arrangement, and a large number of oligonucleotide synthesis companies supply dual labelled probes with the correct dyes to support the 7700 and later instruments. Applied Biosystems, who market TaqMan® probes, recommend that the design of such probes should adhere to the following recommendations for optimum signal generation: the GC content should be in the 30–80% range; the strand with the greater number of Cs rather than Gs should be selected; runs of identical nucleotides should be avoided; there should be no Gs at the 5′ end (which would otherwise partially quench the fluorophore); the probe should have a melting temperature at least 5°C greater than the amplimers′; and finally, the probe sequence should not overlap with the amplimer regions (Gelmini et al., 1997).

The spacing of the dyes may be optimized with respect to Förster distance. Following cleavage, separation of donor from acceptor is maximized, but some spatial effects may also be observed on binding of the complementary strand. Providing the experimental parameters are optimized, the 5′ nuclease system produces a signal to noise ratio that is usually much better than that achievable with other assay methods (Fig. 3.4). It is the only probe system where the dyes are covalently linked before separation by molecular cleavage. The method is very versatile and can be used on a large number of platforms, but there are limitations and these are outlined next.

Applicability can be limited by the target and probe sequence requirements for efficient hydrolysis. PCR amplification may be shown to be highly efficient, but the signal from the hydrolysis probe may be low. Therefore the key design parameter for this type of assay is to choose a probe sequence that will be hydrolysed efficiently and, in general, this is achieved by designing the primers around the probe rather than the converse. In particular, the use of the 5′ nuclease assay for AT rich DNA can be problematic. To obtain a T_m for efficient cleavage of the probe it should be about 5–10°C higher than that of the amplimers. For some sequences this would require the synthesis of very long probes that could only be produced in low yields. The development of the minor groove binding (MGB) probe technology that increases the T_m of such probes (Kutyavin et al., 2000; Salmon et al., 2002) has reduced this problem somewhat, but adds to the complexity and cost of the probe. Similarly the use of locked nucleic acids (LNAs) may be used to improve the probe T_m (Letertre et al., 2003). Difficulties associated with the synthesis of dual label probes and the signal to noise ratios obtained were discussed earlier. We have experienced problems with almost all suppliers, but generally less often with small-scale syntheses from reputable companies. The use of an additional complementary probe labelled with a quenching moiety may be useful in improving the signal to noise ratios of problematic probes (Nurmi et al., 2002).

With many thermal cyclers, heating and cooling may take up a large proportion of the total assay time. Instruments with faster temperature transitions such as the Roche/Idaho LightCyler®, Corbett RotorGene®, Cepheid SmartCycler® and electrically conducting polymer (ECP) cyclers enable overall PCR assay times to be significantly reduced. However, whilst PCR amplification has been shown to proceed at considerable speed on fast systems, signal to noise ratios from the 5′ nuclease process may be impaired without relatively long hold times at the annealing step and the 5′ nuclease assay has been shown to need more time to generate a signal than the PCR per se (Whitcombe et al., 1999a; Lee et al., 2002), as illustrated in Fig. 3.4.

The 5′ nuclease assay has been reported for a large number of applications and is supported by the majority of synthesis companies and suppliers of instruments, such that, for many new users,

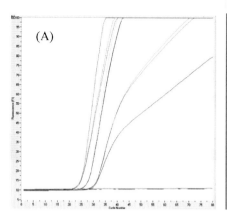
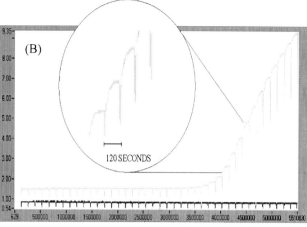

Figure 3.4 Amplification using the 5′ nuclease assay on the LightCycler®. (A) The amplification plot shows 80 cycles (in excess of completion) of thermal cycling of three dilutions (three replicates) of a 10-fold dilution series and three no-template controls on the Roche LightCycler®. The reaction utilized anti-Taq antibody hot start and UNG carry-over protection. The reaction used the primers and probes from the Applied Biosystems Human β-Actin kit which utilizes a FAM-TAMRA internally quenched TaqMan® probe. The amplification plot shows raw data (no analysis/background subtraction) and illustrates the excellent signal: noise ratio achievable with this chemistry. (B) The same reaction amplified on the Idaho LightCycler® carried out by continuous monitoring of a positive and negative sample throughout all stages of amplification. The data shows that ~120 seconds are required each cycle to take the hydrolysis of bound probe to completion (plateau) each cycle. A colour version of this figure is located in the plate section at the back of the book.

the trademark TaqMan® is used as a synonym for real-time PCR. These applications include expression studies, detection of pathogens, allelic discrimination, and genomic applications to name but a few.

Probe library

A recent innovative development in probe technology is 'Probe Library' (Mouritzen et al., 2005). This is a commercial off-the-shelf array of probes. The probes are not specific; rather each probe is pre-designed to bind to a 10- to 15-mer sequence that is reproduced frequently throughout a specific genome. The PCR assay design is carried out by the end-user with a web-based design package. A sequence is entered and the software then designs primers around the probes which exist in the library. There are probe libraries for different genomes; if a genome is not catered for then the closest match may provide a solution.

The probes themselves are dual-labelled probes. The length of each probe is kept deliberately short so that the sequence occurs with a high enough frequency to be useful. The T_m

of the probe is raised, so it can hybridize to the amplicon, by using LNA. Signalling comes from 5–3′ hydrolysis of the probe concomitant with primer extension. The biochemical process is the same as the 5′ nuclease assay, but the probe is not designed for a particular amplicon. The specificity of the reaction has to be derived from the amplimers. When the assay is optimized the end-user is able to carry out a large-scale synthesis of the sequence or purchase it from the supplier. This should allow the assay to be applied on a wide variety of instrument platforms. Additional benefits will be realized from the kit-scale synthesis and the associated quality control that is required for such a product. Problems with dye stoichiometry observed in research scale synthesis of dual labelled probes could be minimized.

Probe library is useful for rapid development of real-time PCR assays. Market literature suggests an assay can be optimized in a few days. The main application will probably be in gene expression studies. A useful feature for this application is that the probes are designed to have similar efficiencies in terms of 5′ nuclease activity. This significantly facilitates the optimization process and eliminates some aspects of experimental error. The authors would not consider this probe system suitable for high-confidence applications such as diagnostics since high nucleic acid concentrations in some samples could lead to a false-positive signal. It may be possible to carry out allele-specific multiplex assays using two or more probes over a polymorphism, but this has yet to be reported.

Direct hybridization methods

Direct hybridization methods use one or more probes that hybridize to the product strands during low temperature stages of each PCR cycle and are released from the product strand during high temperature stages. Unlike the 5′ nuclease assay, the signal is not cumulative, but does increase each cycle as the level of probe binding rises with product accumulation. Since the signal is derived from a hybridization event the signal is proportional to the product concentration, but is also a function of probe binding efficiency. The signal from hybridization assays may be based on direct binding to product or on the competitive formation of different hybrids. Generally, where the hybridization event is simple, the effect of probe binding inefficiency will decrease. Here we describe the different types of hybridization probe in detail.

Linear hybridization probes

Several probe systems using the direct hybridization of linear oligonucleotides fall into this class. These oligonucleotides are designed to have a minimal secondary structure when not bound to target.

Dual hybridization probes were developed to provide a strand-specific chemistry for PCR applications and are the most often reported probes of this class. The term hybridization probe is often used as a synonym for this probe method. Two oligonucleotide probes, each labelled with one of the dyes of a FRET pair, are used. Typically one dye is located at the 3′ end of one of the oligonucleotides and the other dye is located at the 5′ end of the other oligonucleotide, the latter being protected to stop primer extension by incorporation of a phosphate or other blocking group. Both probes are designed to bind to the

non-amplimer region of one of the amplicon strands and to hybridize adjacent to one another such that the dyes are in a *cis* configuration with respect to the nucleic acid strand being probed. As product accumulates both probes bind to the strand and the basis of detection is to monitor the drop in donor fluorescence or the rise in acceptor fluorescence or, in non-multiplexed reactions, to monitor the quotient of acceptor/donor fluorescence. Typically fluorescence is measured at the end of the annealing step and 10–30 seconds is required to generate sufficient signal.

Dual hybridization probes were first described for use in PCR on the Idaho LightCycler® (Cardullo *et al.*, 1988; Wittwer *et al.*, 1997). The two channel optical systems of these machines used fluorescein as the donor moiety and the Amersham dye Cy5 as the acceptor. The later three-channel Roche LightCyclers® use three dyes, two of which are optimized for the LightCycler®, LC640 and LC705, although it is possible to multiplex using alternative dyes with emission wavelengths that correspond to the discreet optical detectors of the LightCycler®. The assay can be performed on other platforms such as the Corbett RotorGene® in either a universal donor arrangement and a universal acceptor arrangement by using a number of dye combinations and a non-fluorescent quenching moiety.

For applications requiring accurate quantification of initial target copy numbers efficient signal generation is required during the annealing step. For this to occur both probes should be designed to bind at approximately the same temperature which should be above that at which the amplimers bind. The probes should also melt off at a temperature commensurate with efficient extension by the *Taq* polymerase. This typically requires the probes to have a melting temperature of 65–75°C. If the probe T_m is too low, insufficient signal will be generated each cycle. If the probe T_m is too high, the probes may clamp the PCR and reduce amplification efficiency. The reporting efficiency of the probes can be adversely affected by probe-probe interactions if there is complementarity in sequence between them. This is important in multiplexed reactions where the concentration and number of probes is increased. Ideally the probes should be designed to bind to one amplicon strand, close to, but not overlapping the reverse amplimer region. The T_m of probes can be calculated using software tools such as the applet found on the Idaho Technology website, or using other commercial packages such as that supplied by Roche for use on the LightCycler®. If the design of probes is carried out as described then the probe efficiency will be high enough to provide accurate quantification of initial target molecules. A commonly observed effect in this and other hybridization assays is a decrease in probe signal in later cycles. This 'hook' effect is a result of increased specific product accumulation and competitive displacement of probes as the product strands anneal upon cooling (Fig. 3.5). This effect does not reduce the ability to accurately quantify initial target numbers.

A less widely reported variation of the dual hybridization probe is one that uses two probes with the dyes in a *trans* configuration with respect to the probed strand (Bernard *et al.*, 1998). In this embodiment one 'free' probe is used in combination with a fluorescent labelled amplimer. The energy from the dye on the amplimer may be transferred to the specific probe during the low temperature steps and the probe system may be similarly applied.

One of the useful features of hybridization probes generally is the ability to carry out a multiplex qualitative analysis through melting point analysis. For example, the same fluorescent dye on two different probes may be used to detect two amplification species in the same reaction providing their respective melting motifs can be resolved by the instrument and software. If the software can (a) determine more than one peak and (b) calculate the relative areas under these peaks using integral analysis, then a relative

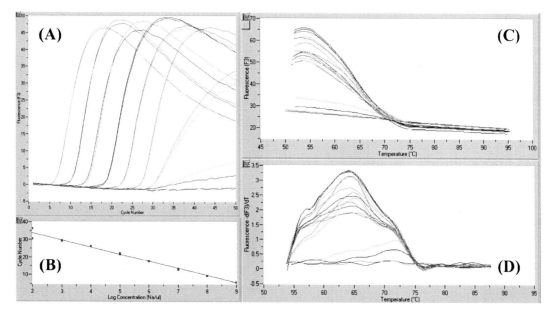

Figure 3.5 Amplification using dual hybridisation probes on the LightCycler®. (A) The amplification plot shows 50 cycles of thermal cycling of replicates of eight dilutions of a 10-fold dilution series of a reference for the target, and no-template controls on the Roche LightCycler®. The amplification shows the data analysed using the background subtraction algorithm. The reaction utilised anti-*Taq* antibody hot-start and UNG carry-over protection. The reaction used the primers and probes for the Human β-globin gene using a 3′ labelled fluorescein donor probe and a 5′ Cy5.5 acceptor probe. The acceptor probe was blocked to prevent 3′ extension using an octendiol moiety. A 5′–3′ exonuclease minus *Taq* polymerase was used to illustrate that the "hook" effect that is observed in linear hybridisation probes as a result of probe displacement by product in later cycles. (B) The derived standard curve from the amplification plot. (C) The melt curve showing probe disassociation from product. (D) The first negative differential of fluorescence with respect to temperature showing the peak at which the maximum rate of probe dissociation occurs. A colour version of this figure is located in the plate section at the back of the book.

quantitative value for each species can be determined within a single reaction. Detection of internal controls using this approach is discussed in Chapter 7.

A probe pair can be used to detect strand variation by designing one of the probes to bind over a site where strand variation may occur. For example, allelic variation is readily detected using this approach. In the case of dual hybridization probes one probe is designed to bind to a conserved (non-variant) sequence adjacent to the sequence variation. This first probe is known as the 'anchor' probe. It is designed to hybridize to the target over a temperature range at which the second 'mutation' probe's T_m will change depending on the specific sequence. The mutation probe can be observed to melt at different temperatures depending on which variant is present. The genotype of a given sample may be determined using this approach. A single melting peak would indicate a homozygous genotype whilst two melting peaks would be obtained from a heterozygous genotype. This is a useful method because a single probe can detect two variants, and this may be multiplexed by fluorescent emission to increase throughput if needed. The approach can also be used to study the allele frequency in a mixed pool for a given population. Using the integral analysis of melting peaks allows a relative value for genetic variations such as single nucleotide polymorphisms (SNPs) to be determined for a given population.

Yin-Yang probes are dual linear probes that work on the basis of competitive hybridization between each other and the PCR product (Li et al., 2002). This principle was first demonstrated for the end-point detection of PCR products in 1989 (Morrison et al., 1989). Two labelled oligonucleotides are used. A fluorophore on the donor/reporter probe is quenched by an acceptor on a second shorter complementary strand. Upon the accumulation of PCR product the acceptor probe is displaced by hybridization of the reporter to the longer complementary sequence on the target amplicon, giving an unquenched signal.

A limitation of these approaches is that they require two oligonucleotide probes and the application of dual hybridization probes may be difficult for applications where there is little or no conserved region for one, let alone two. The empirical optimization of two probes may also take longer than the time taken for just one. With these points in mind, the single linear hybridization probe approaches will now be covered of which ResonSense®, Hybeacons® and Light-Up® probes are examples that have been described for a limited number of applications. The experimental considerations for the design and implementation of these probes is similar to those for dual hybridization probes because of their similar mode of operation.

ResonSense® uses a single linear probe and a DNA binding dye as a FRET pair. Signal is generated when the probe binds to the accumulating product and energy transfer occurs between the dye on the probe and the DNA binding agent in the DNA duplex that is formed (Cardullo et al., 1988; Lee et al., 1999; Taylor et al., 2001). The use of ResonSense® in real-time quantitative PCR has been reported, (Lee et al., 2002). The ResonSense® probe may be labelled with the fluorophore at either end or internally. A 3' probe label will prevent primer extension. If the label is at the 5' end or is internal then an additional 3' blocker molecule will be required. The probe system may be used in either a universal donor arrangement (Lee et al., 2002), or in a universal acceptor arrangement using a non-fluorescent quenching moiety. The broad emission spectrum of the minor groove binding dye SYBR® Gold allows a variety of acceptor dyes to be utilized. As an example of the universal donor arrangement, SYBR® Gold may be used to transfer energy to probes labelled with Cy3, Cy5 or Cy5.5. SYBR® Gold and related dyes have two excitation peaks, one of which is in the UV. The use of a UV excitation source would allow for efficient excitation of the SYBR® Gold whilst minimizing the excitation of much longer wavelength dyes as described. Alternatively, the drop in emissions from probes labelled with donor dyes such as fluorescein can be monitored by using longer wavelength DNA binding acceptor dyes such as SYTO 63. This universal acceptor embodiment provides a method of applying ResonSense® on instruments other than the LightCycler®.

Hybeacons® are linear probes that use a fluorescence moiety attached to an internal nucleotide that exhibits enhanced fluorescence on hybridization to specific target (French *et al.*, 2001). The probe is 3′ protected to prevent primer extension during the course of the PCR. The increase in signal allows for direct quantification of initial target copy numbers and qualitative analysis of closely related sequences using melting point methodology. The uncomplicated structure of the probe and the simple mechanism of signal generation make the approach ideal for multiplexing in strand variation analysis.

Light-Up® probes (Svanik *et al.*, 2000a,b; Isacsson *et al.*, 2000) are peptide nucleic acid (PNA) probes that use a DNA binding agent to report the hybridization of the PNA to a specific DNA sequence. The application of Light-Up® probes for real-time PCR has been described (Wolffs *et al.*, 2001). In this method the probe binds to the target and brings the binding agent spatially close to the probe allowing it to bind with enhanced fluorescence. The dye reported in Light-Up® probes is Thiazole Orange (Lee *et al.*, 1986). PNA has been employed previously in some PCRs as a reagent to clamp amplification (Kyger *et al.*, 1998). The PNA in this application has been shown not to clamp the PCR because the probe target sequence is located away from the amplimer regions.

The molecular beacon and other conformational probes

The use of a conformational change in fluorescent labelled probes to detect nucleic acid hybridization was first reported in 1987 (Heller *et al.*, 1987), and was later proposed for a homogeneous assay, in the context of nucleic acid amplification, in 1995 (Livak *et al.*, 1995; Young, 1996). Molecular beacons were first reported in 1995 (Tyagi *et al.*, 1996, 1998) and have been described in a number of applications including direct 16S RNA detection (Schofield *et al.*, 1997), RNA detection in nucleic acid sequence-based amplification (NASBA) (Leone *et al.*, 1998), and recently in the detection of DNA-binding proteins (Heyduk *et al.*, 2002).

The process of detection is different. The probe is not consumed during the course of the reaction. The molecular beacon's 5′ and 3′ ends are designed to be complementary using a sequence not present in the target so the two fluorescent moieties are brought together in a 'hairpin'

structure for to FRET to occur. Amplification is detected by the sequence-dependent separation of the hairpin as the 'loop' sequence in between binds to its target sequence on the amplicon. The probes may be monitored for certain applications using melting point methodology (Robinson *et al.*, 2000). Poor signalling from molecular beacon probes can be a problem since, as an intramolecular process, the stem–loop folding may be kinetically favoured over probe/target binding which is an intermolecular process (Tsourkas *et al.*, 2003). Nucleic acid analogues have been used to improve template probe hybridization (Ortiz *et al.*, 1998).

An alternative embodiment uses a single fluorophore label in the loop rather than two labels in the stem structure. Hybridization can be followed because the fluorophore, which fluoresces in the ssDNA loop, is quenched when the loop hybridizes to the target (Hwang *et al.*, 2004).

The chemical structure of a molecular beacon may be identical to that of a probe used in the 5′ nuclease assay in that it may be labelled at the 5′ and 3′ ends with a donor and an acceptor moiety respectively, and may also be protected from primer extension at the 3′ end. The difference in mode of operation of a hydrolysis probe and a probe such as a molecular beacon merits some discussion. Although the probe is chemically similar in structure, it is the process of detection that is different. However, when the process is carried out using similar reaction conditions, the signal in either assay will contain both a component derived from the de-conformational change on hybridization and a component derived from any strand-specific hydrolysis by the 5′–3′ exonuclease activity of the enzyme. The relative contribution from either component will depend on the assay design and the enzyme used as some enzymes have different activities. For example the Stoffel fragment of *Taq* polymerase lacks 5′–3′ exonuclease activity. The advantages and application of a molecular beacon to use both processes has been reported (Kong *et al.*, 2002) and the utility of either process for SNP scoring have been discussed (Täpp *et al.*, 2000).

A universal acceptor arrangement is generally used for molecular beacons as it allows the assay to be carried out on all instruments that can analyse the 5′ nuclease format. The original work used EDANS (5-((2-aminoethyl)amino) naphthalene-1-sulphonic acid) as the donor. This is excited in the UV and emits in the blue region of the spectrum. The dark quencher DABCYL ((((4-(dimethylamino)phenyl)azo)benzoic acid) was used as a non-fluorescent acceptor. The use of a dark quencher for molecular beacons is still commonplace; however, the donor moiety is more commonly one that can be used on current instruments, although any FRET dye pair will work and some have been reported (Zhang et al., 2001). The use of different fluorophores for quantitative multiplexing is possible (Vet et al., 1999). A review of the use of such structures for a number of applications is give in Broude et al. (2002).

One application of the hairpin structure has been the development of a generic detection chemistry. This has been available in the past as Sunrise® and is now available as Amplifluor®. A molecular beacon structure (fluorophore plus dark quencher) is synthesized as a 5′ tail on one of the amplimers (Nazarenko et al., 1997). During PCR the Taq polymerase generates the complementary sequence to the probe on the 3′ end of the other amplicon strand. As the two amplicon strands hybridize the hairpin structure is opened out, the donor and acceptor are separated and an increase in fluorescence is observed. A generic Amplifluor® probe called a UniPrimer® has been developed. This is designed to recognize a unique sequence that is tagged onto the 5′ end of one of the amplimers. When the complement to this sequence is synthesized after the second amplification cycle, a target for the UniPrimer is created which can then generate signal as it participates in subsequent rounds of amplification.

Recent developments include the Lux® primer method (Nazarenko et al., 2002). This uses the same principle as Amplifluor®, but only the fluorophore is needed because the acceptor for quenching is provided by bases in the hairpin stem structure. The fluorophore must be two or three nucleotides from the 3′ end attached at the C-5 position of a thymidine base. The primer folds such that the hairpin loop terminates with CG or GC to provide the intrinsic quenching. The dsDNA structure created upon extension

of the complementary sequence allows the dye to fluoresce.

Although the Amplifluor® method provides only generic detection with the probe having no specificity for the internal sequence of the amplicon so that non-specific priming can still generate a signal, it does offer some advantages over DNA target binding probes. The signal to noise ratio is typically better than that obtained from using a DNA binding dye and the probe may be a made as a universal reagent as in the UniPrimer® just described. A generic approach to multiplexing using different fluorophore emissions can be readily implemented. A similar method using a primer incorporated probe sequence for a generic approach using common hydrolysis probes has been reported for single tube genotyping (Whitcombe et al., 1998).

The Eclipse® probe system works on the molecular beacon principle, but additionally uses a minor-groove binder (MGB) moiety to stabilize the binding of the loop to its complementary sequence (Afonina et al., 2002). The probe is chemically similar to the MGB probes that can be used in the 5′ nuclease assay, but is typically labelled with the MGB and dark acceptor on the 5′ end to prevent the probe being hydrolysed. This type of probe can be used for a number applications and melting point analysis can be used to identify sequence variation as described above.

Self-probing amplicons

Self-probing amplicon systems have a probe incorporated into an amplimer at the 5′ end. As the amplimer is extended the probe becomes part of the product strand. The probe has complementarity with a sequence in the newly synthesized region of the amplicon. On cooling after the nascent amplicon strand has been melted away from the template strand at the high-temperature step of PCR, the probe can bind intramolecularly to the complementary sequence to generate a signal. The principle was first reported in 1999 and used a molecular beacon that was connected to the 5′ end of an amplimer via a linker molecule (Newton et al., 1993; Whitcombe et al., 1999a,b). The linker was used to prevent the generation of a complementary strand to the probe sequence at the end of the extension of the reverse strand.

The probe is designed so that the energetics favour binding of the beacon's loop region to the target sequence over the pairing between the stem structures of the beacon. The self-probing intramolecular mechanism has been proven to be the dominant mechanism of signal generation, rather than intermolecular probing of other amplicons (Thelwell et al., 2000). It has also been shown that the cleavage of such probes by the 5′-3′ exonuclease activity of the Taq enzyme does not occur (Thelwell et al., 2000). In the original work the probe had fluorescein as the donor moiety quenched by Methyl Red as a non-fluorescent acceptor. These probes are known commercially as Scorpions® probes.

The main advantage of this method is that the probe is made in the same synthesis as the amplimer so a separate probe is not required. Since the probe becomes linked to the sequence it is probing, the reporting event is a zero order reaction (Saha et al., 2001) and the consequent speed of signal generation has been reported to be much faster than that of the respective native molecular beacon involved in intermolecular probing (Whitcombe et al., 1999a; Thelwell et al., 2000). It is actually more efficient when using short annealing times (Thelwell et al., 2000).

The even stochiometry of a Scorpions® probe with its complement means the amplification signal is directly proportional to the initial amount of target nucleic acid. This has also been demonstrated (Lee et al., 2002) with a ResonSense® probe attached to an amplimer, a probe system that has been termed Angler®. These types of assays will be of use on faster thermal cyclers as a means of improving throughput.

Other variants of the Scorpions® theme include the duplex Scorpion® primer method (Solinos et al., 2001) where the amplimer is labelled with a linear probe that has specificity for the amplicon strand that will be synthesized on the 3′ end of the amplimer. The probe contains one dye of a FRET pair. The other dye is on a separate oligonucleotide which has complementarity with the probe. Until priming by the amplimer occurs these can interact. When the amplimer has been extended the probe is able to bind to the newly generated complement in a more favoured intramolecular reaction. The probe therefore becomes separated from the complementary oligonucleotide and the consequent change in the FRET signal from the dyes indicates that amplification has occurred.

IntraTaq® is the self-probing embodiment of the 5′ nuclease assay. It has been shown to be more efficient in terms of the signal to noise ratio and the speed of the reaction (Solinas, et al., 2002). In theory it should be possible to link any probe system to an amplimer in a similar fashion and one would expect a similar improvement in signalling in terms of speed for assays based on hybridization. One advantage of the self-probing amplicon that is not immediately apparent is that the probe is better controlled for, compared to a separate probe, since it is synthesized on the same molecule (Lee et al., 2002). The main disadvantages of such probes are that synthesis can be complex and therefore relatively costly and the supply base is limited to manufacturers with the technical ability to produce the more complicated formats.

Intellectual property

Differences in patent law, such as between the US and Europe, mean that the protection afforded to intellectual property (IP) shows significant regional inconsistency. In some territories patent protection may not have been sought and there will be no barriers to the use of a technology. Additionally, region by region, granted patents may not provide the same protection since the filings may have been examined and opposed on the basis of different prior art. At times the validity of a patent may be challenged so that the protection of the IP may change. So the patent landscape, and the restrictions on the use of real-time PCR process, will inevitably vary depending on when and where the process is to be used. The authors therefore cannot be definitive about the rights an individual or an organization may or may not have to use the technologies that have been discussed above, and nothing in this section is to be taken as legal advice. However, it is hoped that this section will at least provide an introduction to understanding what can be a very complex issue.

By far the largest application area for real-time methods is in the measurement of gene expression. Many researchers will be developing a process for which they would like freedom of use, but they should be aware of the licensing programmes needed to make the technology available and be aware of third party IP and understand how it may be accessed. PCR technology has been a commercial success for many players. The patent landscape is complex and exploitation of new technologies has created a competitive and litigious commercial scene. There have been long and expensive court cases and major players have been forced out of business. Therefore users should consider and further investigate the points addressed here. Given the cost and development times of validated assays, naivety about IP could prove costly.

The history of the development of the PCR process has had a significant effect on the licensing of the core PCR technologies. Cetus Corporation worked with Applied Biosystems (as Perkin-Elmer at the time) on the development of early thermal cycling instrumentation and they protected the associated IP. The process became a valuable tool to the life science researcher and, when Roche acquired the process from Cetus, it was made available through an open licensing programme with researchers who purchased licensed enzymes, reagents and thermal cycling instruments having the right to carry out the process for their research activities. The authors have had contact with many groups who have misunderstood the licensing programme and have assumed that rights of use are automatically obtained for all areas under this programme. (This has been the case with, mostly, academic, government, or non-profit making organizations that may have less of a commercial interest than pharmaceutical and diagnostics companies.) Other activities such as service diagnostics require the end user to either obtain a 'field' licence or purchase licensed diagnostic platforms and reagents. However, the key patents covering the core PCR IP are reaching the point where they will lapse so the requirement for licensing will diminish over the next few years. After this period, licensing will only be a requirement for the use of IP in more recent patent filings. An 'extended licensing' programme is required from the key stakeholders of much IP, which is only now achieving granted patent status, that covers their real-time processes, instruments, and modified enzymes. This should mean a continuation of the open licensing programme for research with additional requirements for some real-time methods.

The requirement for a licence can be broken down into key technology groupings. These are summarized below. The authors recommend that a detailed search of the patent literature and an appraisal of the status of patents identified should be carried out, with professional help, by end users who plan to use a PCR process for application in fields other than pure research.

Enzymes

Taq polymerase and other enzymes that fall within the scope of the *Taq* patents will require the user to have a licence for use until the patents lapse. However, there is at least one commercial enzyme that is claimed to fall outside of the patent. Over the last ten years chemical hot-start methods have become essential to many users. The processes to make and use these are new IP and researchers may require licences to use them.

Instruments

The 'automation patent' covering the process of amplification in an automated process under computer control varies geographically. Therefore the requirement for an instrument to be licensed will depend upon the heating technology it uses and specific embodiments covered by other key patents. Optical instruments are protected in some territories and freedom to use particular machines will vary such that the choice of instrument may change the requirement for licensing in some geographical regions. Some instruments purchased prior to the grant of the associated patents may not be licensed for some real-time methods. Indeed some licensed instruments with this capability were sold with a licence that did not cover their use for optical analysis because of the early status of these patent families at the time.

Probe methods

As is evident, many FRET-based probe methods have been developed. The freedom of use position for different probe formats in real-time PCR varies. Some probe systems, although they actually function by a different process, may well be covered by kit claims in the main real-time patents. For example, in some territories kits comprising probes with two or more fluorogenic labels, 3′ protection and polymerase without the exonuclease activity removed fall within the scope granted patents. Therefore the requirement to research your application and choose a suitable probe method may be invaluable in obtaining the rights of use.

Dye technology

The dye labels on purchased oligonucleotides may have been synthesized with a limited licence. Fortunately there are many dyes generally available for application, including some that are generic versions of popular labels.

Core technology

Many improvements to the PCR portfolio have come from third parties. These have been in the areas of enzyme hot starts, carry over prevention and allele-specific priming methods to name but a few.

Sequences

Many primer probe sequences and genomes are the subject of granted patents. The freedom to use these sequences depends upon local patent law and may require a licence for certain applications.

The choice of technologies that you make for carrying out your real-time method may, therefore, significantly affect the ultimate cost of applying your process and your freedom to do so. There are some useful websites that can help researchers to understand the patent landscape.

Summary and future improvements

There are many chemistries that may be applied effectively to carry out the real-time quantitative detection of PCR amplification. The chemistries differ in their utility and no single solution best suits every application. The optimum assay will depend upon the application, the instrument platform used, the software suites for data analysis, and the target sequence. The specifications of new instruments are improving, generally becoming faster and generally, but not always, more flexible in operation. New real-time PCR detec-

tion methods such as fluorescence polarization, time resolved fluorescence and the application of quantum dot technology have potential, but have yet to be realized commercially. They could allow for improvements in both the level of multiplexing and the signal to noise ratios obtainable compared to current approaches. Such methods could improve the speed and robustness and reduced the costs associated with real-time PCR. The art is progressing such that readers should also look to the open literature for new advances that are rapidly being made.

Acknowledgements

The authors would like to thank the editors for the invitation to submit a revised version for the second edition, and to Enigma Diagnostics Ltd for the financial contribution to the coloured images.

References

Afonina, I.A., Reed, M.W., Lusby, E., Shishkina, I.G. and Belousov, Y.S. 2002. Minor groove binder-conjugated DNA probes for quantitative DNA detection by hybridisation-triggered fluorescence. BioTechniques 32, 940–949.

Bassam, B.J., Allen, T., Flood, S., Stevens, J., Wyatt, P. and Livak, K.J. 1996. Nucleic acid sequence detection systems: revolutionary automation for monitoring and reporting PCR products. Australas. Biotechnol. 6, 285–294.

Becker A., Reith A., Napiwotzki, J. and Kadenbach B. 1996. A quantitative method of determining initial amounts of DNA by polymerase chain reaction cycle titration using digital imaging and a novel DNA stain. Anal. Biochem. 237, 204–207.

Bernard, P.S., Lay, M.J. and Wittwer C.T. 1998. Integrated amplification and detection of the C677T point mutation in the methylenetrahydrofolate reductase gene by fluorescence resonance energy transfer and probe melting curves. Anal. Biochem. 255, 101–107.

Broude, N.E. 2002. Stem-loop oligonucleotides: a robust tool for molecular biology and biotechnology. Trends Biotechnol. 20, 249–256.

Cardullo, R.A., Agrawal, S., Flores, C., Zamecnik, P.C. and Wolf, D.E. 1988. Detection of nucleic acid hybridisation by non-radiative fluorescence resonance energy transfer. Proc. Natl. Acad. Sci. USA 85, 8790–8794.

Clegg, R.M. 1992. Fluorescence resonance energy transfer and nucleic acids. In: Methods in Enzymology. D.M.J. Lilley and J.E. Dahlber, eds. Academic Press, New York 211, 353–388.

Dubertret, B., Calame, M. and Libchaber, A. 2001. Single-mismatch detection using gold-quenched fluorescent oligonucleotides. Nat. Biotechnol. 19, 365–370.

French, D.J., Archard, C.L., Brown, T. and McDowell, D.G. 2001. Hybeacon® probes: a new tool for DNA sequence detection and allele discrimination. Mol. Cell. Probes 15, 363–374.

Gelfand, D.H. 1989. *Taq* DNA polymerase. PCR Theory, principles, and application to DNA amplification. In: PCR Technology. Erlich, H.A. ed. Stockton Press, New York, pp. 17–22.

Gelmini, S., Orlando, C., Sestini, R., Vona, G., Pinzani, P., Ruocco, L. and Pazzagli, M. 1997. Quantitative polymerase chain reaction-based homogeneous assay with fluorogenic probes to measure c-erB-2 oncogene amplification. Clin. Chem. 43, 752–758.

Heller, M.J. and Jablonski, E.J. 1987. Fluorescent stokes shift probes for polynucleotide hybridisation assays. European patent application 86116652.8.

Heyduk, T. and Heyduk E. 2002. Molecular beacons for detecting DNA binding proteins. Nat. Biotechnol. 20, 171–176.

Higuchi, R., Dollinger, G., Walsh, P.S. and Griffith, R. 1992. Simultaneous amplification and detection of specific DNA sequences. Biotechnology. 10, 413–417

Higuchi, R., Fockler, C., Dollinger, G. and Watson, R. 1993. Kinetic PCR analysis: real-time monitoring of DNA amplification reactions. Biotechnology 11, 1026–1030.

Holland, P.M., Abramson, R.D., Watson, R. and Gelfand, D.H. 1991. Detection of specific polymerase chain reaction product by utilising the 5′-3′ exonuclease activity of *Thermus aquaticus* DNA polymerase. Proc. Natl. Acad. Sci. USA 88, 7276–7280.

Hwang, G.T., Seo, Y. J. and Kim, B.H. 2004. A highly discriminating quencher-free molecular Beacon for probing DNA. J. Am. Chem. Soc. 126, 6528–6529.

Isacsson, J., Cao, H., Ohlsson, L., Nordgren, S., Svanik, N., Westman, G., Kubista, M., Sjöback, R. and Sehlstedt, U. 2000. Rapid and specific detection of PCR products using light-up probes. Mol. Cell. Probes 14, 321–328.

Ju, J., Ruan, C., Fuller, C.W., Glazer, A.A. and Mathies, R.A. 1995. Fluorescence energy transfer dye-labeled primers from DNA sequencing and analysis. Proc. Natl. Acad. Sci. USA 92, 4347–4351.

Kong, D.M, Gu, L., Shen H.X. and Mi, H.F. 2002. A modified molecular beacon combining the properties of TaqMan probe. Chem. Commun. 8, 854–855.

Koshkin, A.A., Rajwanshi, V.K. and Wengel, J. 1998. Novel convenient syntheses of LNA [2.2.1] bicyclo nucleosides. Tetrahedron Lett. 39, 4381–4384.

Kutyavin, I.V., Afonina, I.A., Mills, A., Gorn, V.V., Lukhtanov, E.A., Belousov, E.S., Singer, M.J., Walburger, D.K., Lokhov, S.G., Gall, A.A., Dempcy, R., Reed, M.W., Meyer, R.B. and Hedgpeth, J. 2000. 3′-minor groove binder-DNA probes increase sequence specificity at PCR extension temperatures. Nucleic Acids Res. 28, 655–661.

Kyger, E.M., Krevolin, M.D. and Powell, M.J. 1998. Detection of the hereditary hemochromatosis gene mutation by real-time fluorescence polymerase chain reaction and peptide nucleic acid clamping. Anal. Biochem. 260, 142–148.

Lee, L.G., Chen, C.H. and Chiu, L.A. 1986. Thiazole orange: a new dye for reticulocyte analysis. Cytometry 7, 508–517.

Lee, L.G., Connell, C.R. and Bloch, W. 1993. Allelic discrimination by nick-translation PCR with fluorogenic probes. Nucleic Acids Res. 21, 3761–3766.

Lee, M.A., Brightwell, G, Leslie, D., Bird, H. and Hamilton, A. 1999. Fluorescent detection techniques for real-time multiplex strand specific detection of *Bacillus anthracis* using rapid PCR. J. Appl. Microbiol. 87, 218–223.

Lee, L.G, Livak, K.J., Mullah, B., Graham, R.J., Vinayak, R.S. and Woudenberg T.M. 1999. Seven-colour, homogeneous detection of six PCR products. Biotechniques 27, 342–349.

Lee, M.A., Siddle, S.L. and Hunter, R.P. 2002. ResonSense®: simple linear probes for quantitative homogeneous rapid polymerase chain reaction. Anal. Chim. Acta 457, 61–70.

Leone, G., Schijndel, H., Gemen, B., Kramer, F.R. and Schoen, C D. 1998. Molecular beacon probes combined with amplification by NASBA enable homogenous, real-time detection of RNA. Nucleic Acids Res. 26, 2150–2155.

Letertre, C., Perelle, S., Dilasser, F., Arar, K. and Fach, P. 2003. Evaluation of the performance of LNA and MGB probes in 5′-nuclease PCR assays. Mol. Cell. Probes 17, 307–3111.

Li, Q., Luan, G, Guo, Q. and Liang, J. 2002. A new class of homogeneous nucleic acid probes based on specific displacement hybridisation. Nucleic Acids Res. 30, e5.

Livak, K.J., Flood, S.A.J., Marmaro, J., Giusti, W. and Deetz, K. 1995. Oligonucleotides with fluorescent dyes at opposite ends provide a quenched probe system useful for detecting PCR product and nucleic acid hybridisation. PCR Methods Appl. 4, 357–362.

Matayoshi, E.D., Wang, G.T., Krafft, G.A. and Erickson, J. 1990. Novel fluorogenic substrates for assaying retroviral proteases by resonance energy transfer. Science 247, 954–958.

Morrison, L.E., Halder, T.C. and Stols, L.M. 1989. Solution-phase detection of polynucleotides using interacting fluorescent labels and competitive hybridisation. Anal. Biochem. 183, 231–244.

Morrison, T.B., Weis, J.J. and Wittwer, C.T. 1998. Quantification of low-copy transcripts by continuous SYBR® Green I monitoring during amplification. Biotechniques 24, 954–962.

Mouritzen, P., Noerholm, M., Nielson, P.S., Jacobsen, N., Lomholt, C, Pfundheller, H.M. and Tolstrup, N. 2005. ProbeLibrary: A new method for faster design and execution of quantitative real-time PCR. Nat. Methods 2, 313–316.

Mullis, K.B. and Faloona, F.A. 1987. Specific synthesis of DNA *in vitro* via a polymerase catalyzed chain reaction. Methods Enzymol. 155, 335–350.

Nazarenko, I.A., Bhatnagar, S.K. and Hohman, R.J. 1997. A closed tube format for amplification and detection of DNA based on energy transfer. Nucleic Acids Res. 25, 2516–2512.

Newton, C. R., Graham, A., Heptinstall, L.E., Powell, S.J, Summers, C., Kalsheker, N., Smith, J.C. and

Markham, A.F. 1989. Analysis of any point mutation in DNA. The amplification refractory mutation system (ARMS). Nucleic Acids Res. *17*, 2503–2516.

Nazarenko, I., Pires, R., Lowe, B., Obaidy, M. and Rashtchian. 2002. Effect of primary and secondary structure of oligodeoxyribonucleotides on the fluorescent properties of conjugated dyes. Nucleic Acids Res. *30*, 2089–2195.

Newton, C.R., Holland D., Heptinstall, L.E., Hodgson, I., Edge, M.D., Markham, A.F. and Mclean, M.J. 1993. The production of PCR products with 5′ single-stranded tails using primers that incorporate novel phosphoramidite intermediates. Nucleic Acids Res. *21*, 1155–1162.

Nurmi, J., Wikman, M., Karp, M. and Lövgren, T. 2002. High-performance real-time quantitative RT-PCR using lanthanide probes and a dual temperature hybridization assay. Anal. Chem. *74*, 3525–32.

Ogura, M. and Mitsuhahi, M. 1994. Screening method for a large quantity of polymerase chain reaction products by measuring YOYO-1 fluorescence on 96-well polypropylene plates. Anal. Biochem. *218*, 458–459.

Ortiz, E., Estrada, G. and Lizardi, P.M. 1998. PNA molecular beacons for rapid detection of PCR amplicons. Mol. Cell. Probes *12*, 219–226.

Ririe, K.M., Rasmussen, P.R. and Wittwer, C.T. 1997. Product differentiation by analysis of DNA melting curves during the polymerase chain reaction. Anal. Biochem. *245*, 154–160.

Robinson, J.K., Mueller, R., Filippone, L. 2000. New molecular beacon technology. Am. Lab. *32*,30–24.

Saha, B. K., Tian, B. and Bucy, R.P. 2001. Quantitation of HIV-1 by real-time PCR with a unique fluorogenic probe. J. Virol. Methods. *93*, 33–42.

Saiki, R.K., Scharf, S., Faloona, F.A., Mullis, K.B., Horn, G.T., Erlich, H.A. and Arnheim, N. 1985. Enzymatic amplification of beta-globin genomic sequences and restriction site analysis for diagnosis of sickle cell anemia. Science *230*, 1350–1354.

Salmon, M.A., Vendrame, M., Kummert, J. and Lepoivre, P. 2002. Detection of apple chlorotic leaf spot virus using a 5′ nuclease assay with a fluorescent 3′ minor groove-DNA binding probe. J. Virol. Methods *104*, 99–106.

Schofield, P. Pell, A.N. and Krause, D.O. 1997. Molecular Beacons: trial of a fluorescence-based solution hybridization technique for ecological studies with ruminal bacterial. App. Environ. Microbiol. *63*, 1143–1147.

Selvin, P.R. 1995. Fluorescence resonance energy transfer. Methods Enzymol. *246*, 300–334.

Solinas, A., Brown, L.J., McKeen, C., Mellor, J.M., Nicol, J.T.G, Thelwell, N. and Brown, T. 2001. Duplex scorpion primers in SNP analysis and FRET applications. Nucleic Acids Res. *20*,1–9.

Solinas, A., Thelwell, N. and Brown, T. 2002. Intramolecular TaqMan probes for genetic analysis. Chem. Commun. *19*, 2272–2273.

Svanik, N., Westman, G., Wang, D. and Kubista, M. 2000a. Light-Up probes: Thiazole orange-conjugated peptide nucleic acid for detection of target nucleic acid in homogeneous solution. Anal. Biochem. *281*, 26–35.

Svanik, N., Nygren, J. Westman, G. and Kubista, M. 2000b. Free-probe fluorescence of light-up probes. J. Am. Chem. Soc. *123*, 803–809.

Taylor, M.J., Hughes, M.S. Skuce, R.A. and Neill, S.D. 2001. Detection of *Mycobacterium bovis* in bovine clinical specimens using real-time fluorescence and fluorescence energy transfer probe rapid-cycle PCR. J. Clin. Microbiol. *39*, 1272–1278.

Täpp, I., Malmberg, L., Rennel, E., Wik, M. and Syvänen, A.-C. 2000. Homogeneous scoring of single nucleotide polymorphisms: comparison of the 5′-nuclease TaqMan® assay and molecular beacon probes. Biotechniques *28*, 732–738.

Thelwell, N., Millington, S., Solinas, A., Booth, J. and Brown, T. 2000. Mode of action and application of scorpion primers to mutation detection. Nucleic Acids Res. *28*, 3752–3761.

Tsourkas, A., Behlke, M.A., Rose D. and Gang, B. 2003. Hybridisation kinetics and thermodynamics of molecular beacons. Nucleic Acids Res. 31, 1319–1330.

Tyagi, S. and Kramer, F.R. 1996. Molecular Beacons: probes that fluoresce upon hybridisation. Nat. Biotechnol. *14*, 303–308.

Tyagi, S., Bratu, D.P. and Kramer, F.R. 1997. Multicolour molecular beacons for allele discrimination. Nat. Biotechnol. *16*, 49–53.

Vet, J.A.M., Majithia, A.R., Marras, S.A.E, Tyagi, S., Dube, S., Poiesz, B.J. and Kramer, F.R. 1999. Multiplex detection of four pathogenic retroviruses using molecular beacons. Proc. Natl. Acad. Sci. USA *96*, 6394–6399.

Whitcombe, D., Brownie, J., Gillard, H.L., McKechnie, D., Theaker, J., Newton, C.R. and Little, S. 1998. A homogeneous fluorescence assay for real-time PCR amplicons: its application to real-time, single tube genotyping. Clin. Chem. *44*, 918–923.

Whitcombe, D., Theaker, J., Guy, S.P., Brown, T. and Little, S. 1999a. Detection of PCR products using self-probing amplicons and fluorescence. Nat. Biotechnol. *17*, 804–807.

Whitcombe, D., Kelly, S., Mann, J., Theaker, J., Jones, C. and Little, S. 1999b. Scorpions™ primers – a novel method for use in single tube genotyping. Am. J. Hum. Genet. *65*, 2333.

Wittwer, C.T, Ririe, K.M., Andrew, R.V., David, D.A, Grundy, R.A. and Balis, U.J. 1997. The LightCycler® – a microvolume multisample fluorimeter with rapid temperature control. Biotechniques *22*,176–181.

Wolffs, P., Knutsson, R., Sjöback, R. and Rådström, P. 2001. PNA-based light-up probes for real-time detection of sequence-specific PCR products. Biotechniques *31*, 766–771.

Young, D. 1996. Viral load quantitation: An integral part of future health management. Australas. Biotechnol. *6*, 295.

Zhang, P., Beck, T. and Tan, W. 2001. Design of a molecular beacon DNA probe with two fluorophores. Angew. Chem. Int. Ed. Engl. *40*, 402–405.

Reference Gene Validation Software for Improved Normalization

4

Jo Vandesompele, Mikael Kubista and Michael W. Pfaffl

Abstract

Real-time PCR is the method of choice for expression analysis of a limited number of genes. The measured gene expression variation between subjects is the sum of the true biological variation and several confounding factors resulting in non-specific variation. The purpose of normalization is to remove the non-biological variation as much as possible. Several normalization strategies have been proposed, but the use of one or more reference genes is currently the preferred way of normalization. While these reference genes constitute the best possible normalizers, a major problem is that these genes have no constant expression under all experimental conditions. The experimenter therefore needs to carefully assess whether a certain reference gene is stably expressed in the experimental system under study. This is not trivial and represents a circular problem. Fortunately, several algorithms and freely available software have been developed to address this problem. This chapter aims to provide an overview of the different concepts.

Introduction

Real-time PCR has become the *de facto* standard for mRNA gene expression analysis of a limited number of genes. Given its large dynamic range of linear quantification, high speed, sensitivity (low template input required) and resolution (small differences can be measured), this method is perfectly suited for validation of microarray expression screening results on an independent and larger sample panel, and for studies of a selected number of candidate genes or pathway constituents in an experimental setup (biopsies,

treated cell cultures or any other sample collection). More recently, real-time PCR has also entered the high throughput gene expression analysis field based on 384-well block thermal cyclers and newer platforms, such as array based devices from Biotrove and Fluidigm that allow parallel gene expression analysis of even higher number of genes and samples (1–48 samples for 48–3072 different genes depending on platform and configuration).

It is important to realize that any measured variation in gene expression between subjects is caused by two sources. On the one hand, there's the true biological variation, explaining the phenotype or underlying the phenomenon under investigation. On the other hand, there are several confounding factors resulting in non-specific variation, including but not limited to template input quantity and quality, yields of the extraction process and the enzymatic reactions (reverse transcription and polymerase chain reaction amplification). One of the major difficulties in obtaining reliable expression patterns is the removal of this experimentally induced non-biological variation from the true biological variation. This can be done through normalization by controlling as many of the confounding variables as possible (next section).

Reference genes as golden standard for normalization

There are several strategies to remove experimentally induced variation, each with their own advantages and considerations (Huggett *et al.*, 2005). While most of these methods cannot completely reduce all sources of variation, it has been shown to be very important to try to control all the sources of variation along the entire workflow of PCR based gene expression analysis. If one does not meticulously try to standardize each step, variation can and will be introduced in your results that cannot be eliminated by applying the final normalization (Stahlberg *et al.*, 2004). It is thus recommended to ensure similar sample size for extraction of RNA and to standardize the amount of RNA for DNase treatment and reverse transcription into cDNA. Furthermore, artificial RNA molecules can be spiked into the sample prior to extraction or to the RNA extract prior to reverse transcription (Gilsbach *et al.*, 2006; Huggett *et al.*, 2005; Smith *et al.*, 2003). This will give an indication of the efficiency of the reverse transcription and qPCR procedures, and will reveal any inhibition. The spike however, may not be extracted with the same yield as compartmentalized natural mRNAs and will not control fully for the final amount of input material in the reaction.

Taking everything into consideration, it has been agreed that the reference gene concept is the currently preferred way of normalizing real-time PCR data (3rd London qPCR symposium, April 2005). Several companies have recognized the problem and provide validated reference gene panels for various organisms (Table 4.1). The reference gene concept is particularly attractive because the reference genes are internal controls that are affected by all sources of variation during

the experimental workflow in the same way as the genes of interest. The reference genes were expressed in the cells, and their mRNAs are present during prelevation, nucleic acid extraction, storage, and any enzymatic processes such as DNase treatment and reverse transcription. Furthermore, PCR-based quantification results for a gene of interest are best normalized using a factor that is measured using the same methodology (i.e. a reference gene's expression is also measured by PCR).

While the use of reference genes for normalization of gene expression levels is certainly the gold standard some new approaches for normalization have recently been developed. Argyropoulos *et al.* have developed a generic normalization method for real-time PCR against total mRNA. During reverse transcription this method incorporates a long tailed sequence to each mRNA. After double-stranded cDNA synthesis the tailed sequence can be quantified and is a measure of the mRNA fraction (Argyropoulos *et al.*, 2006). It remains to be determined what the fraction of mispriming of the tailed oligo-dT primer is to e.g. highly abundant RNA molecules such as the ribosomal RNAs. Talaat *et al.* (2002) and Kanno *et al.* (2006) use a gene expression normalization strategy based on DNA. The former measures the DNA content of a total nucleic acids extract by PCR and uses this as normalization factor. In the latter approach the DNA content is determined spectroscopically,

Table 4.1 Commercial reference gene panels

Company	Targeted genera and number of reference genes	URL
Applied Biosystems	*Homo* (10), *Mus* (2), *Rattus* (2),	http://www.appliedbiosystems.com
Eurogentec	*Homo* (12), *Mus* (2), *Rattus* (1), *Oryctolagus* (1)	http://www.eurogentec.com
Invitrogen	*Homo* (16), *Mus/Rattus* (15), *Drosophila* (2)	http://tools.invitrogen.com/content/sfs/ brochures/711-022292_LUXhskg_brochure.pdf
PrimerDesign	*Homo* (12), *Mus* (12), *Rattus* (12), *Caenorhabditis* (12), *Xenopus* (12), *Arabidopsis* (12), *Ovis* (12), *Danio* (12), Bos (7), Sus (10), Nicotiana (6), Cucumis (6), Solenopsis (6)	http://www.primerdesign.co.uk/genorm_all_ specials.asp
TATAA Biocenter	*Homo* (12), *Mus* (12)	http://www.tataa.com/webshop/Endogenous-Control-Panels/View-all-products.html

and the sample is spiked with a cocktail of artificial RNA molecules that is proportional to the DNA content. These spikes can be used for normalization and also for estimating the transcript copy number per cell. A major problem with this DNA normalization strategy is that current RNA extraction protocols are not designed to co-purify DNA, so the extraction yields may be different between different samples, with the DNA yields being suboptimal. A third alternative to the use of stably expressed reference genes is the quantification of a specific internal control or so called *in situ* calibration (Stahlberg *et al.*, 2003). In these approaches, the investigator does not look for a stably expressed reference but, based on knowledge about the biological system, selects a gene whose expression is correlated or anti-correlated with that of the gene of interest. While this approach does not really allow the comparison of expression levels of a given gene between samples, it is a powerful approach when the expression ratio of two marker genes is significant for disease or reflects a biological phenomenon. A final alternative is currently being successfully used and further explored by Vandesompele and colleagues (unpublished). In this novel approach, repeat sequences that are expressed in the transcriptome are quantified and assumed to reflect the amount of total mRNA. Alu repeats are by far the most abundant repeat sequences in the human genome, and approximately 1500 human genes contain at least one Alu in their 5′ or 3′ untranslated region. The rationale is that differential expression of some genes will not alter the total expressed Alu repeat content. The current downside is that the Alu repeats are primate specific, and hence can only be used to normalize primate samples. It is currently under investigation whether other repeats can be used as references in other organisms.

Software and algorithms for reference gene evaluation and selection

While reference genes have the intrinsic capacity to capture all non-biological variation and as such constitute the best normalizers, a major problem is that there is substantial evidence in the literature that most of the commonly used reference genes are regulated under some circumstances. It

is thus of utmost importance to validate in your own experimental situation whether a candidate reference gene is suitable for normalization. The implications of using an inappropriate reference gene for real-time reverse transcription PCR data normalization is recently demonstrated by Dheda *et al.* (2005). If unrecognized, unexpected changes in reference gene expression can result in erroneous conclusions about real biological effects. In addition, this type of change often remains unnoticed because most experiments only include a single reference gene.

Reporting that reference genes might show variable expression under certain conditions is not really helpful. We therefore need strategies to find proper reference genes, and to implement them in a sensible normalization procedure. However, it is important to realize that the evaluation of expression stability represents a circular problem: how can the expression stability of a candidate be evaluated if no reliable measure is available to normalize the candidate? This circular problem is addressed in the following algorithms and Table 4.2.

geNorm

Vandesompele *et al.* (2002) were the first to quantify the errors associated with the use of a single (non-validated) reference gene, to develop a method to select the most stably expressed reference genes, and to propose the use of multiple reference genes for calculation of a reliable normalization factor.

As indicated above many studies have reported that reference gene expression can vary considerably, but Vandesompele *et al.* (2002) systematically addressed the critical issues of using reference genes, and proposed an adequate workaround for their variable expression. To this purpose, they rigorously measured the expression level of 10 common reference genes in 85 samples from 13 different human tissues. Special attention was paid to select genes that belong to different functional and abundance classes, which significantly reduced the risk that genes are co-regulated. To determine the errors related to the common practice of using only one reference gene for normalization, they defined and calculated the single reference gene normalization error as the ratio of the ratios of two reference genes

Table 4.2 Algorithms and software for evaluation of candidate reference genes

First author	Publication year	Software	Algorithm	Number of citations*	URL
Vandesompele	2002	geNorm	Reference gene ranking based on stepwise elimination of least stable gene; expression stability is defined as average pairwise variation (standard deviation of log transformed ratios) of a given gene with all other candidate reference genes	1128	http://medgen.ugent.be/genorm
Akilesh	2003	General Pattern Recognition	Ranked list of statistically changed genes based on Student's t-tests after a multiple gene normalization	32	–
Haller	2004	–	Statistical equivalence test for demonstration of equal expression of a reference gene between two sample groups	16	–
Pfaffl	2004	BestKeeper	Reference gene ranking based on repeated pairwise correlation and regression analysis of a given gene with all other tested candidate reference genes	126	http://www.gene-quantification.de/bestkeeper.html
Szabo	2004	–	Statistical linear mixed-effects modelling	56	http://www.huntsmancancer.org/publicweb/content/biostat/szabo.html
Andersen	2004	Normfinder	Statistical linear mixed-effects modelling	92	http://www.mdl.dk/publicationsnormfinder.htm
Brunner	2004	–	Statistical analysis of variance (ANOVA)	52	–
Abruzzo	2005	–	Statistical linear mixed-effects modelling	28	http://bioinformatics.mdanderson.org/Supplements/MicroFluidics/index.html
Huang	2006	–	Statistical model of simultaneous confidence intervals and practical equivalence testing	–	–

*According to Google Scholar on 5 July 2008.

in two different samples. These analyses clearly demonstrated that a normalization strategy based on a single (non-validated) reference gene leads to erroneous expression differences of more than 3- and 6-fold in 25% and 10% of the cases, respectively (with sporadic cases showing errors greater than 20-fold). This clearly warrants the search for stably expressed genes and an accurate normalization method.

To evaluate the presumed constant expression level of the tested candidate reference genes, a robust and assumption-free quality parameter was developed based on raw non-normalized expression levels. The underlying principle is that the expression ratio of two proper reference genes should be constant across samples. For each reference gene, the pairwise variation with all other reference genes is calculated as the standard deviation of the logarithmic transformed expression ratios, followed by the calculation of a reference gene stability value (M-value) as the average pairwise variation of a particular reference gene with all other tested candidate reference genes.

To manage the large number of calculations, the authors have written a freely available Visual Basic Application for Microsoft Excel (termed geNorm) that automatically calculates the expression stability values for any number of candidate reference genes in a set of samples (Table 4.2). The software employs an algorithm to rank the candidate reference genes according to their expression stability by a repeated process of stepwise exclusion of the worst scoring reference gene. Clear expression stability differences were apparent upon comparison of the candidate reference genes within and between the different tissue panels, which demonstrates that the choice of a proper reference gene is highly dependent on the tissues or cells under investigation. Because of the rather large single reference gene normalization errors, and the tissue and gene dependent expression stability differences, the authors suggest that the geometric mean of multiple reference genes be calculated as a normalization factor for real-time RT-PCR data. This factor controls for possible outlying values and abundance differences between the different genes

Finally, the authors outlined a strategy to determine the minimal number of reference genes for accurate normalization, by variation analysis of normalization factors calculated for an increasing number of reference genes. It turned out that three stable genes sufficed for samples with relatively low expression variation (homogeneous samples), but that other tissues or cell types required a fourth or fifth reference gene to deal with the observed expression variation. Of course, if one is only interested in 'on versus off' expression, or huge expression differences, there is no need for normalization using three or more stably expressed reference genes. In contrast, to measure small expression differences reliably (e.g. 2- to 3-fold), more accurate normalization based on multiple reference genes is needed.

To validate the accuracy of the proposed RT-PCR normalization method, the authors analysed publicly available microarray data and showed that geometric averaging of carefully selected control genes is equivalent to frequently applied array normalization strategies such as median ratio normalization and sum of intensity normalization. In a second validation experiment, it is shown that normalization using the geNorm selected best reference genes result in better removal of non-biological variation compared to geNorm identified 'unstable' reference genes.

To evaluate the geNorm ranking Gabrielsson *et al.* (2005) incorporated a bootstrap step. The ranking method was bootstrapped by resampling with replacement from the original set of samples. The resampling procedure was repeated 10,000 times. To check the robustness of the ranking procedure with respect to outliers, they also repeated the ranking with trimmed standard deviations, excluding the most outlying 10%, 20%, and 40% of log ratios in the computation of the standard deviation of the pairwise log ratios. The results obtained by the bootstrap procedure were in agreement with the original (one pass) geNorm ranking. Furthermore, the ranking was also robust in that it was essentially unaffected by trimming away 10%, 20%, or 40% of the most outlying log ratios. This again demonstrates that geNorm allows robust selection of stable reference genes.

In summary, the common practice of non-validated single reference gene normalization results in relatively large errors. This is a compelling argument for the use of multiple reference genes. Depending on the observed inherent expression variation of candidate reference genes and the tissue heterogeneity of the samples under investigation, the geometric mean of the three to five most stable reference genes allows reliable normalization. The normalization strategy presented is a prerequisite for accurate RT-PCR expression profiling and provides a first step in determination of the biological significance of subtle expression differences.

BestKeeper

This Microsoft Excel based program was developed by Pfaffl *et al.* (2004) and has many feature similarities with the previously discussed geNorm program (Table 4.2). The main differences are that BestKeeper uses C_t values (instead of relative quantities) as input and employs a different measure of expression stability. The founding principle for identification of stably expressed reference genes is that proper reference genes should display a similar expression pattern. Hence, their expression levels should be highly correlated. As such, BestKeeper calculates a Pearson correlation coefficient for each candidate reference gene pair, along with the probability that the correlation is significant. All highly correlated (and putatively stably expressed) reference genes are then combined into an index value (i.e. normalization factor), by calculating the geometric mean. Then, correlation between each candidate reference gene and the index is calculated, describing the relation between the index and the contributing reference genes by the correlation coefficient, coefficient of determination (r^2) and the *P*-value.

One unique feature of this software is that in addition to reference gene analysis, genes of interest can also be analysed, using the same method. This identifies highly correlated genes, as well as genes that behave similarly to the reference genes, and may be included in the calculation of the normalizing index. Another unique feature is that a sample integrity value is calculated. The underlying rationale is that outlier values might obscure the accuracy of the reference gene evaluation. Hence, an intrinsic variance value of expression for each sample is calculated as the mean squared difference of a given sample's C_t value for each particular gene with this gene's mean C_t across all samples. The intrinsic varia-

tion of a given sample can further be expressed as an efficiency corrected *n*-fold over- or under-expression of a particular reference gene with respect to the mean C_t value of that gene across all samples. If justified, strongly deviating samples due to inefficient sample preparation, incomplete reverse transcription or sample degradation can be removed from the BestKeeper index. Removal is recommended by the authors for a sample with a 3-fold over- or underexpression (compared to the mean expression level).

It is important to note that the Pearson correlation coefficient is only valid for normally distributed values with equal variance. Most often C_t values tend to be normally distributed (because these correspond to logarithms of copy numbers), but this cannot always be ascertained. The authors therefore plan to implement the Spearman rank correlation coefficient, which is distribution-free (does not assume normality of the values) and does not suffer from outlier values as does the Pearson correlation coefficient.

In conclusion, the BestKeeper software allows pairwise correlation analysis for up to ten candidate reference genes, ten genes of interest, and 100 biological samples. In addition, a sample integrity value is calculated, allowing removal of spurious data.

General pattern recognition

The GPR software developed by Akilesh *et al.* (2003) is quite different from the other software and algorithms discussed in this chapter. It will not rank candidate reference genes according to their expression stability, nor identify proper reference genes. In contrast, it is specifically suited for identification of differentially expressed genes between control and experimental samples by normalizing each gene by each possible reference gene (called normalizer in the software), with the simple definition of a reference gene that it should be expressed in both control and experimental samples.

GPR goes through several iterations to compare the change of expression of a gene normalized to every other gene in the set of genes being analysed. GPR takes advantage of biological replicates to extract statistically significant changes in gene expression, making it independent of the fold change between the control and experimental groups. This circumvents the biases inherent to standard microarray and qPCR analysis (whereby a minimal 2-fold change is often considered as significant). GPR is claimed to be superior to standard ANOVA techniques in its ability to better handle PCR dropouts without merging datasets.

GPR is a Microsoft Excel-based software algorithm that outputs a ranked list of statistically changed genes using raw input data (C_t values) comprised of between three and five 96-well or 384-well real-time PCR datasets from both a control and experimental group. GPR compares the datasets from both groups using Excel's built-in Student's *t*-test after multiple gene normalization.

GPR first filters data into overlapping gene and normalizer 'bins'. This filtering process is controlled by a user-defined cycle cut-off (CC) value. The CC is the PCR cycle number above which data is disregarded. After ~36–38 cycles, stochastic amplification of low copy-number targets can lead to large variability in the data. Consequently using the CC eliminates this noisy data. A gene passes through the 'gene filter' if all observations in *both control and* experimental groups fall below the cycle cut-off value. As such, GPR will consider a gene for further analysis if it is well expressed in either control or experimental groups (or both), but will disregard a gene if it not well expressed ('off') in both groups. A gene passes through the 'normalizer filter' if all observations in both control and experimental groups fall below the cycle cut-off value. In other words, GPR will consider a gene as a normalizer only if it is well expressed in both control and experimental groups, but will disregard a gene if it is not well expressed ('off') in either group. This ensures that only genes that have measurable expression levels in both groups are used as normalizers and that genes that may be off (C_t >CC) are not considered as normalizers.

After applying the gene and normalizer filters, GPR proceeds with global pattern recognition. For each dataset (a column of up to 384 C_t values), GPR takes each eligible gene and normalizes it to each eligible normalizer in succession to generate delta-C_t values as follows: delta-C_t (gene) = C_t (gene) − C_t (normalizer). For each gene–normalizer combination, the

delta-C_t values generated for the control and experimental groups are compared by a two-tailed heteroskedastic unpaired Student's *t*-test and a 'hit' is recorded if the p-value from the *t*-test falls below a user-defined P-value (e.g. 0.05). At the end of the normalization routine, GPR tallies the hits for each gene against all eligible normalizers and ranks the genes in descending order of number of hits. An experiment-independent score is obtained by dividing the number of hits for a gene by the total number of eligible normalizers (e.g. 50 hits out of 65 eligible normalizers is a score of 0.769). The genes with the highest scores have changed most significantly in the dataset.

A downside of the GPR software is that you need two groups (with a minimum of three and maximum of five samples per group) to proceed with your analysis, and that you need to study a lot of genes (at least 24 different genes (48 or 96 being better) of which at least half should qualify as normalizers (i.e. expressed in both groups).

Equivalence test for equal expression of a reference gene between two sample groups

In most gene expression studies, the goal is to show that a gene is differentially expressed (e.g. between two different patient groups). For validation of a candidate reference gene, the goal is just the opposite, i.e. to prove that the reference gene is equally expressed in the two groups, or is not influenced upon treatment. It is tempting to simply run a standard statistical test (e.g. the parametric *t*-test or non-parametric Mann–Whitney test) to see if the result is statistically significant at a certain significance level. If the difference is statistically significant, then it is a valid assumption that the gene is differentially expressed (higher in one group than in the other). If the difference is not statistically significant, people often assume that the expression is similar or equivalent in the two groups. However, this assumption is invalid, and leads often to erroneous conclusions. If your test reaches the conclusion of 'no statistically significant difference', it simply means that the current evidence (data) is not sufficiently strong to persuade you that the gene is differentially expressed. It is not the same as saying that the expression levels are the same in the two groups.

In other words, 'the absence of evidence is not evidence of absence (of differential expression)'.

To address the issue of equivalence testing in real-time PCR based gene expression analysis, Haller *et al.* (2004) developed a statistical test for the identification of stably expressed reference genes. The decision of the test depends on the cut-off used for differential expression. The authors suggest to use of a 3-fold change as a cut-off. This means that a gene is considered as equivalently expressed if the expression difference between the two groups is significantly smaller than three. The authors note that the cut-off should be carefully adjusted to the distribution of each experiment. In any case, the fold change of a significantly differentially expressed gene of interest after normalization should be at least the fold change used for the cut-off for evaluation of the reference genes. The input for the equivalence test (as for any parametric test) should be logarithmized values (either C_t values or logarithms of quantities derived from a standard curve).

A downside of the current equivalence test is that two sample groups are not always available (e.g. if one has no prior knowledge of groups, or more than two groups are available). A workaround could be the development of a one-sample equivalence test (by first rescaling log transformed expression levels to a mean of 0, and then testing if the expression levels are equivalent or similar to 0. Another concern is that this test is not entirely assumption-free. The equivalence test assumes equal amounts of input material in the PCR reaction, which poses a circular problem. As stated in the introduction of this chapter, the observed reference gene expression variation is not only due to true biological variation in gene expression, but also to several confounding factors, not taken into account in the test.

Advanced statistical models

Recently, four papers have been published describing the development and application of advanced statistical models for describing the expression stability of candidate reference genes. It is beyond the scope of this chapter to fully explain the underlying mathematics and statistics, but we will illustrate the different concepts (Table 4.2).

Szabo *et al.* (2004) and colleagues developed two models for describing the expression variability of candidate reference genes in either a single tissue type (formula 1) or across different cell types (formula 2). Formula 1 models the expression *yij* of gene *j* in sample *i*, where μ is the overall mean log-expression, *Ti* is the difference of the *i*th sample from the overall average and G*j* is the difference of the *j*th gene from the overall average. The key feature of the model that makes it different from a traditional ANOVA model is that it allows for heteroskedastic errors to account for different variability in the genes. The variability around the gene-specific mean log-expression $\mu + T_i + G_j$ is expressed by the error standard deviation σ*j*. This model was selected from a range of competing models with different error variances.

$$\log (y_{ij}) = \mu + T_i + G_j + \varepsilon_{ij}, \tag{1}$$

where $\varepsilon_{ij} \sim N(0, \sigma^2_j)$. On the basis of this model, the estimate of the variance of the log-average of the expression of the candidate reference genes can be calculated and used for stability ranking (the lower the variance, the more stable the gene), and geometric averaging of the best-performing genes for calculation of a reliable and robust normalization factor. The authors note that their ranking is very similar to the geNorm ranking (see p. 49), and even provide mathematical proof of the equivalency of the intuitive geNorm stability value *M* and their modelled error. However, their algorithm also ranks the best two reference genes (which geNorm does not do, relying on pairs of genes to determine stability).

To evaluate the expression stability within and between different tissues, Szabo *et al.* (2004) developed formula 2 that models the expression of gene *j* in the *i*th sample of tissue-type *k*, where μ denotes the overall mean log-expression, C*k* is the difference of the *k*th tissue type from the overall average, *Ti*(*k*) is the specific effect of the *i*th sample from tissue-type *k*, G*j* is the difference of the *j*th gene from the overall average and (*CG*) *kj* is the tissue-type-specific effect of gene *j*. The variability comes from two sources – the specific gene (σ^2) and the tissue-type ($\zeta^2 k$) – which are assumed to be independent and multiplicative.

$$\log(y_{i(k)j}) = \mu + C_k + T_{i(k)} + G_j + (CG)_{kj} + \varepsilon_{i(k)j} \quad (2),$$

where $\varepsilon_{i(k)j} \sim N(0,\sigma^2_j\zeta^2_k)$. Again, the authors note that their results using model 2 correlate very well with the geNorm ranking of stability in the different tissues tested in (Vandesompele *et al.*, 2002). A major advantage of a model-based approach, however, is that the terms are placed within a solid statistical framework, which allows the algorithm to be generalized to a variety of different experimental conditions.

For practical performance, both models were fitted using the gls routine of the nlme library for the freely available statistical language R (http://www.r-project.org/). A step-by-step description of the procedure as well as the R-script can be found on the authors' website (Table 4.2). Apart from R, many advanced statistical software programs are able to fit the models (e.g. PROC MIXED from SAS).

Claus Andersen and colleagues developed a similar linear mixed effects model, whereby both the overall variation of the candidate reference genes is modelled as well as the variation between sample groups (formula 3) (Andersen *et al.*, 2004). This enables the user to evaluate the systematic error introduced in the final results when using this particular gene.

$$\log(y_{igj}) = \alpha_{ig} + \beta_{gj} + \varepsilon_{igj}, \quad (3)$$

where $\varepsilon_{igj} \sim N(0,\sigma^2_{ig})$. Formula 3 models the log-transformed gene expression $yigj$ for gene i in the jth sample from group g, with αig the general expression level for gene i within group g, βgj the amount of mRNA in sample j from group g, and εigj the random variation caused by biological and experimental factors, with mean zero and variance $\sigma 2ig$. Having estimated the intragroup variation $\sigma 2ig$ and the intergroup variation as defined by the variation in αig ($g = 1, \ldots, G$), these two values are combined into a practical gene expression stability value. The authors note that the validity of their model is related to the number of samples and candidate reference genes analysed, i.e. the more, the better the estimates. The sample set should be at least eight per group and at least three genes (with 5–10 genes recommended). It is a further requirement that the candidate reference genes do not have any prior differential expression between the groups. Special attention was further paid to select candidate reference genes that belong to different functional classes, which significantly reduces the chance that these genes are co-regulated. To accommodate all the calculations, the authors have written a freely available Visual Basic Application for Microsoft Excel, termed NormFinder, which automatically calculates the stability value for all candidate reference genes (Table 4.2). In situations where no single optimal reference gene can be found, the authors suggest to use multiple reference genes. The rationale is that the variation in the average of multiple genes is smaller than the variation in individual genes, and that contributions from reference genes with bias for different groups cancel. A complication is the difficulty of weighting the relative importance of the intragroup and intergroup variations, and it is possible that the equal weights used by NormFinder overestimate the cancellation effect. The number of genes to include in the normalization factor is a trade-off between practical considerations and minimizing the variation in the normalization factor. The optimal number of genes is reached when addition of a further gene leads to a negligible reduction in the average of the gene variance estimates.

The above-described models use fixed effects for genes and samples and an error model accounting for gene-specific variability. Abruzzo *et al.* (2005) and colleagues evaluated several other linear mixed effects models, including models with random effects to account for sample differences. The authors conclude that modified versions of the Szabo and Andersen equation that include either fixed effects (formula 4) or random effects (formula 5) better explain the variability. In formula 4 a fixed effect γij is added that represents different expression levels for gene j in sample i. Formula 5 incorporates a random effect Cij to gene j in sample i. The authors show some preference for the random effects model (formula 5).

$$\log(y_{ijk}) = \mu + \alpha_j + \beta_i + \gamma_{ij} + \varepsilon_{ijk}, \quad (4)$$

where $\varepsilon_{ijk} \sim N(0,\sigma^2_j)$

$$\log(y_{ijk}) = \mu + \alpha_j + C_{ij} + \varepsilon_{ijk}, \quad (5)$$

where $C_{ij} \sim N(0, \sigma^2_j)$ and $\varepsilon_{ijk} \sim N(0, \sigma^2)$. The models were fitted using the gls (formula 4) or lme function (formula 5) in S-Plus (Insightful). Again the authors note that normalization to the geometric mean of the best performing reference genes result in smaller standard deviations for almost all normalized genes compared to any single reference gene normalization.

Huang *et al.* (2006) describe a statistical framework to select a set of reference genes with approximately constant expression ratios in given tissues or cells. The fundamental difference between their method and the equivalence test and the linear mixed effects model described earlier is that their approach identifies genes with relatively constant expressions across tissues while the other method select genes with absolutely constant expressions. In Huang *et al.* (2006), the expression levels of the selected genes may vary across the tissues, but the expression ratios of any two of the genes should remain relatively constant for every tissue. This is done by testing the (lack of) parallelism of the lines connecting the mean expressions in the plots where the y-axis is the log expression level and the x-axis is the tissue type. Logarithmic gene expression levels are modelled so that lack of parallelism can be explicitly defined as parameters that can be estimated from expression data. Furthermore, this method controls the overall error rate by obtaining simultaneous confidence intervals for these parameters, and a practical equivalence value for gene expression levels is proposed based on active control genes instead of arbitrarily picking a value as done by Haller *et al.* (2004). This active control gene is a differentially expressed gene known to be involved in the biological phenomenon under investigation.

Formula 6 models the logarithm of expression level y_{ijr} for the rth observation on the gene i from tissue j. τ_{ij} represents the expected log expressed level of the ith gene in the jth tissue, and ε_{ijr} the experimental error present in the rth observation on the gene i from tissue j (random error with mean zero).

$$\log(y_{ijr}) = \tau_{ij} + \varepsilon_{ijr} \qquad (6)$$

Formula 7 measures the lack of parallelism (i.e. absence of interaction of gene i and gene j in tissue s and k).

$$\theta^{sk}_{ij} = (\tau_{is} + \tau_{js}) - (\tau_{ik} - \tau_{jk}), \text{ with } i < j, s < k \qquad (7)$$

The authors try to find those θ^{sk}_{ij} values with absolute values small enough to allow the corresponding reference genes to be used for normalization of gene expression data from the corresponding tissues. To this purpose, they apply a practical equivalence test. The authors note that the geNorm stability value is similar to their measure of interaction but without statistical justification.

Variance of C_t values

A conceptually simple and intuitive way of evaluating reference gene expression stability is the assessment of the variation of the C_t values for a particular gene across the samples. The higher the variation, the less stably the gene is expressed in the experimental setup. Dheda *et al.* (2004) applied this strategy to identify a reference gene with minimal variability under their experimental conditions. They profiled 13 different candidate reference genes in 28 clinical samples (selected from a range of different ages, sex and ethnicity to maximize variability). The variation of the candidate reference genes was visualized using box plots (indicating the median C_t value, the 25 and 75 percentile, and the range), demonstrating clear differences between the genes. The authors propose a standard deviation of less than 2-fold from the mean expression level of a given gene as a requirement for suitability as a reference gene (equivalent to a standard deviation of one in C_t space, assuming 100% PCR efficiency).

While this method is very simple and will separate the more stable from the less stable genes, it is not entirely assumption-free, as it assumes equal amounts of input material (something which is not trivial to guarantee, and what you finally want to correct for using your normalization procedure, critically touching the circular problem of selecting proper reference genes). In any case, the variation of the selected (best) reference gene defines the resolution of the final assay (quantification of the gene of interest). This resolution is dependent on the desired measurement, e.g. one log variation in reference gene is acceptable to reliably detect a two log difference in target gene expression.

ANOVA

Brunner *et al.* (2004) used single-factor ANOVA and linear regression analysis to examine variation among tissues and RT-PCR experiments. C_t values were analysed in Microsoft Excel using single-factor ANOVA and regression analysis in the Analysis ToolPak. Assumptions concerning homogeneity of variance and normality (two requirements to use a parametric ANOVA procedure) were evaluated from inspection of residuals (the difference between an observed value and overall mean for all genes) from the ANOVA. The level and significance of the difference between gene expression levels in different samples are evaluated by Fisher's *F*-statistic (between-tissue-sample mean square divided by the error mean square) assuming the three replicate PCR reactions approximated variance between fully independent observations. The authors further outline their procedure for data analysis to evaluate candidate reference genes.

In a gene quantification experiment of ten candidate reference genes for different tissues of poplar trees, examination of the distribution of the residual values from ANOVA indicated that assumptions concerning homogeneity of variance and normality of data were adequately met. The ANOVA *F*-test of differences among tissues indicated that five of the 10 candidate reference genes showed significant variation in expression among the tissue samples. The mean expression level for each gene in each tissue sample was regressed against the overall means for the different tissue samples. This overall mean provides an index of RNA quality and quantity for that tissue sample. The slope provides an estimate of the degree to which the gene is sensitive to general expression-promoting conditions, and the residuals (deviation from regression prediction) and mean squared residuals estimate the degree to which expression of a gene varies unpredictably after linear effects are removed.

Based on the slope and the coefficient of variation of the regression, the authors define a

stability index as the product of the slope and coefficient of variation. The genes with the lowest stability index will usually provide the best controls. The authors acknowledge that for some studies, no single gene may be adequate. In these cases, the geometric mean of two or more of the most stable reference genes is proposed.

Principal component analysis and autoscaling

Autoscaling is a data pretreatment process that makes variables of different scales comparable. Each variable is autoscaled separately by subtracting its mean value and dividing by its standard deviation (SD). Such a data standardization strategy is perfectly suited for drawing statistically reliable conclusions starting from highly variable datasets (Willems *et al.*, 2008). In gene expression analysis people are sometimes only interested in trends in fold changes and therefore either the logarithm of the expression levels or the C_t values should be autoscaled (Kubista *et al.*, 2006). This is done by the following formula (with the bar denoting an average):

$$Ct_{gene\_A}^{autoscale} = \frac{Ct_{gene\_A}^{raw\,data} - \overline{Ct_{gene\_A}^{raw\,data}}}{SD\left(Ct_{gene\_A}^{raw\,data}\right)} \quad (8)$$

The autoscaled expression values for each gene have zero mean and a standard deviation of one. Hence, in any analysis the genes will be treated as equally important. Whether this is a good assumption or not is up to the researcher. Essentially, analysis of autoscaled data will classify samples and genes based on relative changes in expression, while analysis of unscaled (raw) data also accounts for the magnitudes of the changes in expression. Typically, one does both analyses, since the two classifications may reveal different relations between samples and/or genes (Leung and Cavalieri, 2003).

Table 4.3 shows some data from a study of yeast metabolism (Elbing *et al.*, 2004). Wild-type yeast was grown with ethanol as the carbon source. At time zero glucose was added and the expression of 18 genes was measured as function of time over 60 minutes. The experiment was

Table 4.3 Expression of genes in yeast measured in duplicate as function of time (minutes) after addition of glucose to a yeast culture grown in ethanol

Time	ACTB	ACTB	ADH1	ADH1	ADH2	ADH2	HSP	HSP
C_t values								
0	15.5	15.3	18.5	18.6	16.1	16.1	17.1	17.0
1	15.2	15.2	17.6	18.0	16.7	16.8	16.4	16.2
5	14.9	14.7	16.1	16.2	16.9	16.8	15.8	15.7
10	15.1	15.0	15.5	15.2	18.0	18.1	16.0	15.8
15	15.3	15.3	14.7	14.6	19.3	19.5	16.6	16.2
20	15.6	15.5	14.3	14.6	20.7	20.8	17.4	17.5
30	15.7	15.6	14.7	14.5	21.0	21.2	19.4	19.3
60	16.1	15.7	15.3	15.4	22.3	22.1	22.3	22.3
Autoscaled C_t values								
0	0.197	0.038	1.782	1.698	−1.209	−1.232	−0.239	−0.220
1	−0.591	−0.268	1.180	1.322	−0.948	−0.927	−0.557	−0.573
5	−1.379	−1.796	0.176	0.196	−0.861	−0.927	−0.830	−0.793
10	−0.853	−0.879	−0.226	−0.430	−0.381	−0.360	−0.739	−0.749
15	−0.328	0.038	−0.762	−0.806	0.185	0.251	−0.466	−0.573
20	0.460	0.650	−1.029	−0.806	0.795	0.818	−0.102	0.000
30	0.722	0.956	−0.762	−0.868	0.926	0.992	0.807	0.793
60	1.773	1.261	−0.360	−0.305	1.493	1.385	2.126	2.115

repeated once resulting in duplicate expression profiles. All assays were highly optimized and all genes were significantly expressed, so there was no need to correct for primer-dimers (Chapter 5).

A powerful approach to classify genes and samples based on expression profiles is Principal Component Analysis (PCA) (Chapter 5) (Gower, 1971). Fig. 4.1 shows the raw expression profiles of the yeast genes classified in PC1 vs. PC2 and PC2 vs. PC3 scatterplots. The genes are represented by different symbols based on their functions. The classification is based on unscaled C_t values and hence, the overall magnitudes of the changes in the genes' expression levels are important. The mean expression is reflected by PC1, which is the most significant PC, while variations in expression profiles are contained in the subsequent PCs. Inspecting the PC2 vs. PC3 scatterplots we see that the glycolytic and the glucogenetic genes form two clusters that reflect the common biological functions of its members. Also the candidate reference genes form a cluster.

The next step is to autoscale the data. This gives the same weight to all the genes and PCA will classify them based on their relative changes in expressions. The clusters of the glucogenetic and glycolytic genes in the PC1 vs. PC2 scatterplot are very tight, and also the candidate reference genes form a neat cluster (Fig. 4.2). There is no need to inspect higher order PCs, which are less informative, when we obtain such nice classification with the two main PCs. In fact, it is quite common that the PC2 vs. PC3 scatterplot is the most informative for classification of unscaled data, while the PC1 vs. PC2 scatterplot is the most informative for autoscaled data. The reason is that autoscaling removes the average response of all genes, which often is not particularly selective, and is picked up by PC1 of the unscaled data.

So far we have not normalized the expression levels of the genes of interest to any reference genes. In fact, the candidate reference genes were also classified by the PCA. This is a powerful approach to test the performance of candidate reference genes before selecting those that will

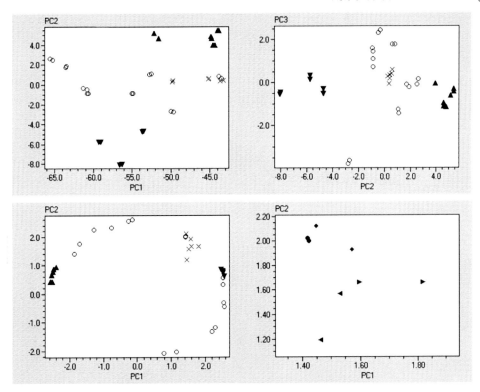

Figure 4.1 Classification of yeast genes based on PCA presented as scatterplots. Top panels show classification of unscaled expression profiles (left: PC1 vs. PC2; right: PC2 vs. PC3) and bottom panels of autoscaled data (left: PC1 vs. PC2; right enlarged section of PC1 vs. PC2). Glucogenetic (▼), glycolytic (▲), reference candidates (×), HSP (●), and other (O) genes. In the enlarged section beta-actin (♦), PDA (▶), IPPI (◀).

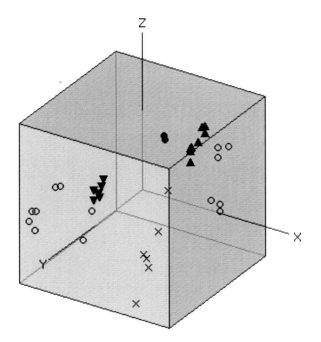

Figure 4.2 PC1 vs. PC2 vs. PC3 scatterplot of autoscaled data. Glucogenetic (▼), glycolytic (▲), reference candidates (×), HSP (●), and other (O) genes.

be used for normalization. The right panel in Fig. 4.1 is an enlarged section of the PC1 vs. PC2 scatterplot of the autoscaled data showing the candidate reference genes. The three candidates show similar spread of the duplicate samples, indicating they have similar stabilities. Comparing the spread of replicates in scatterplots is instrumental in assessing the expression stability of genes, but for stringent comparison the number of biological replicates should be larger. We also see that the three candidate reference genes cluster around a common centre point, with none of them deviating substantially, suggesting that the responses of the reference genes are not biased. Interestingly, the heat shock protein (HSP) is located among the candidate reference genes, suggesting that HSP expression may be invariant in the study. The very tight repeats suggest HSP expression is more stable than that of the three reference gene candidates, which could make it the preferred choice for normalization. However, before jumping to that conclusion we must inspect the variation in the magnitude of HSP expression. In Fig. 4.1 we see that the C_t of HSP ranges from 15.7 to 22.3, which corresponds to $2^{(22.3-15.7)} = 100$-fold variation. This cannot reflect variations in extraction and reverse transcription yields among the samples. The variation in beta-actin expression, for example, is only about one C_t, which is 2-fold. Indeed, if we locate HSP in the PC2 vs. PC3 plot of the unscaled data (Fig. 4.1), we see it is very remote

from the reference gene candidates. In fact, it also separates from the reference genes in a PC2 vs. PC3 (not shown) and a PC1 vs. PC2 vs. PC3 (Fig. 4.2) scatterplot of the autoscaled data. The reason HSP is not distinguished from the reference gene candidates in lower PC dimensions is that its expression profile is different from both the glycolytic and glucogenetic genes as well as of the other regulated genes, and the variation accounted for by the first two PCs (90% based on the eigenvalues) is not sufficient to explain its deviant behaviour.

From the above we conclude that the three candidate reference genes are the best normalizers for this yeast study. We therefore convert all C_t values to relative quantities (copy numbers) and normalize with the expression of the reference genes. The conversion requires we assume values for PCR efficiency and assay sensitivity (Chapter 5). PCA is not particularly sensitive to these parameters, and we assume 90% efficiency for all genes. The copy numbers of the genes of interest are then normalized with the geometric mean of the expression levels of the three reference genes, and the data are converted back to log2 scale. Finally, the data are analysed by PCA (Fig. 4.3). The PC2 vs. PC3 scatterplot of unscaled data clusters the genes based on their overall changes in expression, while the PC1 vs. PC2 scatterplot of autoscaled data clusters the genes based on their relative changes in expression. From the scatterplots we identify five groups:

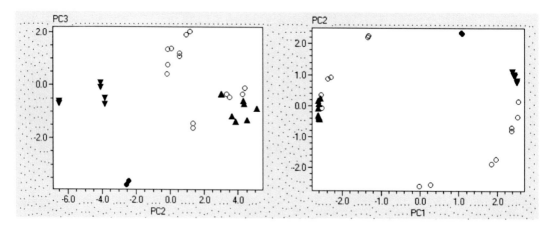

Figure 4.3 Scatterplots of gene expression profiles normalized to the expression of reference genes. PC2 vs. PC3 unscaled (left) and PC1 vs. PC2 autoscaled (right) data. Glucogenetic (▼), glycolytic (▲), HSP (●), and other (O) genes.

three glucogenetic genes, four glycolytic genes that form a cluster with two of the other genes, a group of other genes, HSP, and one other gene. The autoscaled normalized expression profiles are also shown in Fig. 4.4.

Clearly, a procedure based on classification of autoscaled and unscaled expression data by PCA to identify suitable reference genes, followed by normalizing the expression values of the genes of interest with the expression of the validated reference genes for more detailed classification by PCA is very powerful. The entire analysis can be performed using dedicated software such as GenEx from MultiD Analyses (http://www.multid.se). de Kok *et al.* (2005) used PCA in a recent study to evaluate 13 candidate reference genes.

How important is normalization with reference genes? Comparing the scatterplots of the normalized data in Fig. 4.3 with the unnormalized data in the top right (unscaled) and bottom left (autoscaled) panels in Fig. 4.2 we find only minor differences. In fact, autoscaling *per se* is often sufficient for classification of expression profiles. In this study, we measured changes in the expressions of yeast genes as a function of

time, and we did not expect important variations in overall expression levels, extraction and reverse transcription yields. Here, normalization with reference genes, if poorly chosen, could in fact damage the data by adding noise due to large random variations in expression levels or due to systematic variation (e.g. if HSP had been chosen as normalizer). For other data, such as clinical samples from different individuals, possibly from complex tissues and varying disease state the overall expression level and the extraction and reverse transcription efficiencies may differ substantially among samples making normalization to reference genes or *in situ* calibration critical (Stahlberg *et al.*, 2003). Nevertheless, PCA combined with autoscaling is powerful for selection and validation of candidate reference genes and also for the classification of the normalized data.

Confirmation of stable expression in qBase

A typical candidate reference gene selection experiment evaluates between five and ten genes, the more genes studied, the higher the chances of finding stably expressed genes. Once it has been determined which genes and how many are

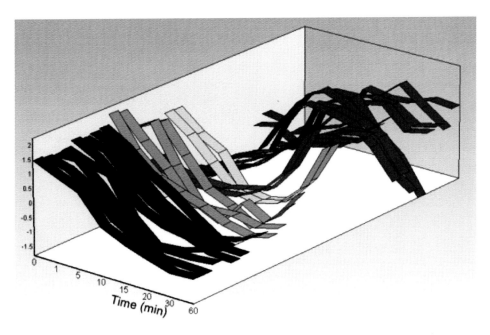

Figure 4.4 Autoscaled logarithmic expression profiles of genes of interest normalized to the expression of reference genes: glycolysis genes (lines 1–8), two other genes with similar profiles (9–12), HSP (13–14), one other gene (15–16), glycolytic genes (17–22), and one group of other genes (23–30).

required for accurate and reliable normalization, this information can be used for future experiments, as long as no significant changes in the experimental setup have been introduced. For example, as it has been determined that *HPRT1*, *GAPD* and *YWHAZ* are the most stable control genes for short-term cultured human fibroblasts, these genes can be used for normalization of all future fibroblast samples, as long as culture conditions, harvesting procedures etc. are kept the same. Nevertheless, it is important to assess the expression stability of the previously selected reference genes in each new experiment (e.g. when culturing the cells again, or extract RNA from new (similar) biopsies, or even when synthesizing new cDNA). This need not be done by re-evaluation all 10 candidate reference genes, but by assessing the performance of the previously selected and validated reference genes. This is automatically done in the qBase software (Hellemans *et al.*, 2007)_(http://medgen.ugent.be/qbase). qBase is a freely available Microsoft Excel based application for management and analysis of real-time PCR data. The program uses a proven delta-C_t relative quantification model with PCR efficiency correction and multiple reference gene normalization along with error-propagation. For each new experiment, qBase aids in the validation of the selected reference genes and normalization process by means of two reference gene quality evaluation parameters. These can only be calculated if more than one reference gene is measured. The coefficient of variation represents the variation of the normalized relative quantities of a reference gene across all samples. Ideally, the variation after normalization is nil. Hence, lower CV values denote higher stability. The M-value is the gene expression stability parameter as calculated by geNorm. The lower the M-value, the more stably expressed is the reference gene. In Table 4.4 these reference gene quality parameters

are calculated for three stably expressed reference genes in a cancer cell line panel.

As mentioned above the geometric mean of multiple (stably expressed) reference genes is a robust and accurate normalization factor. Furthermore, inspecting these normalization factors allows you to inspect possible experimental problems (similar to the integrity index value of the BestKeeper software). qBase displays the calculated normalization factor (geometric mean of indicated reference genes) for each sample along with its standard deviation both in tabular form and in a histogram (Fig. 4.5). Using approximately equal amounts of equal quality input material and proper reference genes, the normalization factor values should be similar for all samples. High variability of the normalization factors indicates large differences in starting material quantity or quality, or a problem with one of the reference gene (either not stably expressed, or not adequately measured). A variation of 2- to 3-fold is generally seen which is acceptable (this is the non-biological variation that you want to remove). Any higher variation should be treated with care.

qBase has been phased out and is now available as a professional real-time PCR data-analysis software qBase*Plus* from Biogazelle (http://www.biogazelle.com).

Conclusion

Several strategies have recently been developed to evaluate candidate reference genes for their suitability as normalizing genes in real-time PCR gene expression quantification experiment. The methods range from simple and intuitive to advanced and some are available as software programs or as scripts to facilitate the evaluation and use by any interested readers. Scientists often ask what method they should use to address the issue of reference gene expression stability in their

Table 4.4 Reference gene quality evaluation in qBase

Gene	Coefficient of variation	geNorm *M*-value
GAPDH	29.75%	0.7781
SDHA	38.66%	0.9219
UBC	30.18%	0.8264
Mean	32.86%	0.8421

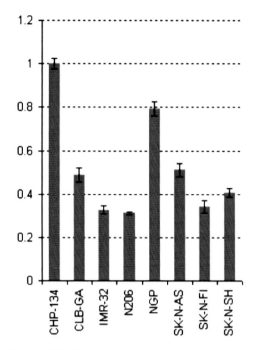

Figure 4.5 qBase normalization factor histogram (geometric mean of three stably expressed reference genes) for indication of possible experimental problems.

experimental setup? There is no clear answer. If there is one lesson to learn from this chapter, it is that every scientist should at least validate their reference gene(s), the actual method used is less critical although it should be reported.

In our laboratory, we have analysed several reference gene data sets over the years (including those from collaborators or people struggling to interpret their data). We applied several of the above mentioned algorithms and software programs, and in almost every case, we obtained highly similar rankings. In a recent study from Willems *et al.* (2006), the authors analysed their reference gene expression data using geNorm and NormFinder and came to the same conclusions, namely that no large differences (except for a few occasional shifts of one or two positions in the obtained rankings) between geNorm and NormFinder were observed.

References

Abruzzo, L.V., Lee, K.Y., Fuller, A., Silverman, A., Keating, M.J., Medeiros, L.J. and Coombes, K.R. 2005. Validation of oligonucleotide microarray data using microfluidic low-density arrays: a new statistical method to normalize real-time RT-PCR data. Biotechniques *38*, 785–792.

Akilesh, S., Shaffer, D.J. and Roopenian, D. 2003. Customized molecular phenotyping by quantitative gene expression and pattern recognition analysis. Genome Res. *13*, 1719–1727.

Andersen, C.L., Jensen, J.L. and Orntoft, T.F. 2004. Normalization of real-time quantitative reverse transcription-PCR data: a model-based variance estimation approach to identify genes suited for normalization, applied to bladder and colon cancer data sets. Cancer Res. *64*, 5245–5250.

Argyropoulos, D., Psallida, C. and Spyropoulos, C.G. 2006. Generic normalization method for real-time PCR. Application for the analysis of the mannanase gene expressed in germinating tomato seed. FEBS J. *273*, 770–777.

Brunner, A.M., Yakovlev, I.A. and Strauss, S.H. 2004. Validating internal controls for quantitative plant gene expression studies. BMC Plant Biol. *4*, 14.

de Kok, J.B., Roelofs, R.W., Giesendorf, B.A., Pennings, J.L., Waas, E.T., Feuth, T., Swinkels, D.W. and Span, P.N. 2005. Normalization of gene expression measurements in tumor tissues: comparison of 13 endogenous control genes. Lab. Invest. *85*, 154–159.

Dheda, K., Huggett, J.F., Bustin, S.A., Johnson, M.A., Rook, G. and Zumla, A. 2004. Validation of housekeeping genes for normalizing RNA expression in real-time PCR. Biotechniques *37*, 112–114, 116, 118–119.

Dheda, K., Huggett, J.F., Chang, J.S., Kim, L.U., Bustin, S.A., Johnson, M.A., Rook, G.A. and Zumla, A. 2005. The implications of using an inappropriate reference gene for real-time reverse transcription PCR data normalization. Anal. Biochem. *344*, 141–143.

Elbing, K., Stahlberg, A., Hohmann, S. and Gustafsson, L. 2004. Transcriptional responses to glucose at different glycolytic rates in *Saccharomyces cerevisiae*. Eur. J. Biochem. *271*, 4855–4864.

Gabrielsson, B. G., Olofsson, L. E., Sjogren, A., Jernas, M., Elander, A., Lonn, M., Rudemo, M. and Carlsson, L. M. 2005. Evaluation of reference genes for studies of gene expression in human adipose tissue. Obes. Res. *13*, 649–652.

Gilsbach, R., Kouta, M., Bonisch, H. and Bruss, M. 2006. Comparison of in vitro and in vivo reference genes for internal standardization of real-time PCR data. Biotechniques *40*, 173–177.

Gower, J. (1971). Statistical methods of comparing different multivariate analyses of the same data, In Mathematics in the Archaeological and Historical Sciences, F. Hodson, D. Kendall, and P. Tautu, eds. (Edinburgh: Edinburgh University Press), pp. 138–149.

Haller, F., Kulle, B., Schwager, S., Gunawan, B., von Heydebreck, A., Sultmann, H. and Fuzesi, L. 2004. Equivalence test in quantitative reverse transcription polymerase chain reaction: confirmation of reference genes suitable for normalization. Anal. Biochem. *335*, 1–9.

Hellemans, J., Martier, G., De Paepe, A., Speleman, F. and Vandesompele, J. 2007. qBase relative quantification framework and software for management and automated analysis of real-time quantitative PCR data. Genome Biol. *8*, R19.

Huang, Y., Hsu, J.C., Peruggia, M. and Scott, A.A. 2006. Statistical selection of maintenance genes for normalization of gene expressions. Stat. Appl. Genet. Mol. Biol. 5, Article 4.

Huggett, J., Dheda, K., Bustin, S. and Zumla, A. 2005. Real-time RT-PCR normalisation; strategies and considerations. Genes Immun. 6, 279–284.

Kanno, J., Aisaki, K., Igarashi, K., Nakatsu, N., Ono, A., Kodama, Y. and Nagao, T. 2006. 'Per cell' normalization method for mRNA measurement by quantitative PCR and microarrays. BMC Genomics 7, 64.

Kubista, M., Andrade, J.M., Bengtsson, M., Forootan, A., Jonak, J., Lind, K., Sindelka, R., Sjoback, R., Sjogreen, B., Strombom, L., *et al.* 2006. The real-time polymerase chain reaction. Mol. Aspects Med. 27, 95–125.

Leung, Y.F. and Cavalieri, D. 2003. Fundamentals of cDNA microarray data analysis. Trends Genet. *19*, 649–659.

Pfaffl, M.W., Tichopad, A., Prgomet, C. and Neuvians, T.P. 2004. Determination of stable housekeeping genes, differentially regulated target genes and sample integrity: BestKeeper – Excel-based tool using pairwise correlations. Biotechnol. Lett. 26, 509–515.

Smith, R.D., Brown, B., Ikonomi, P. and Schechter, A.N. 2003. Exogenous reference RNA for normalization of real-time quantitative PCR. Biotechniques *34*, 88–91.

Stahlberg, A., Aman, P., Ridell, B., Mostad, P. and Kubista, M. 2003. Quantitative real-time PCR method for detection of B-lymphocyte monoclonality by comparison of kappa and lambda immunoglobulin light chain expression. Clin. Chem. 49, 51–59.

Stahlberg, A., Hakansson, J., Xian, X., Semb, H. and Kubista, M. 2004. Properties of the reverse transcription reaction in mRNA quantification. Clin. Chem. 50, 509–515.

Szabo, A., Perou, C.M., Karaca, M., Perreard, L., Quackenbush, J.F. and Bernard, P.S. 2004. Statistical modeling for selecting housekeeper genes. Genome Biol. 5, R59.

Talaat, A.M., Howard, S.T., Hale, W.T., Lyons, R., Garner, H. and Johnston, S.A. 2002. Genomic DNA standards for gene expression profiling in *Mycobacterium tuberculosis*. Nucleic Acids Res. 30, e104.

Vandesompele, J., De Preter, K., Pattyn, F., Poppe, B., Van Roy, N., De Paepe, A. and Speleman, F. 2002. Accurate normalization of real-time quantitative RT-PCR data by geometric averaging of multiple internal control genes. Genome Biol. 3, RESEARCH0034.

Willems E., Leyns, L. and Vandesompele, J. 2008. Standardization of real-time PCR gene expression data from independent biological replicates. Anal. Biochem. 379, 127–129.

Willems, E., Mateizel, I., Kemp, C., Cauffman, G., Sermon, K. and Leyns, L. 2006. Selection of reference genes in mouse embryos and in differentiating human and mouse ES cells. Int. J. Dev. Biol. 50, 627–635.

Data Analysis Software

5

Michael W. Pfaffl, Jo Vandesompele and Mikael Kubista

Abstract

Quantitative real-time RT-PCR (qRT-PCR) is widely and increasingly used in any kind of mRNA quantification, because of its high sensitivity, good reproducibility and wide dynamic quantification range. While qRT-PCR has a tremendous potential for analytical and quantitative applications, a comprehensive understanding of its underlying principles is important. Beside the classical RT-PCR parameters, e.g. primer design, RNA quality, RT and polymerase performances, the fidelity of the quantification process is highly dependent on a valid data analysis. This review will cover all aspects of data acquisition (trueness, reproducibility, and robustness), potentials in data modification and will focus particularly on relative quantification methods. Furthermore, useful bioinformatic, biostatistical as well as multidimensional expression software tools will be presented.

Introduction

Real-time qPCR has a tremendous potential for analytical applications in quantitative DNA and RNA analysis. To tap the full potential, a comprehensive understanding of its underlying quantification principles is important. For most researchers, high reaction fidelity in the nucleic acid (NA) quantification procedure performed is key to a quick result and a biological answer. This is associated with highly standardized pre-analytical steps, like tissue sampling and storage, NA extraction and storage, NA quantity and quality control, and through an optimized RT and/or PCR performance in terms of specificity, sensitivity, reproducibility, and robustness (Bustin, 2004). However, we should not forget that post-qPCR data processing can influence or even change the final result. Events such as qPCR data generation, acquisition, evaluation, calculation and statistical analysis are essential to interpret the biological significance of an experiment.

Data analysis in general is the act of transforming and interpreting data with the aim of extracting useful information and drawing correct conclusions. Depending on the type of data and the biological question, qPCR data analysis might include curve fitting algorithms, data processing, selecting or discarding certain data subsets based on specific pre-set criteria, transformation of logarithm quantification cycle (Cq) values to relative quantities, normalization, rescaling, and a final statistical test of the derived qPCR data. The ultimate goal for qPCR and qRT-PCR is to get a meaningful biological answer at the initial DNA or RNA level, respectively. To understand qPCR data analysis in detail, we have to split the procedure in to various levels, which will be discussed below.

Levels of data analyses in qPCR

On the first level, qPCR data analysis takes place within one raw fluorescence acquisition point. Multiple fluorescence measurements are averaged to a final raw fluorescence reading value per cycle. How many data points are averaged and how the averaging algorithm works is often hidden in the quantification software. Some instrument manufactures, like Bio-Rad (Hercules, CA,

USA), promote the software option of inspecting single cycle fluorescence readings to optimize the polymerase elongation time length to prevent primer-dimers or unwanted and unspecific PCR by-products, or to evaluate reaction kinetics.

The next level of data evaluation is performed within one sample qPCR run, where multiple single fluorescence data points are analysed (mostly between 30 and 50), measured at each cycle and constituting the so-called amplification plot. Most software applications do curve smoothing to show a 'perfect' real-time qPCR history in the amplification plot. We have to realize that the 'real' fluorescence readings, generated in the reaction cup are noisy and rough, and not as 'smooth and beautiful' shaped as shown in the plot. To retrieve the maximal information out of the shape of the amplification plot, rather than a single 'quantification point', more advanced fitting procedures must be applied. In the recent literature various models have been described, either using single (Liu *et al.*, 2002a; Larionov *et al.*, 2005; Ma *et al.*, 2006; Goll *et al.*, 2006) or multiple algorithms fitted to the amplification plot (Tichopad *et al.*, 2003, Wilhelm *et al.*, 2003). More useful information can be generated from these mathematical model variables concerning the background, fluorescence increase and plateau phase (Pfaffl, 2004). Valuable conclusions on qPCR fidelity can be drawn from this data. From the heights of the background fluorescence we can observe if too much reporter dye (SYBR Green I or labelled probe) is present in the reaction setup, whether primer or probe mismatches occur in early cycles and how comparable the samples are. The fluorescence increase gives essential information about amplification fidelity and PCR efficiency. Plateau height informs about the total amount of generated PCR product (Liu *et al.*, 2002b; Tichopad *et al.*, 2004). Some early software applications (LightCycler Software 2001) used amplitude normalization to generate 'identical plateau levels' and unify plateau height (Larionov *et al.*, 2005). The amplitude normalization is based on the suggestion that in an ideal PCR procedure the outcome is determined by the initial available PCR resources. This assumption is valid for ideal PCR but in practice it may not be true. To improve any quantification

procedure the amplitude normalization can be implemented for each individual factor analysed (Larionov *et al.*, 2005).

Further curve smoothing algorithms are applied in melting curve analysis, to test the qPCR product-specificity and to show whether amplimer-dimers, splice variants, or unspecific PCR products have been formed. New approaches to increase the sensitivity of melting curve analysis are the application of innovative saturation fluorescence dyes, e.g. using LC Green (Wittwer *et al.*, 2003) in the LightCycler (Roche Diagnostics, Mannheim, Germany) or Rotor-Gene 6000 (Corbett Life Science, Sydney, Australia) and the use of 'High Resolution Melt' (HRM) curve analysis (Corbett Life Science). These methods are designed to give better-specificity testing and reliable genotyping.

Background subtraction

A prerequisite for cycle threshold (C_t) or crossing point (CP) determination is background fluorescence subtraction. However, accurate measurement of the level of background fluorescence can be a challenge. While a stable and constant background is ideal for subtraction, often the background appears noisy, with a rising or decreasing fluorescence level. Real-time PCR reactions with significant background fluorescence variations occur, caused by drift-ups and drift-downs over the course of the reaction (Wilhelm *et al.*, 2003; Larionov *et al.*, 2005). Averaging over a drifting background will result in an overestimation of variance and thus increase the threshold level (Livak, 1997; Rasmussen, 2001). Key questions surround how the software deals with the background data. Are the data shown real raw fluorescence data or are they already manipulated, e.g. through an additional ROX adjustment or amplitude normalization? Has curve smoothing been applied to the fluorescence data? Which kind of fluorescence background correction and/or subtraction was applied on the hardware?

Most real time platforms show pre-adjusted fluorescence data and in consequence preadjusted CP values. After doing an automatic background correction the CP values are determined by various methods, at a idealized 'constant level' of fluorescence. These constant threshold methods assume that all samples have the same synthesized DNA concentration at the threshold fluorescence. In the recent literature it has been reported that the effect of ROX correction, applied to either an early or late raw fluorescence data, resulted in higher intra-assay variation (up to 35%) and in pre-biased results, which clearly shift the biological answer (Goll *et al.*, 2006).

Background subtraction is a common step in PCR data processing. Often it requires operators involvement to choose between several available options, e.g. subtraction of a minimal value throughout the run, subtraction of an average over a certain pre-platform defined cycle range, or the assumption of different kinds of 'background trends'. To avoid operator involvement we always subtract the minimal value observed in the run. This option has a clear interpretation and works well. It is important that the baseline subtraction is performed after smoothing. So the noise potentially affecting minimal values has already been reduced before baseline subtraction (Larionov *et al.*, 2005).

The real challenge lies in comparing various experimental biological samples. It is not always straightforward to define a constant background for all samples within one run, and in bigger studies between different real-time qPCR runs. These sample-to-sample and run-to-run differences in variance and absolute fluorescence values are leading to the development of a new user friendly CP acquisition modus. Several mathematical models are established to determine the amplification rate, using four parametric logistic or sigmoidal models (Tichopad *et al.*, 2003; Liu *et al.* 2002a,b). These mathematical fit models can also be consulted to determine the optimal CP (Tichopad *et al.*, 2003). Comparable algorithms are already implemented in the LightCycler Software (Roche Diagnostics 2001), and in the Rotor-Gene 3000 and 6000 software (Corbett Life Science). Both types of platform algorithms are more or less 'independent' of the background level, calculate on the basis of raw fluorescence and implement the background data in the CP determination modus (Tichopad *et al.*, 2003; Wilhelm *et al.*, 2003).

Determination of the quantification point

The most important event in qPCR data analysis is to decide at which point of the amplification curve to take the one and only 'quantification point'. This point is well known as cycle threshold (C_t), crossing point (CP) or take-off-point (TOP), depending on the platform used and the analysis software. The importance of the model and algorithm used to get the right 'quantification point' (CP value) is often underestimated (Pfaffl, 2004). All further real-time qPCR data processing events are based on the CP. CP value determination should be done on raw florescence reading data, in a highly standardized reproducible way and independent of any interfering side effects (e.g. ROX adjustment or amplitude normalization). In real-time PCR data analysis, various approaches are used to generate the 'quantification point'. First and widely distributed is the 'cycle threshold method' (C_t method), second is the 'second derivative maximum method' (SDM), and more recently the method of 'non-linear regression analysis' (NLR) and the 'CalQPlex' algorithm (Eppendorf, Germany) was introduced.

Cycle threshold method

For most researchers the C_t method is currently the gold standard. Some real-time cycler software packages offer curve-smoothing and normalization, but the basic C_t method algorithm remains unchanged (Goll *et al.*, 2006). Threshold fluorescence is calculated from the initial cycles, and in each reaction the C_t value is defined by the fractional cycle at which the fluorescence intensity equals the prior set threshold fluorescence. This method is based on an assumption of 'equal' PCR efficiency in all reactions, and accuracy may be compromised if this condition is not met.

The threshold level can be calculated by fitting the intersecting line upon the ten-times value of ground fluorescence standard deviation. This background acquisition mode can be easily automated and is therefore stable and robust (Livak *et al.*, 1997). In the 'fit point method' the user has to discard the uninformative background points, exclude the plateau values by entering the number of log-linear points, and then fit a log-line to the linear portion of the amplification curve. These log-lines are extrapolated back to a common threshold line and the intersection of the two lines provides the C_t value. The strength of this method is that it is extremely robust. The weakness is that it is not easily automated and so requires a lot of user interaction, which can be arbitrary (Rasmussen, 2001; LightCycler Software, 2001).

Sample-to-sample changes in PCR efficiency, caused by RT and PCR inhibitors or enhancers, can question the C_t method and the derived results. As such, the shapes of fluorescence amplification curves differ due to the background level (noisy, constant or increasing), the take off point (early or late), the steepness (good or bad PCR efficiency), the change-over to the plateau phase (quick or steady), and in the appearance of the PCR plateau (constant, in- or decreasing trend) (Tichopad *et al.*, 2003, 2004). PCR amplification efficiency has the largest impact on amplification plot history. Therefore determination of the threshold level, the threshold cycle and in consequence the accuracy of the quantification results are strongly influenced by the amplification efficiency (Pfaffl, 2004).

Second derivative maximum method

Applying the 'second derivative maximum method' the quantification point is automatically identified and measured at the maximum acceleration of fluorescence (Rasmussen, 2001; Tichopad *et al.*, 2003). The exact mathematical algorithm applied in the LightCycler software (Roche Diagnostics) is still unpublished, but is comparable to a logistic or polynomial fit (Tichopad *et al.*, 2003). Corbett Life Science cyclers use a 'comparative quantification' method, where the TOP is calculated on the basis of a sigmoidal model. Both algorithms, the sigmoidal and polynomial curve models, work well with high significance ($P < 0.001$) and coefficient of correlation ($r > 0.99$), as determined in various qPCR experiments with different biological questions (Liu and Saint, 2002a,b; Tichopad *et al.*, 2003, 2004; Rutledge, 2004). Sigmoidal exponential curve fitting was shown to be the most precise method and increases the accuracy and precision of the CP measurements (Wilhelm *et al.*, 2003).

Non-linear regression analysis

Non-linear regression analysis (NLR) has been suggested as an alternative to the C_t method for absolute quantitation (Goll *et al.*, 2006). The advantages of NLR are that the individual sample efficiency is simulated by the model and that absolute quantitation is possible without a standard curve, releasing reaction wells for unknown samples. NLR can be fully automated and may be a powerful tool for analysis of fluorescence data from qPCR experiments. The unfavourable signal to noise ratio of the probe-based assays does not impair NLR analysis. The versatility of NLR depends on the precision needed but if it can be applied, this analysis method may save both time and resources in the laboratory. Further work is needed to improve the precision of the fluorescence copy number conversion factor in order to reduce the bias of NLR observed in this study. However, it is indeed possible to obtain absolute quantitation from real-time qPCR data without a standard curve. In an optimized assay, however, the C_t method remains the gold standard due to the inherent errors of the multiple estimates used in NLR (Goll *et al.*, 2006).

Validity of the crossing point

After the CP value has been determined by one of the above methods, the data point generated has to be validated. Several questions arise and these may be addressed using the assay standard, references and/or controls: is the generated CP data point valuable and reliable?; is the CP sufficiently different from my defined negative standard or non-template-control (NTC) value?; is the outcome comparable with the positive control CP?; is my data point within my defined quantification range, or is the data point out of range so that it must be discarded?.

Here a strict comparison and decision process between analysed CP values and given standards, references and controls is essential and assumed. Especially in so-called 'absolute' quantitative qPCR applications where the quantified NA amount will result in further decision making. A valid and reliable NA quantification is a prerequisite, e.g. in clinical diagnostics, quantitative microbiology and in analysis of genetically modified organisms (GMO) (Burns *et al.*, 2004; Paoletti and Mazzara, 2005). In an 'absolute' quantification approach, the PCR signal is related to input copy number using a calibration curve (Bustin, 2000; Pfaffl and Hageleit, 2001; Fronhoffs et al., 2002). The calibration curve itself can be derived from diluted PCR products, recombinant DNA or RNA, linearized plasmids, or spiked tissue samples. The reliability of such a an 'absolute' real-time RT-PCR assay depends on the condition of 'identical' amplification and reverse transcription (if applicable) efficiencies for both the native mRNA target and the target RNA or DNA used in the calibration curve (Souaze et al., 1996; Pfaffl, 2001). The efficiency evaluation remains one of the essential considerations in qPCR and efficiency correction is necessary in real-time PCR based gene quantification (Rasmussen, 2001; Tichopad et al., 2003).

The synonym 'absolute' quantification is misleading, because the quantification is in fact performed 'relative' to the calibration curve. Clearly, all quantifications should be considered relative (Bustin, 2004). mRNA copy numbers should generally be corrected for differences in input and quality, using various standardization methods, such as mass of tissue, amount of total RNA or DNA (e.g. a verified internal standard), a defined amount of cells, or compared to a reference gene copy number (e.g. ribosomal RNA, or other commonly used reference genes). The current gold standard for gene expression analysis is the use of multiple carefully selected reference genes (Vandesompele et al., 2002). The 'absolute' quantification strategy using various calibration curves and applications has been summarized elsewhere in detail (Pfaffl and Hageleit; 2001; Donald et al., 2005; Lai et al., 2005).

Minimum performance requirements in genetically modified organism (GMO) analysis

The use of a validated method of analysis for 'absolute' quantification of GMOs is required for authorization of GM food or feed products within the European Union (EU). Various EU Commission Regulations (EC) 1829/2003, 641/2004, 882/2004, and a publication by the European Network of GMO Laboratories (ENGL) describe the validation procedure step-by-step. The assessment of method performance and its suitability for GMO testing consists of two independent steps. The purpose of the first phase is to verify that the method performance fulfils the requirements to enter the validation process. Acceptance criteria, like applicability, practicability,-specificity, dynamic range, accuracy, amplification efficiency, correlation of the calibration curve, reproducibility, limit of detection (LOD), limit of quantification (LOQ), and robustness are evaluated and strict cut off criteria are set. The accuracy of the qPCR method must be within $\pm 25\%$, the standard deviation should not exceed $\pm 25\%$ over the whole range of quantification, and the real-time PCR amplification efficiency must be in the range of 90–110% (Paoletti and Mazzara, 2005). If all criteria fulfil the minimal criteria, an inter-laboratory collaborative trial is launched to validate the method in various countries. In the second phase evaluation of method performance characteristics must be done at the end of the validation study, by analysis of collaborative trial results. The purpose of the second phase is to verify that the method performance, as demonstrated by the results of the collaborative trial, confirms the characteristics indicated at the moment of submission in a multi-laboratory setting, and fulfils the requirements to be considered fit for regulatory purpose (Paoletti and Mazzara, 2005).

The GMO validation and strict exclusion criteria represent a good example and basis for further validation strategies in other areas of NA quantification. In clinical diagnostics, quantitative virology and microbiology such stringent and international accepted assay validation premises are urgently needed. They should be established in the near future and disseminated to analytical laboratories as 'quantitative PCR Good Laboratory Practice' (qPCR GLP).

Relative qPCR data analysis

Besides the 'absolute' gene quantification strategy, the relative expression strategy is commonly used by the academic research community. Relative or comparative gene expression analysis compares a gene-of-interest (GOI) in relation to a reference-gene (RG). Today various mathematical models are published and established to calculate the relative expression ratio (R), based on the comparison of the distinct cycle differences, called delta-CP model. In most applications the relative quantification is based on the 'delta-CP' (ΔCP,

Wittwer *et al.*, 2001) or 'delta-delta-CP' (ΔΔCP) values, described by Livak and Schmittgen (2001).

Screening the recent literature applying relative quantification and using the ΔΔCP method (equation 1), the majority of researchers assume an optimal doubling during each PCR cycle with an amplification rate of 100% (i.e. base of exponential amplification of 2) (Livak, 1997, Livak and Schmittgen, 2001).

$$R = 2^{-[\Delta CP_{sample} - \Delta CP_{control}]} \tag{1}$$

This is acceptable for a first approximation of the rough expression ratio (R) between the expressed genes. To obtain reliable relative expression data, more advanced and efficiency (E)-corrected models are published. The assessment of the sample-specific amplification efficiencies must be carried out for each gene and tissue analysed. Models are based on one analysed sample (equations 2 and 3) (Pfaffl, 2001; LightCycler Relative Quantification Software, 2001) or on a group of various samples (equation 4) (Pfaffl *et al.*, 2002).

$$R = \frac{(E_{target})^{\Delta CP_{target}(control-sample)}}{(E_{ref})^{\Delta CP_{ref}(control-sample)}} \tag{2}$$

$$R = \frac{(E_{ref})^{CP_{sample}}}{(E_{target})^{CP_{sample}}} \div \frac{(E_{ref})^{CP_{calibrator}}}{(E_{target})^{CP_{calibrator}}} \tag{3}$$

$$R = \frac{(E_{target})^{\Delta CP_{target}(MEAN\ control-MEAN\ sample)}}{(E_{ref})^{\Delta CP_{ref}(MEAN\ control-MEAN\ sample)}} \tag{4}$$

To be more stable and reliable when applying relative quantification strategy, new approaches using multiple reference genes were introduced (Vandesompele *et al.*, 2002). To base the relative quantification not only on one reference-gene, multi reference-gene normalization software was developed, e.g. Relative Expression Software Tool (REST-384; Pfaffl *et al.*, 2002), qBase (Hellemans *et al.*, 2007), and qPCR-DAMS (Jin *et al.*, 2006). These software applications apply efficiency corrected calculation models, based on multiple sample and on multiple reference-genes

(RG index or normalization factor) consisting at least of three stably expressed reference-genes, as shown in equation 5 (Pfaffl *et al.*, 2004).

$$R = \frac{(E_{target})^{\Delta CP_{target}(MEAN\ control-MEAN\ sample)}}{(E_{RG\ index})^{\Delta CP_{RG\ index}(MEAN\ control-MEAN\ sample)}} \tag{5}$$

Some helpful software tools and Excel spreadsheets (Microsoft Corporation, USA) are available to correct for qPCR efficiency differences. The LightCycler Relative Expression Software (2001), Q-Gene (Muller *et al.* 2002), qBase (Hellemans *et al.*, 2006), SoFar (Wilhelm *et al.*, 2003), DART (Peirson *et al.*, 2003), qPCR-DAMS (Jin *et al.*, 2007) and various REST software applications (Pfaffl *et al.* 2002) allow the evaluation of amplification efficiency plots. In most of the applications a triplicate determination of real-time PCR efficiency for every sample is recommended. Therefore, PCR efficiency corrections should be included in the relative quantification procedure. Therefore a stand-alone software tool, named KEST 2008, was developed to implement the single-run PCR efficiency correction in relative quantification models (Pfaffl, Technical University Munich and Corbert Life Science; download: http://rest. gene-quantification.info/).

Complex qPCR data analysis

The next data analysis level is the comparison between different qPCR runs. The difference between organisms and/or treatment groups, e.g. treatment groups, sex, age, or along a time course, is the real biological question. Various non-treated control or reference treated groups must be implemented in the gene expression comparison. Further multiple inter-run calibrators have to be included in each run and on each plate to allow correction between different runs and repeats. The final goal is to make several runs fully comparable. At this point complex qPCR analysis software and databases are necessary, like qBase or qPCR-DAMS, to deal with large amounts of real-time data.

The next focus is on multiple target genes or multiple transcript analyses within one biological sample, and how to make them comparable.

Commonly this is done by using normalization strategies based on one or more reference-genes (Vandesompele *et al.*, 2002; Pfaffl *et al.*, 2004a). The quality of normalized quantitative expression data cannot be better than the quality of the normalizer itself. Any variation in the normalizer will obscure real changes and produce artefactual changes (Bustin *et al.*, 2002, 2005). These normalizing strategies are summarized and described in detail in published reviews (Huggett *et al.*, 2005, Wong and Medrano, 2005) or in Chapter 4.

Comparison between biological samples represent the next level of complexity. qPCR expression data within different organs, tissue, or types of cell culture can be analysed, using multi dimensional data analysis software applications, e.g. GenEx software. The results can be combined to give multi-tissue expression results or to provide a complex system biological overview. This represents a very high level of complexity and a lot of data conversion, corrections and normalization must be applied to make the real biological expression events visible and comparable.

The flowchart in real-time qPCR data analysis

The following flowchart of the key steps in the qPCR data analysis process, are necessary to give reliable and meaningful biological conclusions. Most of the steps are hidden in fixed software algorithms or models created and defined by the real-time PCR platform manufacturers. But some of the key steps are open and can be influenced by the researcher. The essential key steps are summarized below to indicate how they are influenced by given software applications.

(1) Scope of qPCR application

The researcher has to define the applicability, practicability and type of quantification (absolute or relative). One or more internal or external control samples, negative- and positive-controls, non-template controls (NTC), calibrators and/or reference samples, must be defined. At this stage the scientist has to define minimal and maximal detection limits, minimal and maximal quantification limits, quantification ranges and set unique characteristics for further data evaluation. These data can only be recruited from earlier experiments using identical chemistry,

consumables and real-time PCR platforms. As an example the highly defined 'Minimum Performance Requirements for Analytical Methods of GMO Testing' can be mentioned as a gold standard for test validation (Paoletti and Mazzara, 2005).

(2) Data transfer and profiling

When the raw data is exported from the real-time cycler optical unit and imported to the calculation unit, the system software should (automatically) verify the completeness and integrity of transferred data sets, e.g. missing data or possible outliers. At this stage the auditor cannot perform any relevancy checks on the data sets. Missing or discarded data points must be reported according to the set of qPCR GLP regulations.

(3) Initial data conversion

The fluorescence data are modified and then presented for further calculations. These modifications are done automatically and software specific and cannot be influenced by the researcher. The following initial data conversions are possible: averaging from multiple measurements, corrections with other fluorescence channels, e.g. ROX correction, curve smoothing, amplitude normalization, deletion of outliers, and often an automatic fluorescence background adjustment.

(4) Data analysis

The data sets should be analysed according to the needs of the investigator and the given biological question, e.g. for further quality and integrity tests. Further data processing can be done to the 'raw fluorescence data':

- Data point evaluation, deletion of outliers, management of missing data, amplification plot curve smoothing, and amplitude normalization can be performed.
- A reliable and reproducible background subtraction should be performed, automatically or manually, to correct for differing background fluorescence levels. At this level a lot of different methods and algorithms can be used, mostly pre-set by the platform manufactures (C_t method, SDM, NLR).
- The threshold line should be drawn, automatically or manually, to determine the right

cycle threshold value (C_t, CP, or TOP).

• On some platforms mathematical models, e.g. sigmoidal or logistic models, help the researcher to find the 'right quantification point' and calculate the amplification efficiency in each analytical sample (sigmoidal fit, logistic fit, SDM, NLR, CalQPlex).

(5) Verification of CP data

At this stage the CP data must be verified according to the minimal requirements set for the experiment. These are as discussed in '(1) Scope of qPCR application'. At this point the melting curve analysis may be checked to verify the correct melting point, purity and integrity of amplified real-time PCR product.

(6) Advanced data analysis

Once the CP has been calculated more advanced data analysis can be performed. This depends upon the experimental setup or the biological question. Either an absolute or relative/comparative quantification strategy can be chosen. Data can be analysed using various macros, which will be discussed later in this chapter (qBase, REST, GenEx, qGene, SoFar, etc.). CP data can be further converted by normalization using one or more reference genes or internal, exogenous or artificial controls. Various software applications are distributed within the qPCR community, e.g. geNorm, BestKeeper, REST, Normfinder, Global Pattern Recognition, etc. Advantages and disadvantages of such applications are discussed in Chapter 4.

(7) Statistical data analysis

The statistical analysis of quantitative real-time PCR data is currently a topic with seven secrets! The group comparison has to be done with robust statistical methods, to get scientifically proven and reliable results. However, after several steps of normalization, correction, and adjustment using multiple references and calibrators, which statistical tool is reliable? In the literature only a few publications deal with the statistical analysis of expression data generated by real-time PCR and how they might be analysed in a reliable and statistically correct way (Pfaffl *et al.*, 2002; Muller *et al.*, 2002; Peirson *et al.*, 2003; Yuan *et al.*, 2006; Gilsbach *et al.*, 2006).

(8) Reporting and documentation

Research findings have to be summarized and reported. Results have to be documented and the type of report most suitable to describe the outcome of the applied experiment must be selected. Numbers of molecules, masses or molarities are reported in an absolute quantification, or the fold regulation in a relative quantification. Research findings have to be documented in the correct format for qualified scientific reports, original papers, spreadsheets, or flowcharts, to name a few. Clear and reliable documentation is essential to support the researchers experimental results, independent of whether they are positive or negative, significant or non-significant.

(9) Data visualization

A very important final part of data analysis is the clear data visualization and presentation in a scientific article. The reader should clearly see the experimental result, and the statistical significance without any additional confusing data. Graphical visualization, plot interpretation and the physiological conclusion should be straight forward. Error bars should always be provided.

qPCR data analysis software applications

At the time of writing qPCR analysis software has tended to be eclipsed by the production of new real-time PCR platforms and detection formats. Hardware and chemistry have developed much faster than detection and analysis software. However, we are now at the beginning of a new post qPCR data processing area. The challenge is the development of simple but reliable gene expression analysis and quantification software. The development of 'one-fits-all' software is the goal and this would appear to be the optimal solution. However, can we implement various detection chemistries bearing varying background levels and fluorescence acquisition modes in one analytical software? Optimized analysis models for each real-time platform and for each chemistry might be more effective.

In research and in clinical diagnostics, real-time qRT-PCR is the method of choice for expression profiling. However, accurate and straightforward mathematical and statistical analysis of qPCR data and management of growing data sets have become the major hurdles to effective implementation. Nowadays up to 96-well applications are the standard in research, but in the near future high-throughput 384-well applications will generate huge amounts of qPCR data. qPCR data needs to be grouped, standardized, normalized, and documented by intelligent software applications (Hellemans *et al.*, 2007). Real-time qPCR data should be analysed according to automated statistical methods, e.g. kinetic outlier detection (KOD), to detect outliers and samples with dissimilar efficiencies (Bar *et al.*, 2003; Burns *et al.* 2005). Often the statistical data analysis is performed on the basis of classical parametric tests, such as analysis of variance (ANOVA) or *t*-tests. Parametric tests depend on assumptions, such as normality of C_t value distributions, the validity of which is often unknown to the researchers (Pfaffl *et al.*, 2002; Muller *et al.*, 2002). In relative gene expression analysis, the quantities are derived from ratios where variances can be high. Furthermore, normal distributions might not be expected as such. Logarithmic transformation is a prerequisite for gene expression analysis. It is unclear how a parametric test could be optimally constructed (Pfaffl *et al.*, 2002). New analysis formats, using new algorithms for data analysis and ideas for reliable statistical analysis of real-time qPCR data are urgently required.

To show the state-of-the art in qPCR analysis software the available applications are briefly described:

LightCycler Relative Quantification Software

The first commercially available software for relative quantification was the LightCycler Relative Quantification Software (Roche Diagnostics, 2001). It allows calculation and comparison of the relative quantification results for triplicates of a target versus a calibrator gene. Target genes are corrected via a reference-gene and calculated on the basis of the median value CP of the performed reaction triplets. In newer software versions more repeats can be compared. Real-time PCR efficiency correction is possible within the software and is calculated from the calibration curve slope. A given correction factor and a multiplication factor, which are provided in the

product-specific applications by Roche Diagnostics (2001), have to be entered in the calculation process. Data output is done via a relative expression value, but the software application still lacks a reliable statistical analysis or confidence range of the analysed data (Relative Quantification Software 4.0, Roche Diagnostics, 2004).

REST

The Relative Expression Software Tool (REST) is Microsoft Excel-based and programmed in the Visual Basic Application (VBA), to compare several expressed genes. The basic version compares two treatment groups, with multiple data points in sample group versus control group, and calculates the relative expression ratio between them. The mathematical model used is published and based on the mean CP deviation between sample and control group of target genes, normalized by the mean crossing point deviation of one reference gene, as shown in equation 4. Furthermore, an efficiency correction can be performed, either based on the dilution method or an optimal efficiency of $E = 2.0$ is assumed. The big advantage of REST is the subsequent statistical test of the analysed CP values by a pairwise fixed reallocation randomization test (Pfaffl *et al.*, 2002). Permutation or randomization tests are a useful alternative to better known parametric tests for analysing experimental data (Manly, 1997; Horgan and Rouault, 2000). They have the advantage of making no distributional assumptions about the data, while remaining as powerful as conventional statistical tests. Randomization tests are based on what we know to be true: that treatments were randomly allocated. The randomization test repeatedly and randomly reallocates, at least 2000 times, the observed CP values to the two groups and notes the apparent effect on CP difference each time. The proportion of times random allocation produces a more significant result is reported. The REST software package makes full use of the advantages of a randomization test. In the applied two-sided pairwise fixed reallocation randomization test for each sample, the CP values for reference genes and target genes are jointly reallocated to control and sample groups (= pairwise fixed reallocation), and the expression ratios are calculated on the basis of the mean values. In practice, it is impractical to examine all possible allocations of data to treatment groups, and a random sample is drawn. If at least 2000 or more randomizations are taken, a good estimate of *P*-value (standard error < 0.005 at *P* = 0.05) is obtained. Randomization tests with a pairwise reallocation are seen as the most appropriate approach for this type of application. New REST versions were developed and released in 2005, calculating a geometrically averaged RG index, according to earlier publications (Vandesompele *et al.*, 2002, Pfaffl *et al.*, 2004) analysing 15 target and reference genes (REST-384). Specialized REST-Multiple Condition Solver versions can compare six treatment group with one non-treated control (REST-MCS). During the last years two new REST tools, REST 2005 and REST 2008, were developed by Pfaffl and co-workers at the Technical University Munich and Corbert Life Science (Sydney, Australia). Beside the multiple reference gene normalization, in REST 2008 the single run-efficiency was implemented as a new tool. A further improved statistical testing method on the basis of a bootstrapping method is applied (http://rest.gene-quantification.info/).

Q-Gene

Q-Gene manages and expedites the planning, performance, and evaluation of quantitative real-time PCR experiments (Muller *et al.*, 2002). The software is able to perform a statistical test on the real-time C_t data. Efficiency correction according to the dilution method is also possible. The Q-Gene software application is a tool to cope with complex quantitative real-time PCR experiments at a high-throughput scale (96-well and 384-well format) and considerably expedites and rationalizes the experimental setup, data analysis, and data management while ensuring the highest reproducibility. The expression results are presented by graphical presentation. The Q-Gene Statistics Add-In is a collection of several VBA programs for the rapid and menu-guided performance of frequently used parametric and non-parametric statistical tests. To assess the level of significance between any two groups expression values, it is possible to perform a paired or an unpaired Student's test, a Mann–Whitney *U*-test, or Wilcoxon signed-rank test. In addition, the Pearson's correlation

analysis can be applied between two matched groups of expression values. Furthermore, all statistical programs calculate the mean values of both groups analysed and their difference in per cent (Muller *et al.*, 2002).

qBase

The comprehensive software application qBase was recently developed as a generalized solution to accommodate virtually all relative quantification setups (Hellemans *et al.*, 2007). qBase is an Excel based tool for the management and automatic analysis of real-time quantitative PCR data (http://medgen.ugent.be/qbase/). It employs a proven delta-C_t quantification model with efficiency correction, multiple reference gene normalization and accurate error propagation along all calculations. The qBase browser allows data storage and annotation while keeping track of all real-time PCR runs by hierarchically organizing data into projects, experiments, and runs. It is compatible with the export files from many currently available PCR instruments and provides easy access to all data types (both raw and processed). The qBase analyser contains an easy plate editor, performs quality control, converts CP values into normalized and rescaled quantities with proper error propagation, and displays results in both tabulated and graphical forms. A big advantage of the program is that it does not limit the number of samples or genes and replicates, and allows data from multiple runs to be combined and processed together (Hellemans *et al.*, 2007). The possibility to use up to five reference genes allows reliable and robust normalization of gene expression levels, on the basis of the *geNorm* normalization procedure (Vandesompele *et al.*, 2002). qBase allows easy exchange of data between users, and exports tabulated data for further statistical analyses using dedicated software. qBase has been phased out and is now available as professional real-time PCR data analysis software qBase*Plus* from Biogazelle (http://www.biogazelle.com).

SoFar

The algorithms implemented in SoFar (distributed by Metralabs, Germany) allow fully automatic analysis of real-time PCR data obtained with a LightCycler (Roche Diagnostics) instrument. The software yields results with considerably increased precision and accuracy of real-time quantification. This is achieved mainly by the correction of amplification-independent fluorescence signal trends and a robust fit of the exponential phase of the signal curves. The melting curve data are corrected for signal changes not due to the melting process and are smoothed by fitting cubic splines. Therefore, sensitivity, resolution, and accuracy of melting curve analyses are improved (Wilhelm *et al.*, 2003).

qCalculator

The qCalculator is a VBA-based program that enables the calculation of relative expression data derived from quantitative real-time PCR experiments (Gilsbach *et al.*, 2006). qCalculator enables a flexible calculation of 32 samples (in triplicate) for 20 experiments based on up to 10 reference genes. Each gene can act as target or reference gene without rearrangement of the data, as in REST-384 software. Calculation of the relative expression is done with efficiency correction (according to Pfaffl 2001), or without, on the basis of the delta-delta-C_t efficiency correction approach (Livak and Schmittgen, 2001). Samples for relative comparison can be freely selected and changed at any point of the analysis. Efficiency calculation and absolute quantification is done on the basis of standards curves. Extreme values like outliers can be excluded at any point of the qCalculator analysis procedure. Any changes in the parameters set leads to the immediate recalculation of data. qCalculator includes an automatic statistical comparison of different choices of reference genes. Data output is shown as arithmetic mean (mean ± s.e.m.) of *n* experiments. The two-tailed Student's t-test can be used to compare results under the precondition that their standard error of mean (s.e.m.) values do not differ significantly. The *F*-test is used to verify this requirement. GraphPad Prism 4.0 (GraphPad Software, San Diego, USA) is used to test the normal distribution of the results using the Kolmogorov–Smirnov test (KS test) and to calculate the associated *P* values according to Dallal and Wilkinson (1986). Results and error propagation is based on log2-values and are summarized and graphically displayed. The qCalculator manual and software can be downloaded from the homepage: www.pharmakologie. uni-bonn.de/frames/index_fr.htm

Dart-PCR

Dart-PCR (data analysis for real-time PCR) provides a simple means of analysing real-time PCR experiments from the raw fluorescence data (Peirson *et al.*, 2003). It allows an automatic calculation of amplification kinetics, as well as performing the subsequent calculations for relative quantification and calculation of assay variability. Amplification efficiencies are also tested to detect anomalous samples within groups (outlier detection) and differences between experimental groups (amplification equivalence). Data handling is simplified by automating all calculations in an Excel worksheet and this enables the rapid calculation of C_t values, amplification rate and resulting starting values, along with the associated error, from raw data. Differences in amplification efficiency are assessed using one-way analysis of variance (ANOVA), based upon the null hypotheses, that amplification rate is comparable within sample groups (outlier detection) and that amplification efficiency is comparable between sample groups (amplification equivalence) (Peirson *et al.*, 2003).

Gene Expression Macro

The Gene Expression Macro (Bio-Rad) is a simple tool for calculating relative expression values from real-time PCR data generated by the iCycler, iQ or MyiQ systems (Bio-Rad). The gene Expression Macro runs using Excel and contains specialized data analysis functions. The use of this macro can save valuable time by employing standard methods of relative gene expression analysis in predesigned, easy-to-use spreadsheets. The calculations in the spreadsheet are outlined in the methodology tab of the macro. They are derived from the algorithms outlined by geNorm (Vandesompele *et al.*, 2002). The calculations allow the user to analyse results using multiple reference genes. This often results in more consistent results and more confidence in the conclusions drawn from experiments. As demonstrated in the example calculations provided on the on Gene Expression Macro web page the algorithm does take PCR efficiencies of each assay into account. In general, all of the methods included build upon on the delta-delta-C_t calculation (Livak and Schmittgen, 2001) and assume optimal doubling in each cycle and PCR efficiencies of two. Further an efficiency correction according to Pfaffl (2001) can be implemented in the calculation process. To exactly reproduce the results you would obtain using the 'Pfaffl model' you must define a control sample (calibrator sample in the publication). The Gene Expression Macro will express all values relative to this sample. The control sample is given a value of 1. Relative expression data output is done in a graph using mean expression values and standard deviations. No statistical tool for qPCR data verification is included in the Gene Expression Macro.

qPCR DAMS

Recently a new databank named qPCR-DAMS (Quantitative PCR Data Analysis and Management System) was developed (Jin *et al.*, 2006). It is a database management system implemented on MS Access 2003 (Microsoft Corporation) and VBA. The database tool works with integrated mathematical procedures and is designed to analyse, manage, and store relative and absolute quantitative real-time PCR data. The system consists of five independent blocks: three blocks (gene, plate, and experiment) for inputting, storing and describing raw data, and two blocks (view data and process data) for checking, evaluating, and processing data. Users are allowed to choose among four basic outputs: (1) ratio relative quantification, (2) absolute level, (3) normalized absolute expression, and (4) ratio absolute quantification. A further two advanced options are available within the software package: (5) multiple reference relative quantification and (6) multiple references absolute quantification. The coefficient of variation is monitored at each step during data processing and the accuracy is further improved by an easy data tracking and display system. Unfortunately, qPCR-DAMS lacks a statistical tool to verify the qPCR data. In summary, qPCR-DAMS is a handy and easy-to-use tool to host, manage, evaluate, process and store data from both relative and absolute quantitative real-time PCR. The qPCR-DAMS software and online manual are available free for academic use at http://www.cvm.okstate.edu/research/Facilities/LungBiologyLab/.

Expression profiling

In most studies the expression of a key marker gene in test samples is compared. A number of biological repeats are taken of each sample, and then robust statistical methods are used to infer if there is significant difference between the populations from which the samples where drawn. One may also compare a test sample with standards to estimate, for example, the viral load of a patient for clinical decisions. The approach, however, is only useful when there is a dominant marker gene whose expression signifies the biological condition or a pathogenic RNA. This is not the case for complex diseases, which are caused by a combination of factors and cannot be monitored through the expression of a single gene. Such diseases can only be studied in a meaningful way by expression profiling. In profiling experiments the expression of many genes, optionally all genes, are measured and the disease condition is identified from the pattern of the expressed genes (D'haeseleer, 2005). In some situations the detailed expression pattern can also have prognostic value. Traditionally expression profiles are measured using microarrays, by which the expression of all genes can be assessed in a single experiment. However, the quality of microarray expression data is usually not good enough for detailed classification and accurate prognosis. Real-time PCR is more sensitive, has a wider dynamic range and better reproducibility. Cost per test is also lower, and samples can be measured for the same money. The drawback is that it is not feasible to measure the expression of all the genes. However, this is rarely limiting, because only a fraction of genes have their expression significantly affected in most disease conditions. The expression of most genes show no or little disease dependence that can be discerned from noise and fluctuations caused by other factors. The preferred approach is to use microarrays to identify potentially informative genes, and then validate them by real-time PCR, before setting up a real-time PCR expression profiling study based on those that pass validation. Measuring the expression of only informative genes opens the possibility of data normalization by autoscaling (see below), which makes analysis very robust (Kubista *et al.*, 2006).

Data pre-processing

Real-time PCR raw data are expressed as C_t values. Raw C_t values are not readily comparable for many reasons and may not accurately reflect the initial number of target molecules in the sample. The C_t value depends on instrument factors (gain, filters, light source), detection chemistry (dye, probe), amplicon (longer amplicons bind more dye), and PCR efficiency, which may vary among assays as well as among samples. The C_t value may also be compromised by primer–dimer formation. Further, C_t values reflect the number of molecular copies of cDNAs in the samples, while we are more interested in their concentration. The raw C_t values require reliable normalization. In regular real-time PCR studies the expression of the reporter gene is normalized with the expression of properly selected reference(s) genes. Such normalization also compensates for non-gene-specific variations in extraction efficiency, reverse transcription yield and PCR inhibition. The proper way to pre-treat real-time PCR data for expression profiling is as follows:

1   Compensate for any variation between runs. If data were collected over several runs, C_t values should be related to a common reference sample included in all runs.
2   Correct any primer–dimer contributions. C_t values due to primer–dimer signals should be set to the same value that is higher than any C_t value reflecting product formation.
3   Correct for variations in PCR efficiency between assays (based on standard curve).
4   Normalize for variations in PCR efficiency between samples (normalize with spike).
5   Normalize to sample amount. The copy numbers should be normalized to the volume of biological material, e.g. serum volume or number of cells or alternatively, normalization to the amount of total RNA.
6   Optionally normalize to reference genes. If there is reason to believe there are sample-to-sample variations in extraction efficiency, reverse transcription or PCR inhibitors, these factors can be accounted for by normalization with a stably expressed gene or genes that are not affected under the conditions studied. Such genes may not always be available in an experimental design (Sindelka et al., 2006). To help identifying suitable reference genes, panels of reference gene candidates have been made available for the most common species (www.tataa.com/referencepanels.htm and http://www.primerdesign.co.uk/geNorm.asp).

7   Calculate relative quantities.
8   Convert to log scale. Regulation of expression typically results in fold changes and data should be analysed in logarithmic scale. By convention base 2 is used.
9   Mean centre. Absolute gene expression levels are difficult to measure and often also to interpret. More robust and reliable are relative expression levels. In expression profiling expression levels are often expressed relative to the mean expression of each gene. This is called mean centring of data. Mean centred data has for every gene average expression of zero. Mean centred data are often used to classify samples.
10 Auto-scale. Often relative changes in expression levels are more informative than absolute changes. These are reflected by autoscaled data. Data are autoscaled by first mean centre them (see above) and then dividing with the standard deviation of the expression of each gene. Hence, auto-scaled gene expression data have mean expression of zero and a standard deviation of 1. Autoscaled data are used to classify genes.

Preprocessing is readily performed using software such as GenEx (www.multid.se) and in many cases the preprocessing scheme can be simplified. If there are no primer–dimer signals, the same PCR efficiency and assay sensitivity can be assumed, and if the samples were based on the same amount of material, steps 2–9 cancel and the measured C_t values can be auto-scaled directly. Another difference between expression profiling and comparisons based on the expression of a single gene is that in expression profiling biological repeats are not averaged. Treating them separately it is possible to compare the variation among different biological repeats to tell whether the samples are different or not.

The scatter plot

The minimum number of genes in an expression profiling experiment is two. For reciprocally regulated marker genes the profile based on just two genes can be very sensitive. A classical example is the test for lymphoma based on measuring the relative expression of the kappa and lambda variants of the immunoglobulin light chain. Each β-cell decides during maturation which of the light chain variants to express. In healthy humans 40% of the cells express the kappa variant while 60% express the lambda variant. Like all cancers, lymphoma is a clonal disease, resulting in a large number of identical ß cells originating from the same original cancer cell. All these cells will express either the kappa or the lambda variant, which will result in a kappa to lambda expression ratio that deviates from 60:40. A convenient way to visualize such data is in a scatter plot. If the samples are not normalized to the same sample amount, negative samples will fall on a straight line corresponding to the 60:40 expression ratio (Fig. 5.1). Below and above that line are positive samples with kappa and lambda clonality, respectively (Stahlberg *et al.*, 2003).

The expression of three genes can be visualized in three-dimensional scatter plots, where again similar samples will be close to each other.

The idea of clustering samples in a space made up of expression levels can be generalized to any number of genes. Spaces of higher dimensionality than three cannot be visualized by conventional means, and a number of powerful clustering tools have been developed to analyse such spaces and to present the information. Some of the most common are described below.

Principal component analysis

The information in a multidimensional space can be collected in a space of lower dimension by means of so called principal components (PC). The PCs are defined as vectors that account for most of the variation in the data in the multidimensional space. The first PC is the best linear fit to the data in the multidimensional space; the second PC is orthogonal to the first and accounts for most of the variation not accounted for by the first PC; the third PC is orthogonal to the first two PCs and accounts for most of the remaining variation etc. Most of the information in the original multidimensional expression space can now be represented in a graph of lower dimensionality using the main PCs as axes.

Fig. 5.2 classifies the *Xenopus laevis* development stages based on the expression of 18 genes. Three clusters are seen. They can be identified

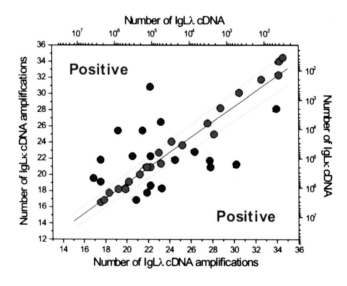

Figure 5.1 Classification of lymphoma samples in a scatter plot based on the expression of kappa and lambda variants of the immunoglobulin light chain. Each symbol represents a sample and its position in the plot is given by the CT values of the immunoglobulin κ and immunoglobulin λ reactions as indicated by the left and bottom axes. The right and top axes indicate the corresponding number of cDNA copies. Negative samples in the plot cluster around a diagonal line.

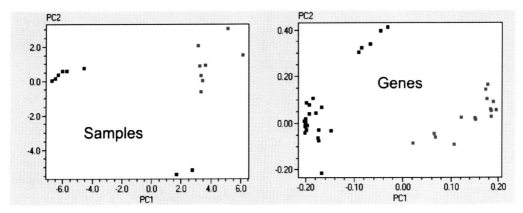

Figure 5.2 Classification of developmental stages (left) and of genes (right) in *Xenopus laevis* by Principal Component Analysis and visualized in a PC1 vs. PC2 plot. Early stages/genes are shown in red, midblastula stages/genes in blue and late stages/genes in green.

as the early 1–8.5 stages, stages 11 and 15 corresponding to the mid-blastula transition, and the late stages 17–44. Fig. 5.2 also shows corresponding classification of the genes, with three main clusters, corresponding to early, mid-blastula, and late genes. The PC1 vs. PC2 plots account for 74% of the information in the entire data set, as estimated from eigenvalues. Adding a third PC increases the amount of information to 82%. This can be viewed in a PC1 vs. PC2. vs. PC3 scatter plots, which indeed reveals some sub-clusters (Fig. 5.3).

Hierarchical clustering

An alternative to PC analysis is hierarchical ag-glomerate clustering. In hierarchical clustering

all information in the data is accounted for, but the data are analysed sequentially, which means that all information is not considered at the same time. The procedure is:

1. In the multidimensional space find the two samples that are closest together and merge them into a cluster.
2. Find and merge the next two closest points, where a point is either an individual sample or a cluster of samples.
3. If more than one sample/cluster remains return to step 2.

The distance between two samples in the multidimensional space is typically calculated as

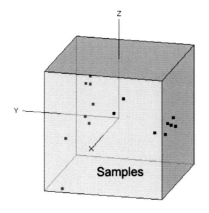

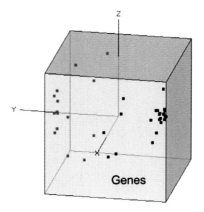

Figure 5.3 Same classification as in Fig. 5.3, visualized in a PC1 vs. PC2 vs. PC3 plot, which accounts for 82% of the variation in the experimental data.

the Euclidian (shortest) distance between them. To calculate distances between groups of samples there are a few options. The distance between the groups can be represented as the distance between the two closest samples, the two farthest samples, or some form of average distance, such as the distance between the centres of gravity of the two groups. Some more advanced methods such as the Ward's algorithm are also available (Ward, 1963). An alternative approach, proposed by Tichopad *et al.* (2006), is to calculate the correlation coefficient of the samples as a dissimilarity measure, and rank the correlation coefficients. Like auto-scaling this also eliminates the effect of expression levels.

Fig. 5.4 shows dendrograms of the *Xenopus laevis* developmental stages and genes, respectively. Both dendrograms show three main clusters, which correspond to the early stages/genes, mid-blastula stages/genes and late stages/genes. The members of the clusters are the same as those identified by the PCA. Also, the dendrograms reveal subclusters.

Self-organizing map

The idea behind the self-organizing map method (SOM; Kohonen, 2001) is to reflect variations in the expression profiles as a collection of cells, each with a representative expression profile, that are arranged to form a map with smooth changes in the profiles. When the expression profiles of the samples are located in the map, similar samples will be found close to each other. The SOM is generated as follows:

1 Initialize the cells in the SOM by giving each a random expression profile with the same genes as used in the experiment.
2 Choose by random one of the samples.
3 Compare the expression profile of the chosen sample with those in the cells of the SOM. The cell with the most similar profile is known as the best matching unit (BMU).
4 The neighbouring cells within a certain radius to BMU are now identified. The radius containing neighbours decreases during the training of the SOM. Initially, it is typically set to the 'radius' of the map, such that all cells are affected, and then it decreases with each time-step.
5 The expression profile of the BMU is adjusted to be similar to that of the selected sample. The expression profile of the neighbouring cells are also adjusted to be similar to the selected sample, but with a weight that decreases with the distance to BMU.
6 The procedure is repeated from step 2 a fixed number of iterations.

SOM trained with the expression profiles to classify samples and a SOM trained with samples to classify genes of *Xenopus laevis* are shown in Fig. 5.5.

Smoothing spline clustering

Gene expression over time is, biologically, a continuous process and as such it can be represented by a continuous function. Individual genes often share expression patterns. The shape of each

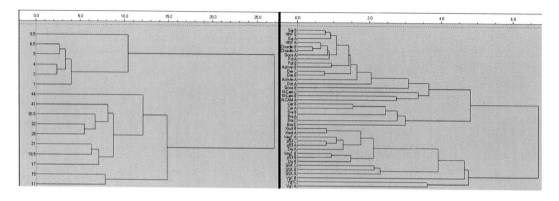

Figure 5.4 Hierarchical clustering of developmental stages (left) and of genes (right) in *Xenopus laevis*. Both dendrograms reveal three main clusters: early stages/genes, midblastula stages/genes, and late stages/genes. The unweighted pair model was used to calculate distances between groups.

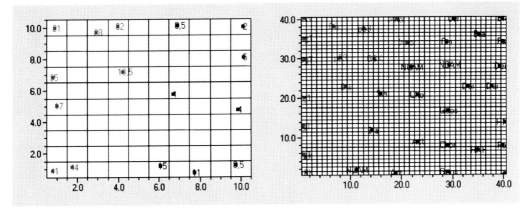

Figure 5.5 Example of a self-organized map (SOM) trained by stages (left) and genes (right) of *Xenopus laevis*. The three groups of stages/genes are seen to cluster.

function, the number of such functions, and the genes that share similar functional forms are typically unknown. Ping Ma *et al.* (2006) developed an approach to reveal related patterns of gene expression and their underlying functions. The method, smoothing spline clustering (SSC), models natural properties of gene expression over time, taking into account natural differences in gene expression within a cluster of similarly expressed genes. Furthermore, SSC provides a visual summary of each cluster's gene expression function and goodness-of-fit by way of a 'mean curve' construct and its associated confidence bands. The approach was used to model gene expression data over the life cycle of *Drosophila melanogaster* and *Caenorhabditis elegans* and revealed 17 and 16 unique patterns of gene expression respectively, in each species. The SSC software application and source code is freely available at the web page http://genemerge.bioteam.net/SSClust.html.

Data analysis in high-throughput screening

Malo *et al.* (2006) are developing statistical tools to detect quality hits in high throughput screens with a high degree of confidence. They examine statistical aspects of data pre-processing, focusing on concerns related to positional effects of wells within plates, choice of hit threshold and the importance of minimizing false-positive and false-negative rates. They argue that replicate measurements are needed to verify assumptions

of current methods and they suggest data analysis strategies when the assumptions are not met.

Global pattern recognition

Akilesh *et al.*, (2003) developed global pattern recognition (GPR) to more reliably evaluate expression changes in real-time PCR data. GPR is inspired by triangulation techniques to determine positional information in cartography and astronomy. It goes through several iterations to compare the change of expression of a gene normalized to every other gene in a real-time PCR array. By comparing the expression of each gene to every other gene in the array, a global pattern is established, and significant changes are identified and ranked. GPR makes use of biological replicates to extract significant changes in gene expression, providing an alternative to relative normalization in real-time PCR experiments.

Conclusion

The successful application of qPCR and post-PCR data processing depends on a clear understanding of the problems. Facilitating data management and providing tools for automatic data analysis are the new goals in qPCR data processing and qPCR application software development. All of the calculation- and statistical-software applications described for this chapter are summarized and discussed at the following web page: http://bioinformatics.gene-quantification.info.

References

Akilesh, S., Shaffer, D.J. and Roopenian, D. 2003. Customized molecular phenotyping by quantitative gene expression and pattern recognition analysis. Genome Res. *13*, 1719–1727.

Bar, T., Stahlberg, A., Muszta, A. and Kubista, M. 2003. Kinetic outlier detection (KOD) in real-time PCR. Nucleic Acids Res. *31*, e105.

Burns, M.J., Nixon, G.J., Foy, C.A. and Harris, N. 2005. Standardisation of data from real-time quantitative PCR methods – evaluation of outliers and comparison of calibration curves. BMC Biotechnol. *7*, 31.

Burns, M.J., Valdivia, H. and Harris, N. 2004. Analysis and interpretation of data from real-time PCR trace detection methods using quantitation of GM soya as a model system. Anal. Bioanal. Chem. *378*, 1616–1623.

Bustin, S.A. 2000. Absolute quantification of mRNA using real-time reverse transcription polymerase chain reaction assays. J. Mol. Endocrinol. *25*, 169–193.

Bustin, S.A. 2002. Quantification of mRNA using real-time RT-PCR. Trends and problems. J. Mol. Endocrinol. *29*, 23–39.

Bustin, S.A. 2004. A–Z of Quantitative PCR, IUL Biotechnology Series, International University Line.

Bustin, S.A., Benes, V., Nolan, T. and Pfaffl, M.W. 2005. Quantitative real-time RT-PCR – a perspective. J. Mol. Endocrinol. 34(3), 597–601.

Dallal, G.E. and Wilkinson. L. (1986) An analytic approximation to the distribution of Lilliefors' test for normality. Am. Stat. *40*, 294–296.

D'Haeseleer P. 2005. How does gene expression clustering work? Nature Biotechnol. *23*, 1499 – 1501.

Donald, C.E., Qureshi, F., Burns, M.J., Holden, M.J., Blasic, J.R. Jr and Woolford, A.J. 2005. An interplatform repeatability study investigating real-time amplification of plasmid DNA. BMC Biotechnol. *5*, 15.

Fronhoffs, S., Totzke, G., Stier, S., Wernert, N., Rothe, M., Bruning, T., Koch, B., Sachinidis, A., Vetter, H. and Ko, Y. 2002. A method for the rapid construction of cRNA standard curves in quantitative real-time reverse transcription polymerase chain reaction. Mol. Cell Probes *16*, 99–110.

Gilsbach, R., Kouta, M., Bönisch, H. and Brüss, M. 2006. A comparison of in vitro and in vivo reference genes for internal standardization of quantitative real-time PCR data. Biotechniques *40*, 173

Goll, R., Olsen, T., Cui, G. and Florholmen, J.R. 2006. Evaluation of absolute quantitation by nonlinear regression in probe-based real-time PCR BMC Bioinformatics *7*, 107

Hellemans, J., Mortier, G., De Paepe, A., Speleman, F. and Vandesompele, J. 2007. qBASE relative quantification framework and software for management and automated analysis of real-time quantitative PCR data. Genome Biol. 8:R19.

Horgan, G.W. and Rouault, J. 2000. Introduction to randomization tests, Biomathematics and Statistics Scotland.

Huggett, J., Dheda, K., Bustin, S.A., Zumla, A. 2005. Real-time RT-PCR normalisation; strategies and considerations. Genes Immun. *6*, 279–284.

Jin, N., He, K., Liu, L. 2006. qPCR-DAMS: a database tool to analyze, manage, and store both relative and absolute quantitative real-time PCR data. Physiol. Genom. *25*, 525–527.

Kohonen, T. (1995), Self-Organizing Maps, Springer Series in Information Sciences, Vol. *30*, Springer, Berlin, 1995. Third Extended Edition.

Kubista, M., Andrade, J. M., Bengtsson, M., Forootan, A., Jonák, J., Lind, K., Sindelka, R., Sjöback, R., Sjögreen, B., Strömbom, L., Stahlberg, A. and Zoric, N. 2006. The real-time polymerase chain reaction. Mol. Aspects Med. *27*, 95–125.

Lai, K.K., Cook, L., Krantz, E.M., Corey, L. and Jerome, K.R. 2005. Calibration curves for real-time PCR. Clin. Chem. *51*, 1132–1136.

Larionov, A., Krause, A. and Miller, W. 2005. A standard curve based method for relative real time PCR data processing. BMC Bioinformatics *6*, 62.

LightCycler Relative Quantification Software, Version 1.0 and 4.0, Roche Diagnostics, 2001 and 2004.

LightCycler Software®, Version 3.5; Roche Diagnostics, 2001.

Liu, W. and Saint, D.A. 2002a. A new quantitative method of real time reverse transcription polymerase chain reaction assay based on simulation of polymerase chain reaction kinetics. Anal. Biochem. *302*, 52–59.

Liu, W. and Saint, D.A. 2002b. Validation of a quantitative method for real time PCR kinetics. Biochem. Biophys. Res. Commun. *294*, 347–353.

Livak, K.J. (1997) ABI Prism 7700 Sequence detection System User Bulletin #2 Relative quantification of gene expression.

Livak, K.J. and Schmittgen, T.D. 2001. Analysis of relative gene expression data using real-time quantitative PCR and the $2^{\wedge}[-\text{delta delta } C(T)]$ Method. Methods *25*, 402–408.

Malo, N., Hanley, J.A., Cerquozzi, S., Pelletier, J. and Nadon, R. 2006. Statistical practice in high-throughput screening data analysis Nature Biotechnol. *24*, 167–175.

Ma, P., Castillo-Davis, C.I., Zhong, W. and Liu, J.S. 2006. A data-driven clustering method for time course gene expression data. Nucleic Acids Res. *34*, 1261–1269.

Manly, B. 1997. Randomization, Bootstrap and Monte Carlo Methods in Biology. Chapman & Hall.

Muller, P.Y., Janovjak, H., Miserez, A.R. and Dobbie, Z. 2002. Processing of gene expression data generated by quantitative real-time RT-PCR. Biotechniques *32*, 1372–1378.

Paoletti, C. and Mazzara, M. 2005. Definition of Minimum Performance Requirements for Analytical Methods of GMO Testing European Network of GMO Laboratories (ENGL) http://gmo-crl.jrc.it

Peirson, S.N., Butler, J.N. and Foster, R.G. 2003. Experimental validation of novel and conventional approaches to quantitative real-time PCR data analysis. Nucleic Acids Res. *31*, e73.

Pfaffl M.W., Tichopad A., Prgomet C., Neuvians T. P. 2004. Determination of stable housekeeping genes, differentially regulated target genes and sample integrity: BestKeeper – Excel-based tool using pairwise correlations. Biotechnol. Lett. 26, 509–515.

Pfaffl, M.W. and Hageleit, M. 2001. Validities of mRNA quantification using recombinant RNA and recombinant DNA external calibration curves in real-time RT-PCR. Biotechnol. Lett. *23*, 275–282.

Pfaffl, M.W., Horgan, G.W. and Dempfle, L. 2002. Relative expression software tool (REST) for group-wise comparison and statistical analysis of relative expression results in real-time PCR. Nucleic Acids Res. *30*, e36.

Pfaffl, M.W. 2001. A new mathematical model for relative quantification in real-time RT-PCR. Nucleic Acids Res. *29*, e45.

Pfaffl, M.W. 2004. Quantification strategies in real-time PCR, pages 87–120 in, A–Z of Quantitative PCR, eds. Bustin SA., IUL Biotechnology Series, International University Line,

Ramakers, C., Ruijter, J.M., Deprez, R.H. and Moorman, A.F. 2003. Assumption-free analysis of quantitative real-time polymerase chain reaction (PCR) data. Neurosci, Lett. *339*, 62- 66.

Rasmussen, R. 2001. Quantification on the LightCycler. In: Meuer, S, Wittwer, C. and Nakagawara, K, eds. Rapid Cycle Real-time PCR, Methods and Applications. Springer Press, Heidelberg.

Regulation (EC) No. 1829/2003. of the European Parliament and Council of 22 September 2003 on genetically modified food and feed.

Regulation (EC) No. 641/2004. of 6 April 2004. on detailed rules for the implementation of Regulation (EC) No. 1829/2003. of the European Parliament and of the Council as regards the application for the authorization of new genetically modified food and feed, the notification of existing products and adventitious or technically unavoidable presence of genetically modified material which has benefited from a favourable risk evaluation.

Regulation (EC) No. 882/2004. of the European Parliament and Council of 29 April 2004 on official controls performed to ensure the verification of compliance with feed and food law, animal health and animal welfare rules.

REST©, REST-XL©, REST-384©, REST-MCS©, REST-RG©; Pfaffl, M.W., Horgan, G.W., Vainshtein, Y., Avery, P. 2002. and 2005, Institute of Physiology, Weihenstephan, Technical University of Munich, Germany; http://rest.gene-quantification.info/.

REST-2005© M.W. Pfaffl and M. Herman 2005, Technical University Munich and Corbett Life Science; http://rest.gene-quantification.info/.

Rutledge, R.G. 2004. Sigmoidal curve-fitting redefines quantitative real-time PCR with the prospective of developing automated high-throughput applications. Nucleic Acids Res. *32*, e178.

Sindelka, R., Ferjentsik, Z., Jonak, J. 2006. Developmental expression profiles of *Xenopus laevis* reference genes. Dev. Dyn. 5, 754–8.

Souaze, F., Ntodou-Thome, A., Tran, C.Y., Rostene, W. and Forgez, P. 1996. Quantitative RT-PCR: limits and accuracy. Biotechniques *21*, 280–285.

Stahlberg, P. Aman, B. Ridell, P. Mostad and Kubista, M. 2003. Quantitative Real-Time PCR Method for Detection of B-Lymphocyte Monoclonality by Comparison of alpha and gamma Immunoglobulin Light Chain Expression. Clin. Chem. *49*, 51–59.

Tichopad, A., Didier, A., Pfaffl, M.W. 2004. Inhibition of real-time RT-PCR quantification due to tissue specific contaminants. Mol. Cell. Probes *18*, 45–50.

Tichopad, A., Dilger, M., Schwarz, G. and Pfaffl, M.W. 2003. Standardised determination of real-time PCR efficiency from a single reaction setup. Nucleic Acids Res. *31*, e122.

Tichopad, A., Pecen, L. and Pfaffl, M.W. 2006. Distribution-insensitive cluster analysis in SAS on real-time PCR gene expression data of steadily expressed genes. Comput. Methods Programs Biomed. *82*, 44–50

Vandesompele, J., De Preter, K., Pattyn, F., Poppe, B., Van Roy, N., De Paepe, A. and Speleman, F. 2002. Accurate normalization of real-time quantitative RT-PCR data by geometric averaging of multiple internal control genes. Genome Biol. 3, 0034.1

Ward, J.H. (1963). Hierarchical Grouping to optimize an objective function. J. Am. Stat. Assoc. *58*, 236–244.

Wilhelm, J., Pingoud, A. and Hahn, M. 2003. *SoFAR* – Validation of an algorithm for automatic quantification of nucleic acid copy numbers by real-time polymerase chain reaction. Anal. Biochem. *317*, 218–225.

Wittwer, C.T., Reed, G.H., Gundry, C.N., Vandersteen, J.G. and Pryor, R.J. 2003. High-resolution genotyping by amplicon melting analysis using LC Green. Clin. Chem. *49*, 853–60.

Wittwer, C.T., Herrmann, M.G., Gundry, C.N. and Elenitoba-Johnson, K.S. 2001. Real-time multiplex PCR assays. Methods *25*, 430–42.

Wong, M.L., Medrano, J.F. 2005. Real-time PCR for mRNA quantitation. Biotechniques *39*, 75–85.

Yuan, J.S., Reed, A., Chen, F., Stewart, C.N. Jr. 2006. Statistical analysis of real-time PCR data. BMC Bioinformatics. *22*(7), 85.

Performing Real-time PCR

Kirstin J. Edwards and Julie M. J. Logan

Abstract

Optimization of the reagents used to perform PCR is critical for reliable and reproducible results. As with any PCR initial time spent on optimization of a real-time assay will be beneficial in the long run. Specificity, sensitivity, efficiency and reproducibility are the important criteria to consider when optimizing an assay and these can be affected by changes in the primer concentration, probe concentration, cycling conditions and buffer composition. An optimized real-time PCR assay will display no test-to-test variation in the crossing threshold or crossing point and only minimal variation in the amount of fluorescence. The analysis of the real-time PCR results is also an important consideration and this differs from the analysis of conventional block-based thermal cycling. Real-time PCR provides information on the cycle at which amplification occurs and on some platforms the melting temperature of the amplicon or probe can be determined.

Optimization of real-time PCR assays

Real-time PCR assays require optimization in order that robust assays are developed which are not affected by normal variations in the target DNA, primer or probe compositions. A robust assay is defined as an assay in which these 'normal' variations cause no effect on the crossing threshold (CT) also known as crossing point (CP) and have only a minimal effect on the observed amount of fluorescence. The important criteria for optimization are specificity, sensitivity, efficiency and reproducibility. It is important to decide before commencing optimization which type of assay is required for a particular application. For example, there is little point in developing a quantitative assay if a simple qualitative assay will be just as informative. Melting curve analysis may also be required for product differentiation and in this case should be considered at the planning stage. If the real-time assay is based on conversion of an existing block-based assay it is important to note that the cycling conditions used for conventional block-based thermal cycling may not always translate easily to a real-time format and so it is important to consider re-optimization of the assay (Teo *et al.*, 2001).

The same principles of optimization apply to assays run on all real-time platforms. The following criteria should be optimized: buffer composition, cycle conditions, magnesium chloride ($MgCl_2$) concentration, primer concentration, probe concentration and template concentration. Commercial mastermixes, which are widely available, simplify the optimization and are convenient. In this chapter all aspects of optimization and analysis of real-time PCR results will be discussed as well as commercially available kits.

PCR mastermix

Commercial mastermixes are available for most of the real-time platforms and although some are marketed for specific instruments and probe formats, they often work equally well on other instruments. Mixes are provided in easy-to-use formats and often contain additional features such as use of dUTP allowing the enzyme uracil-

DNA glycosylase to be used to prevent carry-over contamination. Some commercial mastermixes contain the uracil-DNA glycosylase as well as the dUTP and carry-over contamination never needs to be considered. However, it is important to note that the use of dUTP has been shown to decrease PCR sensitivity. There is little doubt that the use of commercial mastermixes can simplify optimization and promotes uniformity of assay performance over time. However, for some laboratories the cost cannot be justified.

Preparation of in-house mastermixes can be effective and assays can perform well over time if the reaction components are well optimized. The mastermix requires thermostable polymerase, buffer, dNTPs and $MgCl_2$ though the latter will be discussed later in the chapter. The correct amount of polymerase is essential and there can often be a narrow optimal concentration range. Not enough polymerase leads to inefficient amplification, low fluorescence and loss of sensitivity and leads to a high CT value. Too much enzyme also leads to low fluorescence and can contribute to the production of primer dimers and production of other non-specific amplicons. The composition of the core buffer can be essential for some real-time platforms and can affect the T_m of the primers/probes and the performance of the enzyme. For example, when using glass capillaries it is essential that a protein such as bovine serum albumin (BSA) is included in the buffer to prevent the DNA from binding to the glass. The optimal concentration of dNTPs is usually wide and the CT values are not affected by 2–4× increases or decreases; however, the amount of fluorescent signal can be affected. An important consideration is the use of dUTP in the mastermix. *Taq* polymerase preferentially incorporates dATP, dTTP, dGTP or dCTP and although it can incorporate dUTP this is usually less efficient. However, the use of dUTP may be considered essential to prevent carry-over contamination and a balance has to be found between sensitivity and the consequences of contamination. Differences of up to two cycles can be observed between mastermixes containing dUTP and dTTP.

When the assay produces a large quantity of primer dimer or if the sensitivity of the assay is low, a hot-start technique can be applied to increase the stringency of the reaction (D' Aquila *et al.*, 1991; Chou *et al.*, 1992). Hot-start PCR can be achieved by using a commercially available mastermix with an in-built hot-start, or by adding either commercially available anti-*Taq* DNA polymerase antibodies or a modified *Taq* polymerase. When the PCR is performed in glass capillaries, i.e. using the LightCycler, the use of hot-start has been shown to improve performance as it is thought that the binding of *Taq* polymerase and magnesium ions to glass is reduced due to the hot-start mechanism (Teo *et al.*, 2002). Antibody-based hot-start methods can be advantageous as the recommended activation time is only 1–3 minutes compared with 10 minutes for modified *Taq* polymerase. This time difference can be important where fast PCR results are required.

Regardless of whether commercial mastermix or an in-house preparation is used the following components (primers, probes, $MgCl_2$ and template concentration) require optimization.

Primers

Primers should be designed with the aid of appropriate software and attention should be paid to the optimal size of the amplicon produced in the real-time application. Real-time PCR products are usually <500 bp, this may be much smaller than the amplicon size generated on a block-based thermal cycler. The following guidelines should be observed for optimal and accurate primer design:

- The primer 3′ ends should be free from secondary structure, repetitive sequences, palindromes and highly degenerate sequences.
- Forward and reverse primers should not have significant complementary sequences.
- The forward and reverse primers should have equal GC contents, ideally between 40% and 70%.
- The binding sites should not have extensive secondary structure.

The annealing temperature as determined using primer design software should be used as an initial guide, however, it is important to note that this can often vary greatly from the experimentally determined annealing temperature.

Table 6.1 Example of a primer optimization matrix

Reverse/forward	50 nM	300 nM	500 nM	900 nM
50 nM	50/50	300/50	500/50	900/50
300 nM	50/300	300/300	500/300	900/300
500 nM	50/500	300/500	500/500	900/500
900 nM	50/900	300/900	500/900	900/900

One approach, which can be used to determine the true annealing temperature is to synthesize the complementary primer sequence, hybridize the two primers and perform a melting analysis that allows the T_m to be observed under the conditions that will be used in the PCR reaction. The cost of synthesizing another oligonucleotide is often justified by the reduction in time spent repeating the experiment at different annealing temperatures (Teo *et al.*, 2002).

Time should also be spent determining the optimal primer concentration and it is best to try a range of final concentrations from 0.1 μM to 0.5 μM for both primers. For some assays the optimal concentration for the two primers will not be the same, for example, to achieve good melting curves it is sometimes necessary to use asymmetric PCR, where a lower concentration of one primer is used, to increase the amount of the strand complementary to the probe (Lyon *et al.*, 1998; Phillips *et al.*, 2000). This improves the melting curve because increased amplification of the target strand can reduce the competition be-

tween the template strands and template–probe hybridizations. Some manufacturers recommend performing a primer optimization matrix as outlined in Table 6.1. The optimal combination is the one that gives the lowest CT value (Fig. 6.1).

Fluorescent probes

As with the primers, real-time PCR probes are best designed using dedicated software such as Primer Express (Applied Biosystems) for hydrolysis probes, hybridization probe design software from Roche Applied Science or Beacon Designer from Bio-Rad. For any of the probe formats the annealing temperature should be higher than that of the primers in order to ensure that the primers are binding and synthesizing the product before the probe is able to participate in the reaction. It is important that the probes are not able to act in the PCR as primers and to prevent this either a fluorescent molecule or a blocker, for example phosphate, should be placed at the 3′ end of the sequence. Probes complementary to the 3′ termini of the PCR primers should be avoided

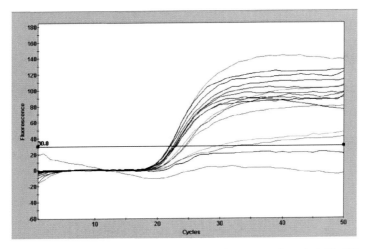

Figure 6.1 Primer optimization experiment. The same sample was analysed at 12 different forward and reverse primer concentrations as detailed in Table 6.1. This experiment was run on the Smart Cycler using the same concentration of hydrolysis probe in each tube. The optimal primer concentration is the one that gives the lowest CT value.

since they may hybridize to the primers. Primer elongation may then occur leading to primer-probe dimers that affect amplification efficiencies (Roche Applied Science). As with the primers it is important to try a range of probe concentrations to determine which is optimal. A wide range of probe concentrations is possible but if the concentration is too low no fluorescent signal will be observed and if the concentration is too high it can lead to a high fluorescent background. The probe concentration should be optimized after optimizing the primer concentration and as with the primer concentration, a range of probe concentrations should be tried between 0.1 μM and 0.5 μM and the concentration which gives the lowest CT values and the highest fluorescent signal should be selected (Fig. 6.2).

SYBR Green I

If the DNA binding dye SYBR Green I is used in the fluorescent detection system it should be optimized for each set of PCR primers. As the amount of SYBR Green I is increased an increase in the amount of fluorescence can be observed. The recommended concentration is a 1:10,000 dilution of the neat stock as supplied by the manufacturer. High concentrations of SYBR Green I have been shown to adversely affect the PCR amplification as they inhibit *Taq* polymerase (Ririe *et al.*, 1997; Wittwer *et al.*,

1997). DNA binding dyes also influence melting temperatures and consequently it is important to use only the optimized level. Some researchers have suggested that SYBR Gold is preferable to SYBR Green I due to its stability during long-term storage (*Lee et al.*, 1999).

Magnesium chloride

A key variable is the magnesium chloride (MgCl$_2$) concentration since Mg^{2+} ions are known to affect both the specificity and the yield of PCR (Oste, 1988). Concentrations that are too low or too high can have deleterious effects on PCR, high concentrations may lead to incomplete denaturation and low yields, whereas levels that are too low reduce the ability of polymerase to extend the primers. High MgCl$_2$ levels also lead to increased production of non-specific products and primer artefacts including primer dimers. This should always be avoided but especially in sensitive quantitative assays. It is recommended that a MgCl$_2$ titration is performed for each primer/probe set. This work is usually best performed after completing optimization of primer and probe concentration. The optimal MgCl$_2$ concentration will generally be the lowest amount that gives the minimum CT value, the highest fluorescent intensity and the steepest slope of the fluorescence/cycle plot (Fig. 6.3). The optimal MgCl$_2$ concentration for

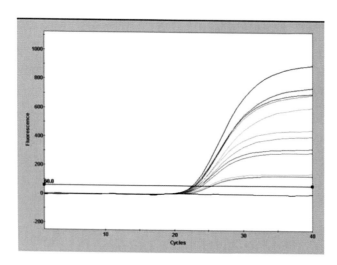

Figure 6.2 Probe optimization experiment. Using two different primer concentrations, five different hydrolysis probe concentrations (250 nM, 200 nM, 150 nM, 100 nM and 50 nM) were run on the Smart Cycler. The optimal probe concentration is the one that gives the lowest CT value and the highest fluorescent signal.

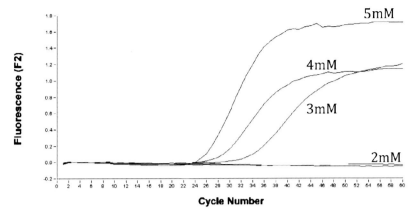

Figure 6.3 MgCl₂ Optimization Experiment. The same sample was analysed at five different MgCl₂ concentrations on the LightCycler (5 mM, 4 mM, 3 mM, 2 mM). Using 5 mM MgCl₂ gave the lowest CT value and the highest fluorescent signal and was therefore the optimal concentration.

DNA assays is usually between 2 and 5 mM and for RNA is usually between 4 and 8 mM. Commercially available real-time PCR buffers are now available where MgCl₂ optimization is not required for example, QuanTitech (Qiagen, Crawley, UK) or FastStart Plus (Roche Applied Science, UK).

Template purity and concentration for optimization

When optimizing assays it is advisable to use a minimum of a high and a low template concentration as too much DNA can inhibit the reaction and too little may be undetectable. For some assays it may not be necessary to use purified DNA or RNA however, if the primers are not

amplifying it is worth repeating the PCR with pure template in the event that PCR inhibitors are present. If the CT is <10 cycles then a higher dilution should be prepared or if the CT is >30 then higher concentrations should be used. If a serial dilution series of template is prepared and amplified then an optimized assay should display an equal number of cycles between amplification of each dilution.

Cycling conditions

The use of optimal cycling conditions are essential for real-time as with any PCR reactions (Rasmussen, 1992). The initial cycling conditions recommended for both hybridization probes and hydrolysis probes are given in Table 6.2. The

Table 6.2 Cycling conditions

	Hybridization probe (performed on Roche LightCycler)	Hydrolysis probe (performed on ABI Sequence Detection System)
Initial denaturation	2–10 min at 95°C	10 min at 95°C
Denaturation	0 s at 95°C	15 s at 95°C
Annealing	5 s at annealing temperature	60 s at 60°C
Extension	Optimal time dependent on amplicon length at 72°C	
Ramp rate	20°C/s or reduce to 2–5°C if annealing temperature < 55°C	
Fluorescent acquisition	At end of annealing step	At end of annealing step
Melting curve	30 s at 55°C Ramp rate 0.1°C/s with acquisition mode on continuous Temperature increased to 80°C	

optimal annealing temperature should be determined as described in the primer section and the optimal extension time should be calculated using the following formula: extension time (s) = amplicon length (bp) ÷ 25. Note: when hot-start techniques are employed the initial denaturation time may need to be increased. If Primer Express is used to design hydrolysis probes then they should all work optimally under the same conditions (Table 6.2).

PCR controls

No template (negative) controls should always be included in real-time runs and where possible appropriate positive controls should also be included. When the assay is quantitative standards of known concentration should be included, although on some platforms it is possible to perform quantitative analysis using a standard curve generated in a previous run thus maximizing the number of samples which can be analysed. Where melting curve analysis is used for product differentiation then it is advisable to include samples that will melt at each of the possible melting temperatures (Logan et al., 2000; Edwards et al., 2001). This allows for any drift in melting temperature that may be observed over time or with different batches of fluorescent probe.

To control for PCR inhibitors an additional inhibition control template should be included in each sample. This inhibition control should be amplified by the same primers as the experimental target but should contain a different probe binding site. When the reaction is significantly affected by inhibitors, their presence is indicated by the failure of both control and experimental target amplification, or by low amplification efficiency. Inhibition may be minimized by diluting the sample, by using an alternative purification technique or by obtaining a different sample. If an inhibition control is not available, negative samples should be analysed for a different target or spiked with a positive control and re-amplified. The use of inhibition controls is particularly important when real-time PCR is being used for clinical diagnostics or other applications where a false-negative result could be critical.

ROX passive reference

Real-time platforms that use Peltier blocks as the heating/cooling mechanism require the use of a reference signal to normalize the fluorescent signal across the block. By using the ROX passive reference all fluorescent signals are normalized leading to more accurate and reproducible results. When the ROX passive reference is used it is important to note that everything in the real-time PCR is relative, which means that if the ROX passive reference changes then the baseline, CT value and amount of fluorescence will also change. When optimizing the passive ROX reference there is a narrow optimal range. If the amount is too high then background noise will be reduced but a low signal will be hard to distinguish from the background and this may lead to false-negative results. If the ROX level is set too low the background noise is increased but a strong signal will be easy to distinguish from the background for high copy number samples, however, weak signals may be lost in the background.

Multiplexed reactions

Many of the instruments now have a greater capacity to run multiplexed reactions and there are more fluorescent dyes available to support this. Multiplexed reactions will require more design and optimization and overall sensitivity may be reduced. But multiplexing is particularly useful when sample material is limited, allowing more data to be generated for each sample, and it can reduce the running costs. The primers and probes are designed as for a single-plex assay except that all the primers and probes in the same reaction should be of similar length, T_m and GC content and oligonucleotides that will interact with each other should be avoided. When optimizing the concentrations of each oligonucleotide it is important with a multiplexed assay to use the lowest primer and probe concentrations possible otherwise there is a higher chance of the oligonucleotides interacting with each other. It may be necessary when running a multiplexed reaction to add additional reaction components to ensure that none of the components become limited. If the multiplexed reactions do not appear to be working well then it is advisable to increase the Taq polymerase and dNTP concentrations by 50–100% and the buffer can be increased from a final concentration of 1x to a concentration of 1.5x. There are also commercially available mastermixes designed specifically for running

Table 6.3 Real-time PCR kits

Kits	Examples of target(s)	Manufacturers/suppliers
Pathogen detection	*Bacillus anthracis*, Cytomegalovirus, HHV6, *Listeria monocytogenes*, *Chlamydia pneumoniae*, enterovirus, norovirus, polyomavirus, *Salmonella* sp., *Escherichia coli* 0157, Parvovirus B19, Hepatitis A, B and C, HSV 1 and 2, Epstein–Barr virus, *Pseudomonas aeruginosa*, *Candida albicans*, *Enterococcus faecalis* and *E. faecium*, *Staphylococcus aureus* and coagulase-negative staphylococci, Dengue virus, HIV-1, Influenza A and B and H5N1, SARS coronavirus, varicella zoster virus, West Nile virus, *Mycobacterium tuberculosis*, *Chlamydia trachomatis*, *Plasmodium* spp., *Mycoplasma pneumoniae*, *Borrelia* spp., group B streptococci, human papillomavirus, *Brucella* spp., *Francisella tularensis*, *Yersinia pestis*, *Cryptosporidium* spp., *Clostridium botulinum*	Alpco Diagnostics AME Bioscience Cepheid GeneOhm Sciences Idaho Technology Minerva Biolabs Qiagen (Artus) Roche Applied Science
Mutation detection	Cytochrome P450 2C19 gene, cytochrome P450 2C8 gene, Glutathione-S-transferase genes, lactase genes, multidrug resistance gene (MDR1), apolipoprotein B gene, human N-actetyltransferase 2 gene, factor V Leiden, factor II (prothrombin), dehydropyrimidine dehydrogenase (DPD) gene, methylenetetrahydrofolate reductase (MTHFR) gene, thiopurine S-methyltransferase (TPMT) gene	Alpco Diagnostics Qiagen (Artus) Roche Applied Science
Genetically modified organism (GMO) detection	GMO maize 35S, GMO soy 35S, GMO Bt-176 maize, genetically modified plants, genetically modified round-up soya	Applied Biosystems Roche Applied Science
Antibiotic resistance screening	Clarithromycin resistance in *Helicobacter pylori*, Methicillin (*mecA*) detection in *Staphylococcus aureus*, vancomycin resistance in *Enterococcus* sp.	Alpco Diagnostics Roche Applied Science
Housekeeping (HK) genes	ALAS mRNA HK genes, b2M mRNA HK genes, HPRT mRNA HK genes, PBGD mRNA HK genes, β-actin, GAPDH, PGK1	Qiagen (Artus) Roche Applied Science
Food testing	*Salmonella* sp., *Listeria* sp., *E. coli* 0157, *Campylobacter* sp., beer screening	Roche Applied Science
Gene expression	Human, mouse, rat, *Arabidopsis*, *Drosophila*, *C. elegans*, canine and Rhesus macaque genes	Applied Biosystems Invitrogen Stratagene

multiplexed reactions. When optimizing a multiplexed reaction each individual reaction should give a similar CP or CT value for the same sample showing that each reaction has a similar efficiency. If one primer/probe set is giving a significantly lower CP/CT value then the concentration of those primers/probes should be reduced to prevent this reaction from affecting the other reactions. If the primers/probes are not working to a similar efficiency then they should be redesigned.

Commercial real-time PCR kits

There are increasing numbers of commercially available reagents for real-time PCR including mastermixes, prepared controls and specific detection kits available. An illustration of the range of kits for specific detection and control kits is detailed in Table 6.3. The use of commercial kits can allow the introduction of a new test more readily since no design or optimization is required, both of which can be lengthy procedures. There is however, a need to evaluate commercial reagents to confirm that the test performs as described by the manufacturers and since there is no design control it is necessary to be aware that there may be limitations of the test, e.g. different sequence types may not be amplified due to the presence of sequence changes occurring in the probe or primer regions. Commercial kits do, however, benefit from reliable manufacturing processes and are usually CE marked which means that the product meets the requirements of all relevant European Union directives. The obvious disadvantage of commercial kits is the increased cost when compared to 'in-house' tests, also commercial tests are only developed where there is market demand for the test, so there will never be commercial tests available for all targets.

Analysis of real-time PCR results

Real-time PCR results can be analysed in a variety of ways depending on the application. Analyses of quantitative standards can be used to generate PCR curves and from these the cycle number at which the fluorescent signal increases above the background fluorescence can be determined. This cycle number can be

used for comparison of results from run-to-run and can be used to generate qualitative positive/negative results. In quantitative assays the standard curve is used to determine the copy number of unknown samples and again this can be compared with other samples. Real-time PCR reactions can also be analysed by melting curve determination when methods including hybridization probes and SYBR Green I intercalation have been used. Melting curves are useful for differentiating primer dimers from specific PCR products (primer dimers usually melt at lower temperatures than specific PCR products) or for differentiating different PCR products in mutation detection. During the initial optimization of real-time PCR assays it is useful to analyse the results by agarose gel electrophoresis in order to correlate product length with melting peaks and to identify any primer artefacts.

Conclusions

Real-time PCR results are dependent on the optimal concentration of each of the reaction constituents as well as on the interrelations of each of these components. The use of commercially available mastermixes, which have already been optimized, provides a solid basis for a robust assay. Optimization of $MgCl_2$, primer and probe concentration is still required to obtain maximum sensitivity and PCR efficiency but can be performed relatively easily. Depending on the platform and probe format employed some optimization of the cycling conditions may be required. When using a well-optimized assay the target will reliably amplify at the same cycle number although variations may be observed in the amount of fluorescence observed as this is more dependent on not only the PCR efficiency but also on variations in the availability of each of the different reaction components.

References
Chou, Q., Russell, M., Birch, D.E., Raymond, J. and Bloch, W. 1992. Prevention of pre-PCR mis-priming and primer dimerization improves low-copy-number amplifications. Nucleic Acid Res. 20, 1717–1723.
D' Aquilia, R.T., Bechtel, L.J., Videler, J.A., Eron J.J., Gocczyca, P. and Kaplan J.C. 1991. Maximizing sensitivity and specificity of PCR by pre-amplification heating. Nucleic Acid Res. 19, 3749.
Edwards, K.J., Kaufmann, M.E. and Saunders, N.A. 2001. Rapid and accurate identification of coagulase-

negative staphylococci by real-time PCR. J. Clin. Microbiol. *39*, 3047–3051.

Lee, M.A., Brightwell, G., Leslie D., Bird, H. and Hamilton A. 1999. Fluorescent detection techniques for real-time multiplex strand specific detection of *Bacillus anthracis* using rapid PCR. J. Appl. Microbiol. *87*, 218–233.

Logan, J.M., Edwards, K.J., Saunders, N.A. and Stanley, J. 2001. Rapid identification of *Campylobacter* spp. by melting peak analysis of biprobes in real-time PCR. J. Clin. Microbiol. *39*, 2227–2232.

Oste, C. (1988). Polymerase chain reaction. BioTechniques 6, 162–167.

Rasmussen, R. 1992. Optimizing rapid cycle DNA amplification reactions. The RapidCyclist. Idaho Technology *1*, 77.

Ririe, K.M., Rasmussen R.P. and Wittwer, C.T. 1997. Product differentiation by analysis of DNA melting curves during the polymerase chain reaction. Anal. Biochem. *245*, 154–160.

Roche Molecular Biochemicals. Optimisation Strategy. Technical Note No. LC 9/2000.

Teo, I.A., Choi, J.W., Morlese, J., Taylor G. and Shaunak S. 2002. LightCycler qPCR optimisation for low copy number target DNA. J. Immunol. Methods *270*, 119–133.

Wittwer, C.T., Herrmann, M.G., Moss A.A. and Rasmussen, R.P. 1997. Continuous fluorescence monitoring of rapid cycle DNA amplification. BioTechniques *22*, 130–138.

Internal and External Controls for Reagent Validation

7

Martin A. Lee, Dario L. Leslie and David J. Squirrell

Abstract

False negatives in PCR can occur from inhibition of one or more of the reaction components by a range of factors. Therefore applications requiring a high level of confidence in the result need to be designed to control for the occurrence of false negatives. While an external, or batch control is often used, the ideal control is one that is included in the reaction cocktail in a multiplex format. Here we discuss the application and development of molecular mimics for use as controls in real-time PCR, and explain a number of concepts and experimental considerations that will aid in the optimization of controlled multiplexed assays.

Introduction

Confidence in assays based on the polymerase chain reaction (PCR) (Saiki *et al.*, 1985; Mullis *et al.*, 1987) may be compromised by the sporadic occurrence of either false-positive or false-negative results. False positives are a common problem and occur mainly as a result of cross-contamination from either positive samples or reaction products. Preventative methods including good laboratory practice, delineated preparation/analysis areas, PCR cabinets with UV treatment, UV air scrubbers, closed tube assays, and uracil glycosylase carry over prevention chemistry (Longo *et al.*, 1990), have effectively eliminated the occurrence of false positives, especially when used in combination. However, there are a number of PCR applications where the avoidance of false-negative results is of equal importance. False negatives occur through failure of one or more of the reagents, the presence of inhibitors, or the failure of the PCR thermal cycling (Rossen *et al.*, 1992; Wilson *et al.*, 1997) and reporting processes. Applications requiring high confidence in the PCR include pathogen detection in clinical diagnosis, food quality control and environmental analysis.

For most molecular tests the use of reference material in a batch test is the only control that can be implemented. However, PCR and other nucleic acid amplification techniques provide not only extremely sensitive detection, but also have the advantage that they can be readily multiplexed to include an internal control (IC). An IC is a second target molecule that can be amplified with, but distinguished from, other products in the same tube. In an ideal assay this should be able to control for all of the reagents in a reaction cocktail, and for variations in machine operation parameters. This may be achieved by the use of a synthetic molecule, a 'molecular mimic' that can be co-amplified using the same set of amplimers as the target species.

Before real-time PCR, approaches for ICs used amplicons with different molecular masses and subsequent analysis using a separation method. For example, the molecular mimic could contain both priming sites, but an internal sequence changed by insertions or deletions (Ursi *et al.*, 1992). The products of amplification could be easily analysed by the use of agarose gel electrophoresis and ethidium bromide staining. There are a large number of such examples reported in the literature. The main drawback of the approach is that the analysis method is time-consuming, the smallest amplicon tends to be amplified more efficiently, and the generation of specific and control product hetero-duplexes (Henley et al., 1996) can be unpredictable and therefore confusing for data interpretation.

Real-time PCR approaches with fluorescence reporting are particularly useful for high-confidence applications because the assays can be carried out in multiplex in the same tube with measurements in situ so the tube never has to be reopened. This closed tube format significantly reduces opportunities for cross-contamination by amplification products and thereby reduces the risk of false positives whilst allowing an IC to control for negatives. Most commercial fluorimeter instruments can co-amplify and detect at least two targets in a quantitative manner. However, increasing the level of multiplexing adds significantly to the complexity of designing and optimizing the assay.

In this chapter we will present various strategies for producing internal controls for different fluorescent chemistries, and discuss some experimental considerations for their optimization and implementation. The experimental considerations described here are based largely on our own experiences and we make some useful suggestions for the development and implementation of controlled assay systems. Competitive homologous internal controls are state-of-the-art for PCR and the main focus of this chapter. Whilst users will find that for many applications the use of a homologous or competitive control is not necessary, the information will be of equal use to those developing heterologous controls and synthetic standards alike. In the literature, the term internal control is often used incorrectly and so we will first discuss some definitions.

Nomenclature

The use of a second amplicon can serve to control, normalize or standardize the result. Indeed the use of a second amplicon as a calibrator to provide for efficiency correction now forms part of the Roche LightCycler® instrument's capability, enabling greater precision and throughput. However, this chapter covers the application of molecular mimics to check reaction outcomes. The strategies described are equally applicable in the development of other internal and external

mimics. In particular the use of mimics as either external or internal standards has advantages. Therefore, here we summarize some nomenclature used to describe the types of second amplicon in Table 7.1. It is easy to see how an internal control using homologous amplimers will add more confidence to a result by confirming that conditions in the PCR are adequate for allowing amplification to occur. However, this only controls for the amplification and, where a nucleic acid reporter probe is also involved, this has to be controlled for as a separate factor.

Strategies for development of internal controls and their analysis

The strategy for detecting an internal control depends upon the instrument's capabilities and the reporting chemistry. The simplest approach is to use a strand-specific probe as the IC with a reporting dye of different emission wavelength to that of the probe for the specific target. This requires that the emission wavelengths of the dyes are either spectrally separated enough to not overlap, or that the emissions can be effectively de-convoluted using the instruments calibration algorithm. For example, on the ABI 7700, a 5' nuclease assay may use FAM as the specific reporter probe and a dye such as VIC, JOE or HEX as the reporter for the control probe. Cycle-by-cycle increases in the emissions of the dyes reports amplification of the respective species. This strategy may be applied to virtually all fluorescent chemistries and instruments.

Since the internal control serves only to report the integrity of the PCR reagents the signal does not necessarily need to be quantitative. The ability to detect the species at the end of thermal cycling should suffice for most applications. Therefore, for assays based on hybridization run on an instrument that can determine melting point (T_m), one can design the control amplicon and/or probe to have a different T_m from that of the specific nucleic acid target to allow discrimination between target and control species. This may be achieved using DNA binding dyes for amplicons with different melting points (Lee et al., 1999), or labelled probes for different sequences. When using specific probes the use of melting point can also include the use of dif-

Table 7.1 Definitions and types of standard, control and reference nucleic acids

	Standard	Control	Reference (Second amplicon type)
Definition / **Type**	A standard is a nucleic acid preparation that has a set or known concentration. Unknown samples may be compared to one or more standards and an absolute value for the unknown can be determined by numerical interpolation/extrapolation from experimentally determined values.	A control is an amplicon added to the PCR and allowed to amplify to verify the integrity of one or more reagent(s) in the cocktail	A nucleic acid target that is used to compare the sample for either a qualitative or relative quantitative analysis
External: amplified in a different well	The standards are amplified separately from the target species in different wells on the same instrument. *	'Batch control' This involves the use of a control in a separate well on the instrument. This approach cannot control for false negatives in different samples. It controls only for successful thermal cycling and the integrity of a common core reaction mix.*	A related nucleic acid that is used for a qualitative comparison of amplification.
Internal: amplified in the same well	**Homologous** — A competitive standard that has the same amplimers as the specific target.	**Homologous** — A competitive control that is included in the same well of the instrument that has the same amplimers as the specific assay and controls for the core reagents and primers, and the thermal cycling.	**Exogenous** — This is an internal reference that is added to the sample and is co-amplified with the target nucleic acid using a second set of amplimers.
	Heterologous — A non-competitive standard with different amplimers to those of the specific target	**Heterologous** — A control that is included in the same well of the instrument but has different amplimers to that of the specific assay that controls for both the thermal cycling and the core reagents.	**Endogenous** — A 'housekeeping gene.' This is usually a gene that is naturally present in the sample at constant levels such that the up- or downregulation of the target species can be relatively quantified using a second set of amplimers

*External standards and controls may also be homologous, or heterologous, to avoid the risk of cross-contamination.

ferent emission wavelength dyes. The merits of strand-specific probes over DNA binding dyes have been discussed in Chapter 3. Whilst DNA binding dyes have some disadvantages as probes in terms of specificity, the same probe is used for detecting the control and the specific target. A polymorphism introduced into the target sequence can generate an IC such that a single hybridization probe can be used for both amplicons with melting points used to differentiate the binding partner. This represents one of the few approaches where the reporter probe can be controlled for as well as the amplification. The probe's function can therefore be better controlled than in approaches with strand-specific probes. Using dual hybridization probes in either quantification and/or melting point analysis it is possible to use one common probe, either the donor or the acceptor, this also provides a better approach for control of the probe. The methods for generating these mimics are now briefly described.

Synthesizing and optimizing molecular mimics for use as internal controls

There are several different approaches using recombinant DNA techniques that can be used to generate a homologous internal control mimic. In each case the objective is to create a synthetic target that can co-amplify with the specific amplicon and that has a 'motif' which may be easily distinguished from that of the specific amplicon. This distinction can be achieved by the use of dyes with different emission wavelengths and/or a shift in the melting point as described above. Several methods have been reported. These include: insertion of the specific target into a vector and insertion/deletion of a sequence between the amplimer regions (Ursi *et al.*, 1992; Zimmerman *et al.*, 1996; Brightwell *et al.*, 1998), the amplification of a generic target from a related sequence (deWit *et al.*, 1993), the use of overlapping and tailed PCRs (Müller *et al.*, 1998; Sachadyn *et al.*, 1998), site-directed mutagenesis (Nash *et al.*, 1995), random misprimed products (Mergenthaler *et al.*, 2004) and polymerase extension of oligonucleotides and subsequent PCR (Rosenstraus *et al.*, 1998). In addition, a complete synthesis of the gene is possible for smaller amplicons (Zimmerman *et al.*, 2000) and

some commercial suppliers will now make the gene and supply it in a vector ready for use. Since the ability to discriminate amplicons by size is not a requirement, the amplicon may be similar, or the same in size, and thus achieve a similar amplification efficiency. Gene synthesis is now available commercially and so the entire sequence may be requested in a choice of vectors. The only constraints are cost and time. However, two main techniques that we use to generate mimics are:

- Direct insertion of amplimer and probe sequences into cloning vectors using recombinant techniques. This can be done by amplifying amplimer/probe sites, or the use of oligonucleotide linkers. In either case the use of restriction overhangs will assist in the correct directional insertion of the fragments.
- Generation of targets using a PCR approach of amplimers tailed with specific amplimer sequences to amplify alternative probe regions. The products contain the specific amplimer sequences incorporated into the PCR product. This product may be subsequently cloned into the desired vector (Fig. 7.1). The use of the additional overhangs with unique restriction sites will assist in directional cloning that may be required for some applications (e.g. sense-specific RT-PCR).

For reverse transcriptase PCR (RT-PCR) a mimic may be made using the methods described and subsequent insertion of the product into a vector with a suitable RNA polymerase promoter. *In vitro* transcription (e.g. with T7 or T3 polymerases) then allows the RNA molecule to be synthesized in quantity. The use of modified nucleotides that make the transcripts resistant to RNase activities could form one strategy that would allow for a more stable RNA control (Capodici *et al.*, 2002). Alternatively, the sequence could be cloned into an RNA virus such as the bacteriophage MS2. Armoured RNA™ is a product available from Ambion Inc. (Pasloske, *et al.*, 1998). This is a method for packaging recombinant RNA into pseudoviral particles using a plasmid system whereby the MS2 bacteriophage coat protein is located downstream

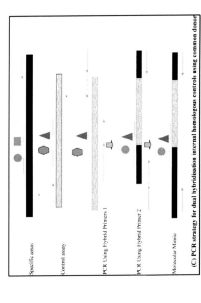

Specific assay

Control assay

PCR Using Hybrid Primers

Molecular Mimic

(a) PCR strategy for generation of 5' nuclease internal homologous controls

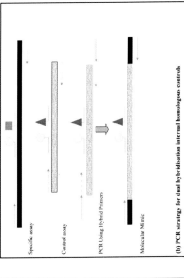

Specific assay

Control assay

PCR Using Hybrid Primers

Molecular Mimic

(b) PCR strategy for dual hybridisation internal homologous controls

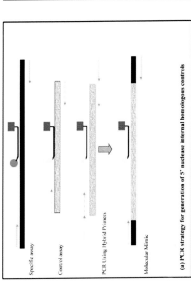

Specific assay

Control assay

PCR Using Hybrid Primers 1

PCR Using Hybrid Primer 2

Molecular Mimic

(C) PCR strategy for dual hybridisation internal homologous controls using common donor

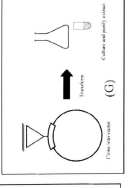

Clone into vector

Transform

Culture and purify extract

(G)

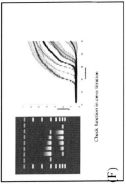

Check function in cross titration

(F)

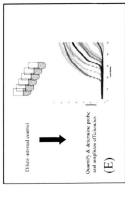

Dilute internal control

Quantify & determine probe and amplicon efficiencies

(E)

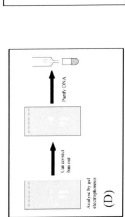

Analyse by gel electrophoresis

Cut correct band out

Purify DNA

(D)

Figure 7.1 Schematic showing strategies for rapidly generating and evaluating competitive internal homologous controls by PCR. (A) For assays such as the 5' nuclease where there is only one probe, (B) For dual hybridization assays using two additional probes, (C) Dual hybridization probes using one common probe for both the control and specific assay, (D) Amplify the IC using PCR and extract the correct band from the gel, (E) Serial dilute the product and determine the amplification and probe efficiencies (Fig. 7.2), (F) Cross-titre the IC with specific target to assess the efficacy of the IC in multiplex PCR before cloning into a suitable vector and (G) Transformation, culture, extraction and use. A colour version of this figure is located in the plate section at the back of the book.

of the promoter and the target gene of interest. RNA encoding the coat protein and the target sequence is transcribed under the control of the inducer. Coat protein is translated and subsequently binds to recombinant RNA to form the product. The packaged molecule is resistant to RNAase digestion yet easily extracted for amplification. The process has been shown to stabilize the RNA compared to *in vitro* transcripts made from plasmid and is an ideal method for RNA IC production. The authors would expect that any recombinant RNA virus containing the control sequence would improve the stability of the nucleic acid over naked RNA.

With any approach there are a number of experimental considerations that should be considered when designing and using the control molecule and these will now be discussed.

Experimental considerations

In designing an experiment the sequence of the specific target and the assay are important considerations. The design of the assay is best facilitated by software tools that can automatically select amplimers and probes and predict (score) amplimer–amplimer and amplimer–probe interactions for the multiplexed assay. Here we highlight some of the considerations that we consider important in achieving a successful IC implementation.

Control sequence

The choice of control sequence is the most important factor since the other considerations are mostly dependent on this. When using nucleic acid probes one useful approach is to use a sequence that has been previously reported/optimized, or that is commercially available for the same assay type. For example, the ABI human β-actin sequence for a 5′ nuclease assay is useful because the probe is efficient and a kit containing it is commercially available. Ideally the probe sequence should not be of the same origin (or endogenous to the sample) as the target for the specific test. For example, it would not be ideal to use the β-actin sequence for a human test. Providing the probe sequence has been shown to work, a sequence from almost any other nucleic acid sequence would be suitable. Ideally, the GC content for the sequence should be similar to that of the specific target.

Amplicon efficiency

Whilst the differences in amplification efficiencies of the target and IC may be compensated for by changing the concentration of the initial target, they should ideally be similar and should be determined empirically. This is most important if the mimic is to be used for competitive quantification since the amplification efficiency can vary significantly between amplicons (Zimmerman *et al.*, 1996). It is useful to include native sequence either side of the amplimer when designing the mimic to ensure that the effects of neighbouring sequence is maintained. The amplification efficiency is best determined using a DNA binding dye such as SYBR®Green I combined with an efficient hot-start to amplify a broad range of concentrations across the dynamic range of the assay. Whilst using a probe directly to do this is satisfactory, checking the amplification efficiency without the probe allows it to be measured directly and is useful in understanding the efficiency of the probe when it is later evaluated. It will avoid dismissing a satisfactory reaction for the sake of iterating the probe design and/ or as a result of a failure in probe synthesis. The efficiency (E) can be calculated by using the slope value for the derived standard curve using (Fig. 7.2).

$$E = 10^{-1/slope}$$

The efficiency of different amplicons for the same (or similar) dilution series can then be directly determined.

Probe (nucleic acid) efficiency

Probe efficiency is important if the signal of the control is to be commensurate with that of the specific probe. The probe efficiency can be affected by a number of factors that are discussed in Chapter 3. The probe sequence to be chosen should be checked for complementarities to other probes, amplimers and amplicons in the reaction. This is most important when the probe melting point approach is used since the signal from a control probe that is too large may swamp a low signal obtained in positives. When assessing potential probes for the mimic sequence it is useful to score the probe efficiency relative to a reference assay. For example, the percentage of signal–noise for both 5′ nuclease assay specific

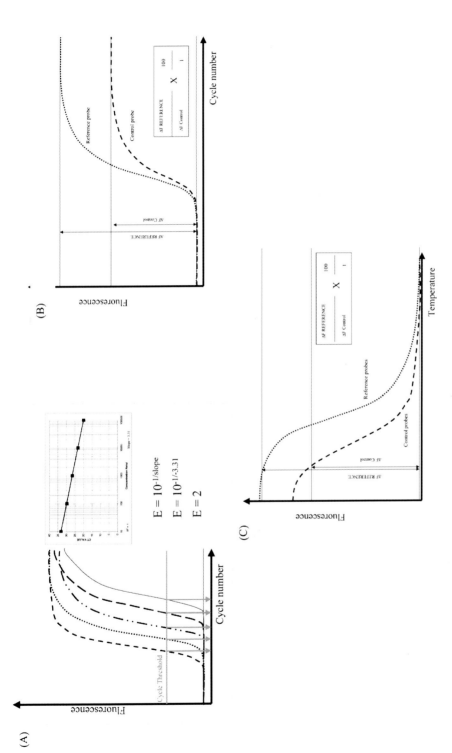

Figure 7.2 Determining and comparing amplification and probe efficiencies of different amplicons. (A) Determining amplification efficiency (E). Ideally, using a generic method such as a DNA binding dye, e.g. SYBR® Green I, the cycle threshold (CT) values are determined for a dilution series of target material over a range of concentrations and the plot of fluorescence intensity against cycle number generates the amplification curves. The derived CT values can be used to generate a standard curve. The slope of the curve can then be used to calculate E for the range tested. This value can be used to compare E with other amplicons using the same or similar dilution series. (B) Determining probe efficiency. Using the 5′ nuclease assay and other assays by comparing the total change in fluorescence (as a function of a reference probe/assay) when the reaction is allowed to go to completion. (C) Using probes based on hybridization by comparing (as a function of a reference probe/assay) the change in fluorescence when the probe is hybridized (at a specific temperature) to when it is un-hybridized (at high temperatures).

and control probes can be compared to that of another assay such as the ABI human β-actin probe. For this to be done the reaction for either probe must be taken to completion (plateaux). These data may be expressed as:

$$\frac{\text{FS value specific assay/control assay at (cycle 50–cycle 1)}}{\text{FS value } \beta\text{-actin (cycle 50–cycle 1)}} \times \frac{100}{1} \quad (1)$$

This provides a comparative means for matching candidate control sequences to given assays. Likewise, the same may be done for comparing dual hybridization probes using:

$$\frac{\text{FS value specific assay/control assay at (95°C–60°C*)}}{\text{FS value } \beta\text{-globin (95°C–60°C*)}} \times \frac{100}{1} \quad (2)$$

This approach may be carried out using any reference probe and is equally applicable to other assay systems to estimate the relative probe efficiency (Fig. 7.2). The quotient used to compare should be based on the values derived from the maximum signal (when the target is in excess of the probe) subtracted from the minimum signal (the background) for each probe. The minimum signal is usually at the start of the amplification or when the probe is dissociated from the target at elevated temperatures.

Control vector

The choice of control vector is important since the vector may affect the stability and amplification efficiency. Plasmid constructs are commonly used and the efficiency of amplification may drop dramatically at lower target numbers (<100 copies per reaction). Linearizing the plasmid by digestion at a unique restriction site in a vector can improve amplification efficiency. The efficiency of closed plasmids is often seen to drop off dramatically at ~100 to 1000 copies. The use of other vectors such as bacteriophage λ vectors and M13 (ssDNA rescue by co-infection with helper phage) may also be useful approaches for generating large amounts of linear IC molecules. PCR product itself is fine for use as a control

although a vector represents the best method of safe storage of the desired sequence. We recommend the initial amplification of control DNA by PCR since the products may be used to evaluate potential mimics prior to the cloning of the molecule into a vector. However, the use of an *in vivo* amplification method for the final synthesis of DNA to be used as control is superior in that individual clones may be readily sequenced, to check their correct identity, and this process avoids the spurious artefacts generated in PCR. These artefacts can be minimized during the evaluation stages by cutting the predicted PCR band out of an agarose gel and using any one of a number of methods for cleaning up the product. These include column purification and/or dialysis, before dilution and evaluation as suitable mimics.

Number of molecules

When using a competitive control the number of molecules to be added into the reaction must be determined empirically so that the required sensitivity of the specific assay is not compromised. The control must be amplified effectively in the absence of specific target. With increasing numbers of specific target molecules in the reaction the control amplification may be outcompeted. We have observed that it is possible to stop this if the T_m of the specific amplicon is higher than that of the control. Reducing the denaturing temperature of the thermal cycle (which is possible on some thermal cyclers) to below that of the specific target, but higher than that of the control, suppresses specific amplification thus allowing the IC to amplify. However, this is not necessary since the amplification of the control only serves to validate the reagents in the absence of the specific target. A cross-titration of specific target to control should be carried out to determine the correct amount of control molecules to include in a reaction. In practice, this amount is routinely determined to be between 10 and 100 copies per reaction if the specific assay is to maintain a sensitivity of 10 or more copies of specific target. Accurate determination of the IC mimic concentration is therefore important. The use of

---

FS, Fluoresence signal
*Or at the temperature where the fluorescence data are to be acquired.

fluorescent dyes and standards is inherently more accurate than ultra-violet optical density (UV OD_{260}) quantification at 260 nm. An alternative method for statistically determining the correct concentration of a stock of internal controls using PCR is provided by Rosenstaus *et al.* (1998). This method describes using a dilution series of the control preparation and performing multiple amplifications. At a given dilution the amplification will become stochastic due to the presence of low or no copies of the template in the reaction in accordance with Poisson's law. The relationship between the average number of internal control copies in a dilution (C) and the probability (P_n) that no molecules exist in a sample of this dilution is given by;

$$C = -\ln(P_n)$$

where P_n is determined by counting the number of negative replicates used to calculate C.

Target nucleic acid

The control must be RNA if it is to be used for controlling the RT step. If a specific primer initiated RT step is used then utilizing the same primers in the mimics will control for both the RT and the PCR and negates the requirement of a second mimic for the PCR step in a one step PCR. Depending on the application, controls at both the RNA and DNA levels may be required in a two-step PCR. The RT efficiency of the mimic and the specific target must be similar if the mimic is to be used as a standard in any application. When the RT–PCR is designed to be sense specific, directional cloning and the choice of vector orientation is important.

Dye/channel assignment

The most efficient dye and/or detector should be used for the specific target assay. This is of importance where a single excitation source is used in a universal acceptor arrangement (Chapter 3), in which case it is usually the shorter wavelength dye that will be most efficiently excited. When a universal donor arrangement is used, it will be the acceptor to which energy is most efficiently transferred from the donor. In either multiplex arrangement, the optimum channel will be dependent on the instrument's specification and the dyes used.

Synthesis

If the IC is to be a synthetic oligonucleotide then the authors would recommend using a different supplier to the company that will be making your amplimers. The risk of cross-contamination by the IC of an amplimer is such as to have been reported numerous times in PCR discussion forums.

Summary and future improvements

The further development of instruments and methods of detection will increase the ability to multiplex PCR to higher levels in a quantitative manner. This introduces new possibilities for controlling for and/or normalizing sample extraction, and/or RT-PCR steps. This will be achieved through a combination of competitive (homologous) and non-competitive (heterologous) internal controls for each process. It is not always possible to control for these processes with one control since each may be subjected to different inhibitory factors. The ability to do this will improve the usefulness of the PCR for both research and diagnostic applications. The ability to control for all such processes will be of most use in emerging applications where remote testing in the non-laboratory environment necessitates the use of internal validation.

Acknowledgements

The authors would like to thank the editors for the invitation to contribute to this text and Andrea Hamilton for proofreading.

References

Brightwell, G., Pearce, M. and Leslie, D. 1998. Development of internal controls for PCR detection of *Bacillus anthracis*. Mol. Cell. Probes *12*, 367–377.

Capodici, J., Kariko, K. and Weissman, D. 2002. Inhibition of HIV-1 infection by small interfering RNA-mediated RNA interference. J. Immunol. *169*, 5196–5201.

deWitt, D., Wooton, M., Allan, B. and Steyn, L. 1993. Simple method for production of internal control DNA for *Mycobacterium tuberculosis* polymerase chain reaction assays. J. Clin. Microbiol. *31*, 2204–2207.

Henley, W.N., Schuebel, K.E. and Nielsen, D.A. 1996. Limitations imposed by heteroduplex formation on quantitative RT-PCR. Biochem. Biophys. Res. Commun. *226*, 113–117.

Lee, M.A., Brightwell, G, Bird, H., Leslie, D. and Hamilton, A. 1999. Fluorescent detection techniques for real-time multiplex strand specific detection of

Bacillus anthracis using rapid PCR. J. Appl. Microbiol. *87*, 218–223.

Longo, M.L., Berninger, M.S. and Harley, J.L. 1990. The use of uracil glycosylase to control carry-over contamination in polymerase chain reactions. Gene *93*, 125–128

Mergenthaler, E., Fodor, M. and Süle, S. 2004. A new method to develop internal controls for the determination of phytoplasma concentration. Acta Phyto. Ent. Hung. *39*, 1–7.

Müller, F.M.C., Schnitzler, N., Coot, O., Kockelkorn, P., Haase, G. and Li, Z. 1998. The rationale and method for constructing internal control DNA used in pertussis polymerase chain reaction. Diag. Microbiol. Infect. Dis. *31*, 517–523.

Mullis, K.B. and Faloona, F.A. 1987. Specific synthesis of DNA *in vitro* via a polymerase catalyzed chain reaction. Methods Enzymol. *155*, 335–351.

Nash, K.A, Klein, J.S. and Inderlied, C.B. 1995. Internal controls as performance monitors and quantitative standards in the detection by polymerase chain reaction of herpes simplex virus and cytomegalovirus in clinical specimens. Mol. Cell. Probes. *9*, 347–356.

Pasloske, B.L., Walkerpeach, C.R., Obermoeller, R.D., Winkler, M. and DuBois, D.B. 1998. Armoured RNA technology for production of ribonuclease-resistant viral RNA controls and standards. J. Clin. Microbiol. *36*, 3590–3594.

Rosenstraus, M., Wang, Z., Chang, S.-Y, DeBonville, D. and Spadoro, J.P. 1998. An internal control for rou-tine diagnostic PCR: Design properties, and effect on clinical performance. J. Clin. Microbiol. *36*, 191–197.

Rossen, L., Norskov, P., Holmstrom, K. and Rasmussen, O.F. 1992. Inhibition of PCR by components of food samples, microbial diagnostic assays and DNA extraction solutions. Int. J. Food Microbiol. *17*, 37–45.

Sachadyn, P. and Kur, J. 1998. The construction and use of a PCR internal control. Mol. Cell. Probes. *12*, 259–262.

Saiki, R.K., Scharf, S., Faloona, F.A., Mullis, K.B., Horn, G.T., Erlich, H.A. and Arnheim, N. 1985. Enzymatic amplification of beta-globin genomic sequences and restriction site analysis for diagnosis of sickle cell anemia. Science *230*, 1350–1354.

Ursi, J-P., Ursi, D., Ieven, M. and Pattyn, S.R. 1992. Utility of an internal control for the polymerase chain reaction. Application to detection of *Mycoplasma pneumoniae* in clinical specimens. APMIS *100*, 635–639.

Wilson, I.G. 1997. Inhibition and facilitation of nucleic acid amplification. App. Environ. Microbiol. *63*, 3741–51.

Zimmerman, K, and Mannhalter, J.W. 1996. Technical aspects of quantitative competitive PCR. Biotechniques *21*, 268–279.

Zimmerman, K., Rieger, M., Groß, P., Turecek, P.L. and Schwarz, H.P. 2000. Sensitive single-stage PCR using custom synthesized internal controls. Biotechniques *28*, 694–702.

Introduction to the Applications of Real-Time PCR

Nick A. Saunders

Abstract

The technique of real-time PCR has features that make its use advantageous in a wide range of applications. A number of examples, covering the main areas of application, are given in the following chapters of this book. In this introduction the important features of these applications are discussed.

Technical areas of opportunity for real-time PCR

The preceding chapters of this book feature technical details of real-time PCR and demonstrate how the concept can be exploited using a wide range of chemistries and instrumentation. The purpose of the final chapters is to indicate the extent of the opportunities for exploitation of real-time amplification.

To summarize the first part of this book, it can be seen that real-time amplification technology has basic properties that are advantageous in a variety of applications. The ability to monitor the progress of reactions in real-time provides data that can be used to calculate the efficiency of the chemical and enzymatic reactions involved in PCR. More importantly, plots of fluorescence against cycle number can be used for template quantification. Instruments capable of real-time fluorescence monitoring can also be exploited to measure nucleic acid duplex melting temperatures and these data can be helpful in assay design, as well as in analysis of the amplification products. The second important feature of real-time technology is the use of sequence-specific probes that

can be incorporated directly into the assay and which are normally responsible for transduction of product accumulation within the reaction tube into changes in fluorescence intensity. The probes can be used to confirm that the amplification product conforms to the expected sequence and can even be applied to distinguish between different amplicons. An important benefit associated with the closed-tube end-point of real-time PCR is that the amplicons are contained within the reaction vessel and so the chance of contamination is minimized.

These properties of real-time machines and chemistries provide two important exploitation opportunities that extend the value of PCR. First, the 'real-time' and 'sequence confirmation' abilities combine to allow the design of rapid and highly automatable protocols for detection of particular target sequences. For example, a real-time PCR for detection of a specific bacterial gene sequence can be used to replace protocols combining classical PCR with one or more of the following additional processes: gel electrophoresis, blotting, purification, restriction endonuclease analysis, sequencing and probing. The second opportunity relies primarily on the 'real-time' property of the technology and is the ability to quantify the initial load of DNA or RNA (for reverse transcription PCR). For quantification it is advantageous but not strictly necessary to use 'sequence confirmation'. Both relative and absolute quantification methods are possible and are widely used.

Applications of real-time PCR

Detection and quantification of pathogens including biothreat agents

The diagnosis of infectious disease represents an important application of nucleic acid amplification methods such as PCR (Speers, 2006). Indeed molecular biology did not figure significantly in the clinical microbiology laboratory until the appearance of the first reported applications of PCR for the detection of HIV-1 (Barinaga, 1988). The detection of nucleic acid components of the infectious agent fitted a growing need for the development of methods that are rapid, specific and sensitive. However, some of the traditional methods of diagnosis such as culture and serology were inexpensive and offered a degree of quantification in addition to having some but not usually all the other required qualities (rapidity, specificity and sensitivity). Consequently, the penetration of nucleic acid amplification based assays into routine use has been selective. Besides the additional expense the adoption of PCR-based detection in the early years had additional disadvantages. Contamination of assays with amplicon from previous reactions was a serious problem causing false positives that was recognized from the first days of the application of PCR to clinical material (Lo et al., 1988). An additional criticism, which came primarily from workers accustomed to culture, was that nucleic acid methods might give positive results due to the detection of nucleic acids from lysed or otherwise non-viable agents. The detection of insignificant quantities of certain agents was also a potential concern.

The introduction of real-time instruments has greatly increased the functionality of nucleic acid amplification based detection by reducing the time required to perform the assays and to obtain a positive identification of the amplicon, by helping to resolve contamination issues and by providing quantitative or semi-quantitative results. Hundreds of real-time PCR assays have been developed over the last few years and many are now available commercially in kit form from manufacturers including Roche, Qiagen and Applied Biosystems, to name but a few (Chapter 6). There has been a steady shift towards the use of commercial kits, especially in frontline laboratories, because of the need to use reagents produced within a quality system that can maintain batch-to-batch consistency. However, many kits are still labelled for research use only. The use of assays and reagents designed and produced in-house remains important especially within specialist laboratories providing reference services. In many cases the demand for in-house tests, although real, is insufficient to warrant economic production by the commercial sector.

Real-time PCRs for agents of infectious disease employ a wide range of chemistries and hardware platforms reflecting the diversity of the sources of target nucleic acids and concerns of the interested laboratories. A significant complication for assays designed to detect and quantify infectious agents is the wide range of sample types and biological entities that have to be accommodated. Samples range from complex foods to blood and may contain significant quantities of materials that may either inhibit the polymerase or, less frequently, sequester other essential reaction components, e.g. magnesium ions. Many infectious agents, especially Gram-positive bacteria for example, are not lysed using standard combinations of chaotropes, detergents and temperature unless the cell wall has been enzymatically pre-treated. In such cases, physical disruption methods such as bead beating or mild sonication may be preferable. The different protocols that are necessary to overcome the sample type diversity are probably a significant reason why no dominant real-time technology has emerged in the field of infectious diseases. Another driver of technology diversity (or hindrance to uniformity) has been the multiplicity of the questions posed. For example, tests for *Staphylococcus aureus* may need to distinguish between MRSA (methicillin-resistant *S. aureus*) and MSSA (methicillin-sensitive *S. aureus*). For MRSA tests, it may be necessary to exclude the possibility that the sample contains an MSSA and the *mec*A gene within the genome of a different organism (i.e. a coagulase-negative staphylococcus). Subsequently, it may be important to determine whether the staphylococci detected carry specific toxins and/or superantigens, which may be relevant in either a food contamination or clinical context. Finally, although not particularly relevant for *S. aureus* strains, real-time

PCR assays can be used to detect SNPs which can be used either as anonymous markers for typing microbial genomes or more frequently as indicators of a relevant phenotype (e.g. antiviral or antibacterial drug resistance).

Three chapters in the applications section of this book are primarily concerned with the detection of infectious agents. This emphasis partly reflects the background of the Editors but is probably justified by the importance and diversity of the field. Chapter 13 describes the application of real-time methods to the detection of bacterial, viral and protozoan pathogens in the clinical setting and Chapter 14 extends this in the context of the particular problem of dealing with fungal infections. Chapter 15 is concerned with the detection of infectious agents that are considered biothreats from either military or civilian (i.e. terrorist) sources. In the case of biothreats the ability to analyse environmental specimens and food is of key importance. Chapter 16 is also partly concerned with the detection and enumeration of micro-organisms in food. These may be involved in food spoilage and/or represent an infection threat.

Detection and quantification of signature sequences in foods

Food authenticity testing represents a considerable challenge to the analyst. Prior to the introduction of PCR the detection of genetically modified components in either raw materials or manufactured foods was considered to be impractical. The introduction of PCR allowed the detection of traces of specific nucleic sequences and led to the development of tests for genetically modified crop plant material based on the presence of vector and transgene sequences. PCR was also applied to the task of determining the species of origin of food material. Many of these tests used alternative methods of quantification since food regulations governing contamination necessarily involve allowable limits rather than absolute proscription. The development and introduction of quantitative real-time PCR methods allowed the replacement of these earlier and less convenient assays.

The use of quantitative real-time PCR for the authentication of foods is now an economically important activity. Laboratories participat-

ing in this field have been required to adopt strict quality standards and to use validated methods. Standardization is therefore more advanced than has been achieved in the field of infectious diseases. Chapter 16 details the use of real-time PCR in the field of food authenticity testing.

Genotyping and haplotyping by SNP analysis

The accurate and rapid determination of the identity of nucleosides at specific loci of a sequence (i.e. SNP analysis) can now be accomplished via a bewildering number of techniques. However, in general the available methods can be classified as being based on either primer extension (e.g. Sanger sequencing, pyrosequencing, allele-specific PCR, etc.) or hybridization (e.g. real-time PCR, oligonucleotide ligation, oligonucleotide microarrays). Real-time PCR SNP detection can be based on either of the core technologies.

Allele specific primers for PCR are designed so that the 3′ terminal base corresponds to the base to be queried, so that extension proceeds efficiently only when the base is paired. Obviously, exploitation of this principle depends upon the use of a polymerase that does not have mismatch repair (3′–5′ exonuclease) activity. One complication is that different mismatched base pairs are extended with rather divergent efficiencies, which also depend upon the local sequence environment (i.e. other bases close to the terminus). However, the minimum difference between mismatch and match that might be anticipated is at least two orders of magnitude and this can be modified by introducing mismatched bases into the primer. Although allele specific primers are useful for SNP detection in real-time PCR assays, it is generally more straightforward to use probes that interrogate the polymorphic base within the PCR amplicon. All imaginable types of real-time probe are probably adaptable for SNP analysis although hydrolysis probes have been around the longest and are the more frequently used. It is a particularly convenient feature of this chemistry to be able to include a hydrolysis probe for each possible allele within each reaction. This gives a straightforward readout of the alleles present. When hybridization probes are used it is usually necessary to perform a melt analysis which may yield more complex results.

The first applications of PCR (Saiki et al., 1985) and real-time PCR (Lee et al., 1993) were the detection of sequence polymorphisms associated with human disease. Two chapters (11 and 17) within this edition are specifically concerned with the detection of SNPs for genotyping. Chapter 17 extends the principle to haplotyping by discussing methods whereby it is possible to determine whether two SNPs are present on the same or different chromosomes. This is relatively simple by real-time PCR when the SNPs are closely spaced a single probe complementary to both target bases can be designed. When the probe has hybridized to the target amplicon melting analysis can be used to determine whether one or both SNPs are present. The principle can also be applied to more widely spaced SNPs (up to 87 bp) by using 'loop-out' probes that are comprised of probes for the two separate target sequences that are joined. When such probes hybridize to their target amplicons they form a structure in which the amplicon region between the probes forms a single-stranded loop. This structure is more thermodynamically stable than the alternative in which the two regions of the probe hybridize to different amplicons and hence it is possible to distinguish between them by melting analysis. PCR assays for haplotyping based on allele specific PCR in which each primers is used to extend a different polymorphic base have also been described. For this method the SNPs need only be spaced within the size range for PCR amplification. Although, dual allele-specific PCR could be done on a real time instrument it would be limited to the relatively short amplicons that can be analysed efficiently.

Real-time PCR-based genotyping assays are generally very robust, versatile, and automatable and are easily standardized. Many tests are commercially available especially for SNPs associated with human disease and antibiotic resistance markers. Large panels of assays are available for research. Over the longer term it remains to be determined whether SNP analysis by real-time PCR can consolidate its position within the research and diagnostic markets against a number of powerful alternative technologies.

RNA quantification

Gene expression analysis is a key area of biological research and has become a very important application for real-time PCR. Prior to the introduction of real-time instruments there were few viable alternatives to Northern blotting for quantification of specific mRNA molecules. Northern blotting has many disadvantages including, that it is at best semi-quantitative, full-length undegraded RNA is needed and the dynamic range and sensitivity is limited. RT qPCR (reverse transcriptase quantitative real-time PCR) does not have these disadvantages and has therefore become the standard in gene expression studies. Recently RT qPCR has assumed even greater importance as the validation method of choice for the vast quantities of microarray data that are now accumulating.

Although RT PCR is a well-established method there are still a number of basic variations in the way it is implemented. These differences can have implications for the real-time quantitative versions of the assays. The reverse transcription reaction can be primed with either random or specific primers. The use of template-specific primers is more efficient and therefore preferable, but random priming can be useful when it is necessary to measure multiple target RNAs, especially when the sample size is limited. Different options are available for linking the reverse transcriptase and DNA polymerase phases of the reaction. The thermostable enzyme *Tth* polymerase has both activities and can be used successfully in RT qPCR although the sensitivity may be lower than for reactions based on enzyme mixtures. Two enzyme reactions (i.e. with both a DNA dependent and an RNA dependent DNA polymerase) can be performed in either one or two tube formats. In the one tube format DNA dependent polymerase and buffer are added to the products of reverse transcription, whereas in the two-tube method an aliquot of the first reaction is added to the second tube containing the fresh reactants. The main advantage of the one enzyme approach over the two enzyme methods is that there is no need to open the tube containing the products of reverse transcription, so the possibility of contamination is minimized.

Absolute quantification of RNA by reference to standard curves constructed using accurately prepared standards comprised of RNA with the same sequence as the target is the ideal output of RT qPCR. However, absolute quantification is difficult to achieve in practice because the quantity of tissue extracted must be accurately determined and subsequently the efficiencies of extraction and transcription must be carefully controlled. Relative quantification versus an internal control mRNA is generally easier to implement and its accuracy is greater. Unfortunately, greater accuracy is accompanied by greater uncertainty because in practice good control genes are very difficult to find.

Two chapters in this section are devoted to RT qPCR. Chapter 9 discusses the methodologies used with particular reference to the quantification of eukaryotic mRNA Chapter 10 deals with the increasingly important problems of quality presented by gene expression microarray data. Real-time PCR has an important place in the validation of these data.

NASBA

This book is mainly devoted to real-time PCR. However, it should not be overlooked that the instruments used for real-time PCR are generally not merely 'black-boxes' tied to a single proprietary chemistry for real-time amplification and fluorescence measurement. This situation has persisted because users have continued to demand some flexibility in the real-time instruments they purchase. It should be possible to perform real-time NASBA on any of the instruments designed for real-time PCR. The only requirements are that the temperature of the reaction vessel is controlled and that repeated fluorescence measurements can be made.

Chapter 12 describes the current state of the art of real-time NASBA. This technique is best suited to the detection of RNA since preliminary steps are required to use DNA targets.

Quantification in NASBA depends upon the comparisons of time to positivity (TTP) values rather than crossing thresholds (CT). There is evidence that the performance of real-time NASBA can equal or even exceed that of real-time PCR in a range of applications. However, since relatively few workers are using NASBA the number of comparisons has been limited. It remains to be determined whether the promise of NASBA, particularly for quantification of RNA templates will be widely realized.

Conclusions

The concluding chapters of this edition detail a wide range of applications for nucleic acid amplification coupled with fluorescent chemistries and instruments for monitoring the reaction in real-time. However, the list is not exhaustive. The target sequences can either be parts of the genome (DNA or RNA) or the transcriptome. The reasons for making the analytical measurements can include detection (e.g. infectious disease), sequence variant detection (e.g. SNP) and gene expression analysis. The methods are powerful, versatile and consequently it seems likely that real-time PCR (and real-time NASBA) will have a long future as a widely exploited and valuable technology.

References

Barinaga, M. 1988. New tests for spotting AIDS. Nature 333, 291.

Lee, L., Connell, C. and Bloch, W. 1993. Allelic discrimination by nick-translation PCR with fluorogenic probes. Nucleic Acids Res. 21, 3761–3766.

Lo, Y.M.D., Mehal, W.Z. and Fleming, K.A. 1988. False positive results and the polymerase chain reaction. Lancet ii, 8612, 679.

Saiki, R.K., Scharf, S., Faloona, F., Mullis, K.B., Horn, G.T., Erlich, H.A. and Arnheim, N. 1985. Enzymatic amplification of beta-globin genomic sequences and restriction site analysis for diagnosis of sickle cell anaemia. Science 230, 1350–1354.

Speers, D.J. 2006. Clinical applications of molecular biology for infectious diseases. Clin. Biochem. Rev. 27, 39–51.

Analysis of mRNA Expression by Real-time PCR

Stephen A. Bustin and Tania Nolan

Abstract

Its conceptual and practical simplicity, capacity for high throughput, and combination of high sensitivity with exacting specificity has made the fluorescence-based real-time reverse transcription polymerase chain reaction (qRT-PCR or RT-qPCR) today's method of choice for the quantification of RNA. The technology continues to evolve rapidly with the introduction of new protocols, enzymes, chemistries and instrumentation and has become the 'Gold Standard' for a huge range of applications in basic research, molecular medicine, and biotechnology. Progress is increasingly associated with an increased appreciation of the limitations associated with this technology and the need for careful experimental design, application and validation.

Introduction

The real-time reverse transcription polymerase chain reaction (qRT-PCR or RT-qPCR) (Gibson et al., 1996) is today's benchmark method for the detection and quantification of RNA targets (Bustin, 2000). Whilst not necessarily more sensitive than conventional RT-PCR (Ozoemena et al., 2004), RT-qPCR assays have several significant advantages (Orlando et al., 1998):

1. They use fluorescent reporter molecules to monitor the production of amplification products during each cycle of the PCR reaction. The combination of the DNA amplification and detection steps into one homogeneous assay obviates the requirement for post-PCR processing, making the assay very convenient (Ke et al., 2000).

2. Their wide dynamic range allows the analysis of samples differing in target abundance by several orders of magnitude (Gerard et al., 1998).

3. There is little inter-assay variation which helps generate reliable and reproducible results (Max et al., 2001; Wall and Edwards, 2002).

4. Fluorescence-based RT-qPCR realizes the inherent quantitative capacity of PCR-based assays (Halford et al., 1999), making it a quantitative rather than a qualitative assay.

This is important, as there is an obvious need for quantitative data in basic research, molecular, forensic and veterinary medicine, as well as in microbiology. Furthermore, the availability of a choice of several robust chemistries, the introduction of improved protocols and the cost-reductions derived from the introduction of more reliable real-time instruments from several manufacturers has resulted in the development of RT-qPCR-based clinical diagnostic assays that hold considerable promise (Bustin and Mueller, 2005).

Nevertheless, reliable and reproducible RT-qPCR assays require meticulous adherence to standard operating procedures for sample acquisition and handling as well as the preparation, quality assessment and storage of RNA. Furthermore, the results obtained from the assay depend on the type of reverse transcriptase and the priming strategy used to generate cDNA and careful optimization of the efficiency of the PCR step. Data analysis is unexpectedly subjective and accurate data interpretation depends on the inclusion of suitable controls, careful analysis of the raw data and the use of appropriate normalization procedures. Proper optimization of pre-analysis (Muller *et al.*, 2004) and assay (Bustin, 2004) conditions, its validation (Afzal *et al.*, 2003) and standardization (Gabert *et al.*, 2003; Niesters, 2001, 2004) are crucial and, for diagnostic laboratories, external quality assurance programmes have been shown to be useful (Marubini *et al.*, 2004). Assay standardization, in particular, is a critical parameter determining the reliability of quantitative data and use of standardized protocols for RNA isolation and RT-qPCR assays can generate fairly comparable data between different laboratories (Vlems *et al.*, 2003). However, in practice it is clear that there is little standardization: a recent survey of protocols and analysis procedures revealed huge variability, even between diagnostic laboratories (Bustin, 2005). Not surprisingly, these differences lead to heterogeneous results (Raggi *et al.*, 2005; Vlems *et al.*, 2002a,b) similar to those observed with conventional PCR assays (Zippelius *et al.*, 2000). Finally, no amount of standardization can correct for the problems encountered when quantitating very low copy number targets (Ladanyi *et al.*, 2001).

Choice of chemistry

The appropriate detection chemistry depends on the application: the use of a double-stranded DNA binding dye, e.g. SYBR Green I, is the most cost-effective chemistry for initial investigations and for primer optimization steps. Legacy assays can be used and new ones are simple to design. Although the primers alone determine the specificity of product detection, some information regarding product size and population can be determined from a melting curve analysis. Other non-probe based chemistries include Promega's Plexor (www.promega.com) and Invitrogen's LUX primer (www.invitrogen.com).

At very low (< 1000 copies) target concentration there is a greater probability of non-specific amplification and problems with primer dimer products becoming more pronounced. In this case the use of a probe to detect amplicons may be preferable. The most widely used is the dual labelled fluorescent (TaqMan) probe, which can be synthesized with added locked nucleic acid (LNA; Sigma Life Science) or minor groove binder (MGB; Exiqon) modifications. LNA is a nucleic acid analogue that contains a 2′-O,4′-C methylene bridge, which restricts the flexibility of the ribofuranose ring and locks the structure into a rigid bicyclic formation, conferring enhanced hybridization performance. MGBs are dsDNA-binding agents that are attached to the 3′ or 5′ of the TaqMan probes and also stabilize the hybridization reaction. In practice, both modifications allow for the design of shorter probes, which increases design flexibility and specificity. This is particularly useful when detecting single nucleotide polymorphisms or when targeting AT rich sequences and when targets are highly structured or repetitive. The recent development of intercalating nucleic acids (Christensen and Pedersen, 2002; Christensen *et al.*, 2004; Filichev *et al.*, 2004) provides additional alternatives for the improvement of assay design.

Amplicons and primers

There is no consensus concerning the best amplicon location on an mRNA molecule. We believe that, where possible, assays should be designed across intron/exon boundaries, selecting the longest possible intron, as this avoids signal generation from amplified DNA. This is essential when using genomic DNA for normalization or when targeting specific splice variants of an mRNA. It is most economical to place the downstream primer across the splice junction, as this allows the use of a single (expensive) probe for the detection of possible splice variants in separate assays and, at the same time, minimizes their amplification in any one assay. However, we have noticed that it is often not possible to design an optimal assay around splice junctions and if the choice is between a poor (i.e. inefficient and/or insensitive) assay designed across a splice junction and an efficient one contained within a single exon, it is better to use the latter. In this case it is advisable to DNase I treat the sample and essential to add a 'no RT' control to determine the contribution of any DNA contamination to the mRNA quantification. For some specialist applications, e.g. the quantification of mRNA in a single fertilized egg, the relative difference in the abundance of DNA and mRNA is such that any DNA will not interfere with accurate quantification of mRNA.

The amplicon length should be as short as possible because this tends to increase amplification efficiency. In addition small amplicons are more tolerant of reaction conditions because they are more likely to be denatured during the 92–95°C step of the PCR, allowing the probes and primers to compete more effectively for binding to their complementary targets. They also require shorter polymerization times to replicate the amplicon, making amplification of genomic DNA contaminants less likely. From a practical perspective it is also easier and cheaper to synthesize a shorter oligonucleotide, if a synthetic sense-strand oligonucleotide is used to generate amplicon-specific standard curves. When it is impossible to find an acceptable short amplicon the design parameters relating to amplicon size should be changed bit-by-bit, until the smallest acceptable amplicon is found. Whenever possible, the amplicon should be around 100bp and no more than 150 bp in length and have a G/C content of 60% or less as it may not denature well during thermal cycling, leading to a less efficient reaction. Furthermore, GC-rich sequences are susceptible to nonspecific interactions that result in nonspecific signal in assays utilizing SYBR Green I dye.

The 3′-terminal position is essential for the control of mispriming during the PCR reaction. Primers with one or more G or C residues at that position will have increased binding efficiency due to the stronger hydrogen bonding of G/C residues. It also helps to improve the efficiency of priming by minimizing any 'breathing' that might occur. However, avoid clamping the 3′ of the primer because ambiguous binding of oligonucleotides to the target site could result in misprimed elongation ('slippage effect'). A reasonable compromise is to aim for two or three G/C in the five 3′ bases of the primer. Similarly, avoid long runs of a single base (i.e. more than three or four, especially G or C) since homopolymeric runs can also cause the 'slippage effect'.

Avoid direct repeats in the target sequence. Stable hybridization to secondary binding sites in repetitive regions result in non-productive binding of primers to nonspecific regions of the sequence so that efficiency of DNA amplification and detection decreases.

Primers must have no intraprimer homology beyond 3′-base pairs, as otherwise primers may form secondary structures that can interfere with annealing to the template.

DNA primers should be between 15 and 25 bases long to maximize specificity, with a G/C content around the 50%. This also contributes to a higher yield of full length product after synthesis and purification.

For primers binding at very AT-rich sequences it is advantageous to substitute one or more of the bases with a locked nucleic acid LNA analogue to reduce the overall primer length while maintaining the T_m.

Avoid primers with secondary structure (i.e., inverted repeats) or with sequence complementarity at the 3′ ends that could form dimers. Due to competition between intermolecular (primer–template, probe–target) hybridization, and intramolecular hybridization, inverse repeats can cause inefficient priming and probing of the target sequence. PCR reaction and/or probe binding can completely fail because of formation of stable hairpins at the binding region, or inside the amplicon in general. Primer-dimers have a negative ΔG value, so primers should be chosen with a value no less negative than -10 kcal/mol.

qPCR assays using hydrolysis (TaqMan) probes are usually carried out as two-step reactions, with a denaturation step followed by a combined annealing/polymerization step, during which fluorescence is measured. Most detection systems rely upon binding of the probe to the template before hybridization of the primers. It is for this reason that the probe T_m should be around 10°C higher than the T_m of the primers, usually in the 68–70°C range. The probe should be designed with the 5′ end as close to the 3′ end of one of the primers on the same strand as possible. This ensures rapid cleavage by the polymerase.

The probe should be no more than 30 nucleotides in length to maximize quenching, with a G/C content of around 50%. If the target sequence is AT-rich, incorporate analogues such as LNA or minor groove binders. Guanine (G) can quench fluorescence and so is avoided as the 5′ base. In this position the G would continue to quench the reporter even after cleavage, resulting in reduced fluorescence values (ΔR), which could result in reduced sensitivity. Any runs of four or more identical nucleotides, especially G, should be avoided because these can cause secondary structure of the probe and reduce hybridization efficiency. Where an alternative sequence cannot be selected disruption of a series of Gs by the substitution of an inosine can significantly improve probe performance.

No oligonucleotide supplier is consistently better than any other and so choice of vendor is best made based on service and support. If using amplicon-specific oligonucleotides to generate standard curves, it is essential to order those oligonucleotides on a different date, so as not to permit any chance of contamination of the primers with that oligonucleotide. It is also advisable to order a number of potential primers for a single target, optimize the assay and then order the probe. Again this helps cut down any contamination that might occur at the manufacturer and ensures that the best assay is selected. Different batches of primer and probe can generate different results, hence it is essential to compare results obtained with any new batch of oligonucleotide. It is also essential to check probe performance, since probe quality varies considerably between batches.

Amplicon oligonucleotides should be PAGE purified, probes should be HPLC purified, whereas primers can be supplied in a standard desalted format. It is essential to aliquot all oligonucleotides, as this prevents contamination of a whole batch and reduces the need to repeatedly freeze/thaw them. Forward and reverse primers can be stored at 10 µM or mixed together after optimization at the required concentrations. Probes should be stored frozen at –70°C both when lyophilized or in solution (at 5 µM). The probe must be protected from light, but long-term stability is variable, with some probes stable for several years, and others hydrolysed after 6 months.

RNA assessment

The quality of the RNA template is an essential determinant of the reproducibility and biological relevance of any subsequent RT-qPCR assay.

Ideally, any starting template should be of the highest quality, it should be free of DNA and there must be no co-purification of potential inhibitors of the RT-step. Furthermore, the original expression profile must be conserved during sample extraction and subsequent processing. The diversity of the different sources of mRNA poses a real practical problem: they include not just relatively simple *in vitro* samples such as tissue culture cells, but *in vivo* biopsies obtained during endoscopy, post-surgery, post-mortem and from archival materials. Naked RNA is extremely susceptible to degradation by endogenous ribonucleases (RNases) that are present in all living cells. Therefore, the key to the successful isolation of high quality RNA and to the reliable and meaningful comparison of qRT-PCR data is to ensure that neither endogenous nor exogenous RNases are introduced during the extraction procedure (Lee *et al.*, 1997). Two related problems are that different mRNAs display differential stability and that mRNA expression profiles can change rapidly after cells or tissue samples have been collected, but before they have been frozen. This implies that quantification of a single transcript or determination of relative transcript levels could be seriously effected by the sampling procedure adopted. This is a particular problem with cells extracted from whole blood (Keilholz *et al.*, 1998) and mRNA profiles can change over several orders of magnitude even in the short time it can take from collecting the blood to processing it in the laboratory. Causes are RNA degradation and the induction of certain genes, e.g. Cox-2, as well as the method of RNA preparation (de Vries *et al.*, 2000). Therefore, appropriate sample acquisition has a major influence on the quality of the RNA and subsequently on any result of qRT-PCR assays (Vlems *et al.*, 2002a) and once a biological sample has been obtained, immediate stabilization of RNA is the most important consideration (Madejon *et al.*, 2000).

The analysis of RNA extracted from archival material constitutes an exception from the highest quality rule. There are now numerous reports that describe the successful extraction and quantification of RNA from formalin-fixed, paraffin-embedded (FFPE) archival materials (Coombs *et al.*, 1999; Lewis *et al.*, 2001; Mizuno *et al.*, 1998). Formalin-fixed tissues are ideal for retrospective clinical studies of disease mechanisms and as the use of PCR technology has become more common in molecular testing, it has enhanced the clinical utility of these tissues. This is important, since such studies have the potential to enable the correlation of molecular findings with the patient's response to treatment and eventual clinical outcome. RT-qPCR assays, with their amplicon lengths of below 100 bp, are ideally placed to amplify the usually degraded RNA from these archival samples whose average size is 200–250 nucleotides. However, care must be taken when interpreting the results obtained from archival material, as gene expression profiles from FFPE samples do not exactly correlate with the profiles generated from the corresponding frozen samples. In a comparative study 64 genes were differentially expressed in matching fresh-frozen normal colon and cancer samples, only 38 were in the corresponding FFPE samples. Furthermore, only 28 of these genes were in common (Bibikova et al., 2004a). Thus, any results obtained using FFPE samples require independent experimental determination (Bibikova et al., 2004b), but also may underestimate or report misleading changes in gene expression patterns.

Just as real-time chemistries have revolutionized the PCR reaction, so the introduction of the Agilent Bioanalyser and Biorad Experion has revolutionized the ability to analyse RNA samples for quality. This technology has become the method of choice for analysing RNA preparations destined to be used in RT-qPCR assays. Both systems use small bench top instruments that integrate sample detection, quantification, and quality assessment. They do this by using a combination of microfluidics, capillary electrophoresis and fluorescent dye that binds to nucleic acid within a RNA chip. Size and mass information is generated by the fluorescence of the nucleic acid molecules as they migrate through the channels of a dedicated chip. The instrument software is used to quantify the amount of unknown RNA samples by comparing their peak areas to the combined area of the six reference RNAs. It has a wide dynamic range and can quantitate as little as 5 ng/μl, although it is most accurate at concentrations above 50 ng/μl. An electropherogram of fluorescence vs. time and a virtual gel image are generated and the software assesses RNA quality by using the

areas under the 28S and 18S rRNA peaks to calculate their ratios. However, it is worth noting that these profiles describe rRNA quality and do not provide direct information about the quality of the mRNA. Accurate delineation of mRNA quality is likely to demand additional quality assessments similar to the 3':5' ratios used during microarray experiments. The analysis of total RNA samples from human colon biopsies is shown in Fig. 9.1.

Although these instruments can be used for quantification, the most accurate quantification is achieved by using either the Nanodrop (www.nanodrop.com) system or, for higher throughput applications, the Ribogreen RNA (www.invitrogen.com) staining dye that binds to RNA and produces a fluorescent signal. The concentration of unknown samples is determined by comparison of the fluorescent signal to that of samples in a standard curve series. This can be used in conjunction with a fluorescent plate reader or some qPCR instruments. Note that it is essential that a single method of RNA quantification is selected and that this is used exclusively throughout a study. It has been clearly demonstrated that different methods of quantification give widely differing results (Bustin, 2005).

RNA integrity

RNA quality encompasses both its purity (absence of protein and DNA contamination, absence of inhibitors) and its integrity. Traditionally RNA quality has been determined by analysis of the A260/A280 ratio and/or analysis of the rRNA bands on agarose gels, an approach updated with the introduction of the Agilent Bioanalyser/BioRad Experion microfluidic capillary electrophoresis systems. A recent report claims that it is possible to calculate a more objective measure of RNA quality by measuring several characteristics of the electropherogram generated by the Agilent 2100, including the fraction of the area in the region of 18S and 28S RNA, the relative height of the 28S peak, the area containing molecules which are defined as having a 'fast' mobility and marker height and use these to calculate and assign an RNA Integrity Number (RIN) to each RNA sample (Schroeder et al., 2006). However, an exhaustive analysis of the influence of RNA integrity on RT-qPCR assay performance comes to a different conclusion

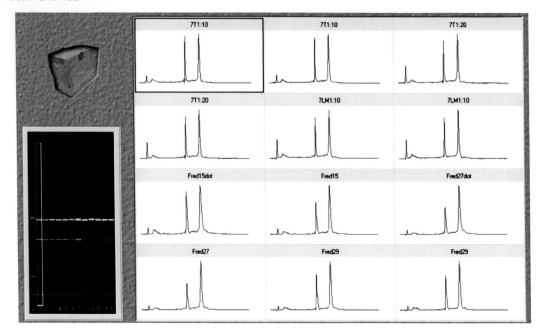

Figure 9.1 Electropherogram generated using the Agilent 2100 Bioanalyser showing high quality RNA preparations with distinct 28S and 18S peak ratios. The amount of small RNA (first peak), which can include 5.8S and 5S ribosomal peaks and transfer RNA, is highly dependent on the preparation protocol.

(Fleige and Pfaffl, 2006). The authors extracted RNA samples extracted from numerous tissue types, subjected them to controlled degradation and analysed them using the Agilent Bioanalyser. The samples had RINs between 10 (apparently intact material) to 4.0 (almost no evidence of rRNA bands). The quantity of individual transcripts in each of these samples was then determined using RT-qPCR assays. In some tissues the quantity of the measured transcripts detected was unaffected by the RIN whereas in others there was a linear relationship and in others a threshold response. Critically, the relationship between transcript quantity and RIN was different for different tissues and there was not a predictable relationship between these factors. The authors conclude that moderately degraded RNA samples can be reliably analysed and quantitated, as long as the amplicons are kept short (<250 bp) and expression is normalized against an internal reference. The discrepancy between the two reports could be due to the relatively poor correlation coefficient (0.52) between RIN and expression values of the reference genes reported by the authors advocating the use of RINs (Schroeder *et al.*, 2006).

Inhibitors

Inhibitory components can result in a significant reduction in the sensitivity and kinetics of qPCR (Cone *et al.*, 1992; Guy *et al.*, 2003; Jiang *et al.*, 2005; Lefevre *et al.*, 2004; Perch-Nielsen *et al.*, 2003; Radstrom *et al.*, 2004; Sunen *et al.*, 2004). Inhibitors may be reagents used during nucleic acid extraction or co-purified components from the biological sample, for example bile salts, urea, heparin or IgG. Blood is well known for containing numerous inhibitors of the PCR reaction (Akane *et al.*, 1994; Al Soud and Radstrom, 2001; Fredricks and Relman, 1998). The presence of inhibitors can produce inaccurate quantitative results and, at worst, a high degree of inhibition will create false-negative results. The most common procedure used to account for any differences in PCR efficiencies between samples is to amplify a reference gene in parallel to the reporter gene and relate their expression levels. However, this approach assumes that the two assays are inhibited to the same degree. The problem is even more pronounced in 'absolute' quantification, where an external calibration curve is used to calculate the number of transcripts in the test samples, an approach that is commonly adopted for quantification of pathogens. Some, or all, of the biological samples may contain inhibitors that are not present in the nucleic acid samples used to construct the calibration curve, leading to an underestimation of the mRNA levels in the test samples (Stahlberg et al., 2005). The increasing interest in extracting nucleic acids from formalin-fixed paraffin-embedded (FFPE) archival material will undoubtedly lead to an exacerbation of this problem. Obviously, such inhibitors are likely to distort any comparative quantitative data. However, a recent survey of practices has revealed that only 6% of researchers test their nucleic acid samples for the presence of inhibitors (Bustin, 2005).

Various methods can be used to assess the presence of inhibitors within biological samples. The PCR efficiency in a test sample can be assessed by serial dilution of the sample (Stahlberg et al., 2003), although this is impossible when using very small amounts of RNA extracted, for example, from single cells or from laser capture microdissected sections. Furthermore, there are mathematical algorithms that provide a measure of PCR efficiency from analysis of the amplification response curves (Liu and Saint, 2002; Ramakers et al., 2003; Tichopad et al., 2003). Internal amplification controls (IAC) that co-purify and co-amplify with the target nucleic acid can detect inhibitors as well as indicate template loss during processing (Pasloske et al., 1998). Another approach utilizes a whole bacterial genome to detect inhibition from clinical samples (Cloud et al., 2003), and we have developed a method, which we have called the 'SPUD' assay, that is suitable for the detection of inhibitors in nucleic acids extracted from all tissue with the exception of potato (Nolan et al., 2006). This is a simple qPCR based assay consisting of predetermined, universal template sequence (SPUD template), primers and the SPUD detecting probe. The sequences have been designed such that they are compatible with studies on all organisms with the exception of potato. In the absence of RNA (cDNA or gDNA) sample template amplification of the SPUD target results in a C_t reflecting optimal reaction conditions. When RNA (cDNA or gDNA) is added any observed deviation from this C_t indicates that the presence of sample has inhibited the qPCR reaction.

Different polymerases react differently to inhibitors and it is worthwhile checking each template preparation for inhibition by testing several polymerases for their efficiency at amplifying the template. Sequence-dependent PCR amplification bias has been observed and results in incorrect or ambiguous quantification (Shanmugam et al., 1993) and can even result in the preferential amplification of one allele over another (Weissensteiner and Lanchbury, 1996). AmpliTaq Gold and the Taq polymerases are totally inhibited in the presence of 0.004% (v/v) blood, whereas HotTub (Tfl) from Thermus flavus, Pwo, rTth, and Tfl DNA polymerases are able to amplify DNA in the presence of 20% (v/v) blood without reduced amplification sensitivity; the DNA polymerase from Thermotoga maritima (Ultma) appears to be the most susceptible to a wide range of PCR inhibitors. HotTub and Tth are the most resistant to the inhibitory effect of K^+ and Na^+ (Al Soud and Radstrom, 1998) and biological samples (Poddar et al., 1998). Thus, the PCR-inhibiting effect of various components in biological samples can, to some extent, be eliminated by the use of the appropriate thermostable DNA polymerase. One thing to bear in mind is that some enzymes are in fact enzyme mixes that combine a highly processive enzyme with a proofreading enzyme to balance fidelity and yield (Barnes, 1994). Finally, contamination of the PCR assay is an ever-present danger (Kwok and Higuchi, 1989).

Reverse transcription

The reverse transcription step is problematic and has been extensively discussed (Bustin and Nolan, 2004; Stahlberg et al., 2004a,b). There are numerous possible approaches to converting RNA into cDNA and each one has advantages and disadvantages which are discussed elsewhere (Bustin and Nolan, 2004). Random or oligo-dT priming both allow a representative pool of cDNA to be produced during a single reaction. However, it has been shown that priming using random hexamer primers does not result in equal efficiencies of reverse transcription for all targets in the sample and that there is not a linear correlation between input target amount and cDNA yield when specific targets are measured (Bustin and Nolan, 2004; Lacey et al., 2005). A recent comparison of the efficiency of RT priming by random primers of varying lengths showed that 15-nucleotide-long random oligonucleotides consistently yielded at least twice the amount of cDNA as random hexamers (Stangegaard et al., 2006). The 15-mers were more efficient at priming, resulting in reverse transcription of >80%, while random hexamers induced reverse transcription of only 40% of the template. Not surprisingly, this resulted in the detection of one order of magnitude more genes in whole transcriptome DNA microarray experiments. However, since the authors did not address the question of linearity it remains to be seen whether this change overcomes that particular limitation of the random hexamer approach.

The choice of oligo dT primers requires the use of undegraded RNA. Even when intact mRNA is used, the cDNA molecules may be truncated, since the RT enzyme cannot proceed efficiently through highly structures regions; accordingly assays should be targeted towards the 3′ end of the transcript. This is an unsuitable choice for experiments that require examination of splice variants, sequences with long 3′ UTR regions or those without polyA sequences. Furthermore, oligo-dT priming is not recommended when using RNA extracted from paraffin tissue sections, since formalin fixation results in the loss of the polyA tails on mRNA (Lewis and Maughan, 2004).

Target-specific primers are the most specific method for converting mRNA into cDNA (Lekanne Deprez et al., 2002) and are the recommended choice when RNA quantity is not a limiting factor. Interestingly, a recent report demonstrates the use of specific primers for the reliable and specific amplification of 72 genes from limiting amounts of RNA using a multiplexed tandem PCR approach (Stanley and Szewczuk, 2005). Nevertheless, as with random priming there may be differences in the efficiencies at which individual RT reactions occur. These variations must be controlled for by reference of unknown samples to a calibrator sample (when using $\Delta\Delta C_t$ analysis) or to standard curves. Specific priming of RNA dilutions results in a linear response of target cDNA yield (Bustin and Nolan, 2004) and so a further advantage of using this priming method is that the efficiency of the combined RT-qPCR reaction can be con-

firmed by analysis of the gradient of the standard curve. The inclusion of a calibrator sample or standard curve on every plate is important since these serve as controls for measuring inter-assay variability that may occur when multiple samples are run on different plates.

Fortunately, whereas the RT step is highly variable, the qPCR segment of the assay is remarkably reproducible when run under optimal conditions, although greater variability is observed between replicates when the assay is run under suboptimal conditions (Hilscher et al., 2005).

Choice of enzymes

The RT-qPCR assay can be performed either as a single RT and PCR enzyme procedure in a single tube or a separate RT and PCR enzyme procedure in a single or a two tube procedure.

Single RT and PCR enzyme

A single enzyme such as Tth polymerase is able to function both as an RNA- and DNA-dependent DNA polymerase and can be used in a single tube without secondary additions to the reaction mix. Its main advantages are the reduced hands-on time and potential for contamination. There are several disadvantages:

1 Since all reagents are added to the reaction tube at the beginning of the reaction, separate optimization of the two reactions is not feasible.
2 The assay can only be carried out using target-specific primers.
3 The assay may be less sensitive, due possibly to the less efficient RT-activity of Tth polymerase (Cusi et al., 1994; Easton et al., 1994).
4 The most thorough study comparing the two procedures found that the one step reaction was characterized by extensive accumulation of primer dimers, which could obscure the true results in quantitative assays (Vandesompele et al., 2002a).

Two enzyme procedures: separate RT and PCR enzymes

Reactions can be either 'uncoupled' or 'coupled'. In the 'uncoupled' alternative the reverse transcriptase synthesizes cDNA in a first tube, under optimal conditions, using random, oligo-dT or sequence-specific primers. An aliquot of the RT reaction is then transferred to another tube (containing the thermostable DNA polymerase, DNA polymerase buffer, and PCR primers) for PCR carried out under optimal conditions for the DNA polymerase. In the 'continuous or coupled' alternative the reverse transcriptase synthesizes cDNA in the presence of high concentrations of dNTPs and either target-specific or oligo-dT primers. Following the RT reaction, PCR buffer (without Mg^{2+}), a thermostable DNA polymerase, and target-specific primers are added and the PCR is performed in the same tube. Interassay variation of two enzyme protocols can be very small when carried out properly, with correlation coefficients ranging between 0.974 and 0.988 (Vandesompele *et al.*, 2002a). The disadvantages of this approach are:

1   In two enzyme/one tube assays a template switching activity of viral RTs can generate artefacts during transcription (Mader *et al.*, 2001).
2   Two enzyme/two tube reactions present additional opportunities for contamination.
3   The RT can inhibit the PCR assay, resulting in an overestimation of amplification efficiency and target quantification (Suslov and Steindler, 2005).

On balance, the increased flexibility and potential for optimization makes the use of a two enzyme protocol preferable to the single enzyme procedure.

PCR

There are two important considerations: first, an absolute requirement for a thermostable polymerase with a 5'–3' nuclease activity when carrying out 5'-nuclease (TaqMan) assays and second, if targeting low copy number targets using SYBR Green I dye chemistry, the use of a hot start *Taq* polymerase, either complexed with an antibody or chemically modified to inhibit polymerase activity at low temperatures. We have used enzymes and kits from all the major suppliers and have found no major, consistent differences in quality but note that antibody inactivated enzymes require a shorter period of activation. Again it is important to use the buffer supplied

with the enzyme, since the manufacturer has optimized it for that enzyme. Some manufacturers include PCR enhancer solutions in their kits. It is important to read the instructions carefully, since they can affect the reaction conditions. For example, if designed for amplification of G/C rich regions, they may lower the DNA melting temperature, thus lowering the primer annealing temperature. Combined reagent mastermixes are increasingly popular. These have the advantage of reducing potential errors when combining components prior to reactions. In many cases they are stabilized with components that will also disrupt secondary structures in templates and so can give improved results. In the case of difficult templates, a sequence-specific effect with specific reagent blends will be observed and so trying a different blend may alleviate problems. They are provided in convenient format and suitable for routine or high throughput assays. The disadvantage is that there is less flexibility for optimization of challenging simplex or multiplex assays. Take care to check the mixture components with particular attention to matching $MgCl_2$ concentration to the detection chemistry used and the presence and concentration of reference dyes such as ROX.

Protocols

PCR optimization

The thermodynamic stability (ΔG) of a duplexed primer/target structure differs for different primers and varies with primer concentration. Therefore, it is important to use primers at concentrations that result in optimal hybridization and priming. Although optimization of PCR reactions used to be an essential part of assay development, recent trends towards high throughput and rapid data reporting have resulted in the elimination of this step in many laboratories. Indeed, some manufacturers claim it is not necessary to optimize primer concentrations when using their particular mastermixes. However, it is clear that the rationale underlying the original recommendations remain valid and that optimization of primer concentration can significantly improve assay detection sensitivity (Fig. 9.2).

Our protocol utilizes SYBR Green I dye to detect amplified products and to visualize

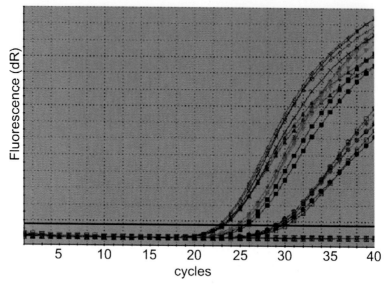

Figure 9.2 The difference in primer binding efficiency effects PCR kinetics. The amplification plots were generated by amplifying the same target using different primer concentrations as described in Table 9.1. The most efficient reaction results from balancing primer binding by optimizing primer concentration. In this example, optimization results in as much as an 8 C_t increase in sensitivity. The same primers in different combinations of forward and reverse between 100 nM and 900 nM were compared in duplicate.

melting of the amplicons after the reaction. Post-amplification melting is performed by gradually increasing the temperature of the reaction and assessing the fluorescence due to amplicon-bound dye at each temperature. As the temperature is increased the fluorescence decreases due to separation of the double strands of the DNA. The temperature at which this occurs depends on amplicon size and sequence. In this way the melting characteristic of primer dimer and specific products can be distinguished since primer dimers are shorter and melt at lower temperature than longer, target amplicon products (Fig. 9.3).

1 Selection of target sample to use during optimization. When using an artificial amplicon or linearized plasmid as the template to determine the PCR efficiency prepare a stock of target sample of around 10^4–10^5 copies amplicon/µl. Test the amplification of the template using 200 nM of each primer to establish that amplification of the sample results in C_t values of between 20 and 30. If the concentration is outside of this range assume that a 10 fold difference in concentration corresponds to 3.323 cycles and adjust the sample concentrations accordingly. Alternatively, if cDNA synthesized from

universal RNA is to be used, prepare six to eight fivefold serial dilutions. Test the amplification of the dilution series of template using 200 nM primers and select the diluted sample that results in Ct values between 20 and 30.

2 Prepare individual working dilutions of forward and reverse primers for optimization matrix. Prepare 20 µl of stocks of 0.5 µM, 1 µM, 3 µM, 6 µM and 9 µM.

3 Prepare a matrix of primer concentrations by adding 2.5 µl of the stock concentration stated in the appropriate column of primer to two qPCR reaction tubes, such that each reaction is prepared in duplicate. HINT: refer to Table 9.1 for the final concentrations of primers. Concentrations are stated as forward primer/reverse primer in nmol in final qPCR reaction.

4 Prepare a mastermix according to Table 9.2A.

5 Add 20 µl mastermix to each primer pair in qPCR tubes. Mix avoiding bubbles and briefly spin reaction tubes to ensure components are in the bottom of tubes.

6 Run in qPCR instrument using three-step protocol as shown in Table 9.2B.

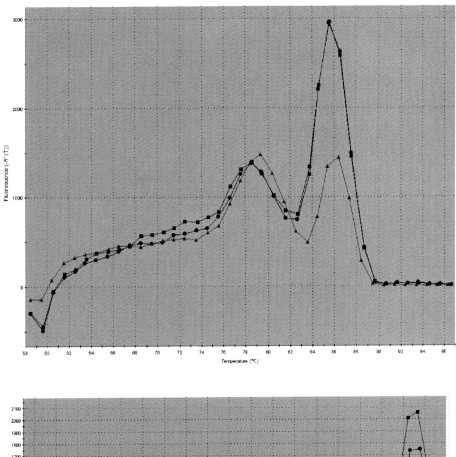

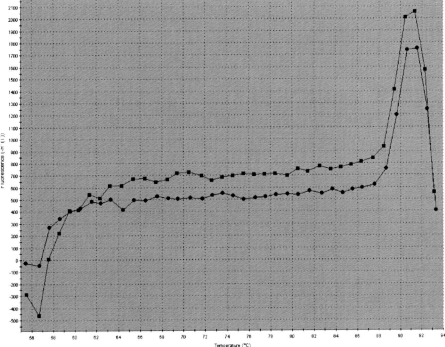

Figure 9.3 (A) SYBR Green I dye melt curve demonstrating the appearance of primer dimers in a poorly designed assay. The target amplicon has a T_m of around 86°C, whereas the primer dimers melt at 79°C. (B) SYBR Green melt curve demonstrating the absence of primer dimers in a well-designed assay. The amplicon T_m is 91°C.

Table 9.1 Forward and reverse primer concentrations for optimization assay

Forward primer → Reverse primer ↓↓	0.5 µM	1 µM	2 µM	3 µM	6 µM
0.5 µM	50/50	100/50	200/50	300/50	600/50
1 µM	50/100	100/100	200/100	300/100	600/100
2 µM	50/200	100/200	200/200	300/200	600/200
3 µM	50/300	100/300	200/300	300/300	600/300
6 µM	50/600	100/600	200/600	300/600	600/600

Table 9.2A qPCR mastermix for primer optimization assay

	Stock concentration	Volume to add for single 25 µl reaction
Water		2.125 µl
SYBR Green I Dye Mastermix (qPCR)	2×	12.5 µl
Reference dye (ROX)*	0.03 µM (optional depending on instrument)	0.375 µl
Sample	10^4 to 10^5 copies or 5 to 10µl diluted cDNA synthesis reaction or alternative template from step 1	5.0 µl

*Note that the given protocol includes the use of the ROX reference dye; if this is not required, adjust water volume accordingly.

Table 9.2B qPCR three-step protocol

1 cycle	Activation	95°C 10 min
40 cycles	Denaturation	95°C 30 s
	Annealing/	60°C 30 s (collect data during each cycle)
	Extension	72°C 30 s
1 cycle	Melt curve	Between 55°C and 95°C (collect data at each temperature)

Table 9.3A qPCR mastermix for determination of assay efficiency and sensitivity

	Stock concentration	Volume (µl) to add for single 25 µl reaction
Water		2.125 µl
SYBR Green I Dye Mastermix (qPCR)	2×	12.5 µl
Reference dye (ROX)*	0.03 µM (optional depending on instrument)	0.375 µl
Primers	As determined during primer optimization process	5 µl

*Note that the given protocol includes the use of the ROX reference dye.

Table 9.3B qPCR three-step protocol

1 cycle	Activation	95°C 10 min
40 cycles	Denaturation	95°C 30 s
	Annealing	60°C 30 s (collect data during each cycle)
	Extension	72°C 30 s
1 cycle	Melt curve	Between 55°C and 95°C (collect data at each temperature)

7   Refer to the data analysis section for general guidelines. Select the primer combination that fulfils the following criteria in the given order of priority:

8   Absence of primer dimers: use the melt curve profile.

9   Lowest C_t: this determines the primer combination resulting in the most sensitive (and efficient) reaction.

10   Highest end point fluorescence (the highest ΔRn). This is likely to signify the highest number of amplicon products being formed.

11   Absence of signal in the NTC.

Determination of assay sensitivity and reproducibility

1   Prepare a 10-fold serial dilution of the template described in step 1 of PCR optimization using a minimum of five samples between 10^7 and 10^3 copies in 5 μl. It is possible to quantify a dynamic range of 10 to 12 logs on most systems (i.e. from 1 to 10^{10} to 10^{12} target copies). Ensure that the range spans the C_t values anticipated for experimental analysis, repeating the dilution series using different concentrations if required.

2   Add 5 μl of each dilution point to qPCR reaction tubes, including each sample in duplicate.

3   Prepare a qPCR mastermix according to Table 9.3A using the primer concentrations determined from the primer concentration optimization process.

4   Add 20 μl mastermix to each sample in qPCR tubes. Mix, avoiding bubbles.

5   Run in qPCR instrument using three-step protocol as shown in Table 9.3B.

6   Follow the steps outlined in the data analysis section.

7   Use the instrument software to plot C_t against log target concentration and determine the slope and R^2 values. Confirm that the slope lies between −3.2 and −3.5 and that the R^2 value is above 0.98, as described for standard curve analysis.

Perform the qPCR as soon as possible after preparing the reaction mixes. Some assays are stable for 12 h if stored at 4°C overnight; however, others are more sensitive to storage. Reactions containing glycerol and DMSO or made from commercial mastermixes can be stored at −20°C.

RT-qPCR assay

Reverse transcription
See flow chart for a summary of the steps (Fig. 9.4).

Variation between RT experiments
When using random primers, the efficiency of the RT reaction may vary between reactions. It also appears that the RT efficiency for different targets within the same reaction is variable (Lacey *et al.*, 2005) and it is important to control for these factors by following the steps below:

•   When possible perform RT reactions on all samples at the same time.

•   Use the same amount of RNA in all reactions.

•   When multiple samples are processed in different batches include a common positive reference sample alongside all batches.

•   Quantify target genes relative to a standard curve including the target copy number in the positive control samples from each batch.

•   Express quantity of target in samples relative to the quantity measured in the control sam-

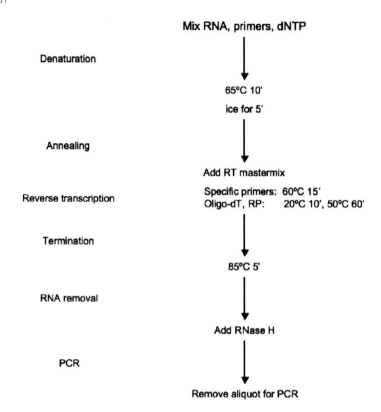

Figure 9.4 Outline of reverse transcription protocol.

ple included in the same RT batch (Lacey *et al.*, 2005).

1   Keep reagents and microfuge tubes on ice. Mix and briefly spin each component and combine the following in a microfuge tube: by combining the reagents in the order listed in Table 9.4A. Note that this protocol is adapted for use with specific primers, random primers or oligo-dT priming. In some cases a combination of oligo dT and random primers may be appropriate. CRITICAL: prepare no-RT controls in duplicate: to prepare each no-RT control, prior to adding RT enzyme to the mastermix, add 18μl mastermix to 7 μl of water. Incubate at 65°C for 10 min, then snap cool on ice for 5 min

2   Add 15 ml of the RT mastermix to the RNA/primer mix. Gently mix the tube contents and briefly spin.

3a *Specific priming*: incubate at 60°C for 15 minutes.

3b *Oligo dT priming or random priming*: incubate at 20°C for 10 min followed by 50°C for 60 min

4   Terminate the reaction at 85°C for 5 minutes, and then place on ice. Collect by brief centrifugation.

5   Optional step: add 1 μl (2 U) of *E. coli* RNase H and incubate at 37°C for 15 minutes.

6   Proceed to qPCR.

Polymerase chain reaction

Option A: qPCR using SYBR Green I dye detection

6a Make a mastermix by adding the reagents in the order shown in Table 9.5A, mix gently by repeatedly pipetting up and down (making sure there are no bubbles), and finally add *Taq* polymerase and mix gently again [buffer, MgCl$_2$, dNTPs, SYBR Green I dye and *Taq* are combined in commercial 2× mastermix preparations].

7a Add 5 μl template to the reaction tubes or wells.

8a Add 20 μl volume of reaction mix to each tube or well containing template. Cap

Table 9.4A RNA/primer mix components

Component	Volume per reaction
Total RNA 1–200 ng (max. 500 ng)	x (maximum 7 µl)
3′ (antisense) primer (2 µM) *or*	1 µl
Random primer (50 ng/µl) *or*	1 µl
Oligo-dT (500 ng/µl)	1 µl
Water	to 10 µl

Table 9.4B RT mastermix components

Component	Volume per reaction
10x RT buffer	2.5 µl
$MgCl_2$ (25 µM)	5 µl
DTT (0.1 M)	2.5 µl
RT Enzyme (200 U/l)	1 µl
Water	to 15 µl

Table 9.5A PCR mastermix components (SYBR Green I dye)

Mastermix	Final concentration	Volume per reaction (µl)
dH_2O		14.125*
10× buffer	1×	2.5
50 mM $MgCl_2$	2.5 mM	1.25
20 mM dNTP	0.8 mM	1
10 µM Primers	Refer to optimization results	0.25
ROX diluted (optional)	Refer to instrument manufacturer's recommendations	0.375*
SYBR Green I diluted 1:5000		0.25
5 U/µl Taq		0.25

*Note that this protocol includes the use of the reference dye ROX; if this is not required, adjust water volume accordingly.

Table 9.5B qPCR three-step protocol

1 cycle	Activation:	95°C	10 min	
40 cycles	Denaturation	95°C	30 s	
	Annealing:	60°C**	30 s	Collect data
	Extension:	72°C	30 s	Collect data

Followed by a melting profile (as below or according to the instrument manufacturer's instructions):

1 cycle	95°C	1 min	
41 cycles	55°C	30 s	Collect data
	Repeat and increase temperature by 1°C per cycle		Collect data

*Annealing temperature is primer design dependent. A two-step protocol with combined annealing and extension steps into a single 62°C incubation for 60sec can be used when primers are being evaluated to be used with a TaqMan probe.

Table 9.6A PCR mastermix components (TaqMan)

Mastermix	Final concentration	Volume per reaction (µl)
dH$_2$O		12.125*
10× buffer	1x	2.5
50 mM MgCl$_2$	5mM	2.5
20 mM dNTP (final)	0.8mM	1
10 µM Primers	Refer to optimization results	0.25
ROX diluted	Refer to instrument manufacturer's recommendations	0.375*
5 µM Probe	200 nM	1
5 U/µl Taq		0.25

*Note that the given protocol includes the use of the reference dye ROX. For alternative volumes or to exclude, adjust water volume accordingly.

carefully, spin briefly to ensure there are no bubbles, and place the tubes, strips, rotor or 96/384-well plate into the real time thermal cycler.

9a Perform a three-step PCR reaction according to the thermal profile in Table 9.5B.

Option B: qPCR using dual-labelled probe detection

6b Make a mastermix by adding the reagents in the order shown in Table 9.6A, mix gently by repeatedly pipetting up and down (making sure there are no bubbles), and finally add *Taq* polymerase and mix gently again.

7b Add 5 µl template to the reaction tubes or wells.

8b Add 20 µl of reaction mix to each tube or well containing template. Cap carefully, spin briefly to ensure there are no bubbles, and place the tubes, strips, rotor or 96/384-well plate into the real time thermal cycler.

9b Perform a two-step PCR reaction according to the thermal profile outlined in Table 9.6B.

Note: reactions containing hydrolysis probes are usually performed using a two-step PCR profile. During this reaction the double-stranded template is melted at 95°C and the annealing and extension steps both occur during a single incubation at 62°C. Although this is suboptimal for amplification by *Taq* polymerase it is believed to encourage more efficient cleavage of the internal probe, resulting in maximum fluorescent signal per cycle. Experiments including other probe based chemistries, e.g. Molecular Beacons, Scorpions, can be run under these conditions but a three-step protocol is optimal (Table 9.6B).

Data analysis

The analysis of qPCR data can be highly subjective, since C$_t$ values can be altered by changes to baseline and threshold levels. Although software default options can assuage some subjectivity, it is also true that these settings are not always appropriate. Clearly, it is important to understand the analysis procedure in order to be able to generate reliable data. The following steps constitute a basic data analysis:

Table 9.6B qPCR two-step protocol

1 cycle	Activation	95°C	10 min	
40 cycles	Denaturation	95°C	30 s	
	Annealing/ extension	60°C	1 min	Collect data

1 Examine amplification plots of standard or reference samples and identify abnormal plots. A normal amplification plot is defined as consisting of a horizontal baseline region, a log phase of amplification, then a linear amplification phase, followed by a plateau. Note samples that record a C_t, but where the amplification plot clearly does not show the appropriate characteristics (Fig. 9.5), these must be subjected to further investigation.

2 The threshold settings are subjective. Select a setting that is within the logarithmic region of amplification. This is best viewed as the linear region of the plot of log normalized fluorescence vs. C_t (log dRn vs. C_t). The precise setting should be within a range where all the amplification plots to be analysed are parallel, hence the setting represents cycles during which the amplification efficiencies are equivalent for each reaction to be compared. Select the threshold setting where the replicates are closest.

3 Check slope is between −3.2 and −3.5 and reproducibility of replicates $R^2 > 0.98$.

4 Check controls for expected results.

5 Check melt curve profiles for SYBR Green I dye detection and ensure that products are specific.

6 Where possible, ensure that the C_t of the positive control (or dilutions of standard curve) is within 1 C_t of that recorded during a previous experiment and that the initial and final fluorescence levels are similar.

7 Ensure that all data points to be recorded are within the dynamic range defined by the standard curve

8 Examine replicates. All replicates should well be within 0.5 C_t of each other. At low C_t the tolerance should be lower than at high C_t. Above cycle 35 the variability will be greater and quantification may be unreliable.

9 Ensure a minimum difference of 5 C_t between any NTC signal and a sample data point (Bustin and Nolan, 2004) (Fig. 9.6). When using SYBR Green I dye take care with C_t values when primer dimer products are evident in the sample after melt curve analysis.

In general the higher the fluorescence released per cycle the more sensitive and accurate the measurement.

Normalization

All RT-qPCR assay results are subject to variability caused by technical as well as biological variation. It is essential that technical variability is kept at a minimum so as to optimize the chances of identifying biologically relevant changes in mRNA levels. Consequently, data normalization is an essential part of a meaningful RT-qPCR assay and requires standardization. Unfortunately, normalization is a rather problematic area and, as yet, there is no universally accepted method for data normalization that accounts for all variables encountered during the course of a RT-qPCR experiment. All proposed methods represent a compromise and the selection depends upon the study aims and acceptable tolerance (Huggett *et al.*, 2005).

Normalization against input RNA amount is feasible and, if validated, perfectly acceptable (Bustin, 2002). However, there are problems when comparing tissues that are proliferating at different rates, since they will contain different amounts of rRNA and the mRNA/rRNA ratios are likely to be different. Furthermore, differences in cDNA synthesis are not taken into account and in some situations (e.g. embryology) may generate irrelevant results.

Normalizing to an endogenous reference (not 'housekeeping') gene or genes is the most common method for internally controlling for error in RT-qPCR and is currently the preferred option, as long as it is carried out correctly (Goossens *et al.*, 2005). The inevitable consequence of inappropriate normalization is the acquisition of biologically irrelevant data (Dheda *et al.*, 2005; Tricarico *et al.*, 2002). When using the endogenous reference strategy target gene levels are expressed relative to those of internal reference genes and so all of the steps required to obtain the final PCR measurement are controlled for. The procedure is simplified as both the gene of interest and the reference gene are measured using real-time RT-PCR. However, it is essential that reference gene expression in the target tissue is carefully analysed (Perez-Novo *et al.*, 2005)

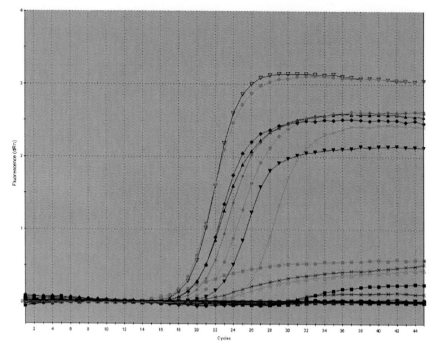

Figure 9.5 Example of problematic amplification reaction resulting in inappropriate C_t values. Clearly, there is a problem with the amplification of four of the unknown samples. This is often due to the presence of inhibitors in the assay, and a 10- or 100-fold dilution of the RNA results in efficient amplification and accurate C_t determination.

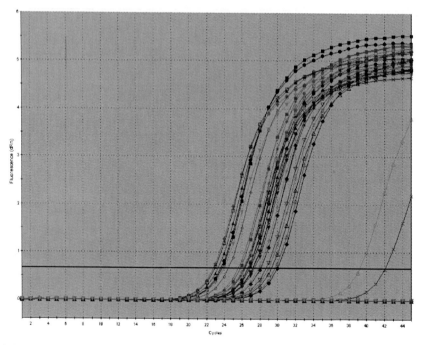

Figure 9.6 Analysis of a SYBR Green I reaction with positive NTC. In this example, the replicate NTCs are recording C_ts of 39.5 and 42. Since the lowest C_t generated by the unknowns is at 30, i.e. the ΔC_t is >5, it is acceptable to use these data for quantification. The positive C_ts are caused by primer dimers, and if the aim of the experiment was to quantitate very low abundance targets, it may be necessary to redesign the primers.

and the minimum variability is determined and reported (Vandesompele *et al.*, 2002b). The current gold standard combines the evaluation of a panel of several reference genes together with a method for selecting reference genes with the most stable expression, e.g. GeNorm medgen. ugent.be/~jvdesomp/genorm/or Bestkeeper (Pfaffl *et al.*, 2004).

Normalization against gDNA is another option (Kanno *et al.*, 2006), particularly when performing RT-qPCR analyses on RNA obtained using kits such as the Invitrogen CellsDirect kit. The main problems here are (1) differential stability of DNA and RNA may distort quantification, (2) the sample cannot be DNase-treated, (3) whilst being an internal control, there is no equivalent RT step and (4) high relative concentrations of gDNA may inhibit the RT-PCR. Finally, it is also possible to normalize against area dissected when using laser capture microdissection, or cell number when extracting RNA from nucleated blood cells.

Standard curve

Analysis of the data from the standard curve can provide a substantial amount of information about the assay. For this reason, assays should be initially validated on a serial dilution of high-quality template even in situations where data analysis will not require reference to a standard curve. The template material used to generate a standard curve should accurately reflect the sample complexity and this means using a total RNA preparation, either from the tissue sample under investigation or from a commercially supplied reference RNA, or spiking a known amount of cDNA, plasmid DNA or oligonucleotides into a tRNA solution. Each concentration should be amplified in triplicate to allow a determination of reproducibility. The standard curve is constructed from a measure of C_t against log template quantity. It is realistic to expect a linear dynamic range of at least six logs with highly reproducible quantification, with more than nine logs feasible (Fig. 9.7). This defines the working dynamic range for the assay. One measure of assay efficiency is made by comparison of the relative C_t values for subsequent dilutions of sample. The efficiency of the reaction can be calculated by the equation: $E = 10^{(-1/\text{slope})} - 1$. The efficiency

of the PCR should as close to 100% as possible, corresponding to a doubling of the target amplicon at each cycle. Using this measure an assay of 100% efficiency will result in a standard curve with a gradient of -3.323 (also see http://www. gene-quantification.de/efficiency.html). An optimized assay will result in a standard curve with a slope between -3.2 and -3.5. Reproducibility of the replicate reactions also reflects assay stability, with R^2 values of 0.98 or above being indicative of a stable and reliable assay. The intercept on the C_t axis indicates the C_t at which a single unit of template concentration would be detected and is therefore an indication of the sensitivity of the assay.

Fig. 9.8 illustrates a typical result obtained for amplicons detected using SYBR Green I. Note that the slopes of all amplification plots are identical, indicating that the amplification efficiencies of every sample are the same. One of the plateaus shows a significantly reduced ΔR_n value, but this clearly does not interfere with the ability to quantitate accurately from that sample. Also note that the NTCs are showing positive C_ts of 44.2 and 45, respectively. This phenomenon is fairly common with SYBR Green I assays and is due to the formation of primer dimers in the absence of template. In this example it would be acceptable to increase the threshold line, which would result in negative NTCs (Bustin and Nolan, 2004). However, we would suggest that the presence of the primer dimers is reported, together with the ΔC_t between the highest C_t recorded by an unknown and the NTC.

Fig. 9.9 illustrates a typical result for amplicons detected using TaqMan chemistry. The replicates show a good standard of pipetting and all of the amplification plots have the same slopes, indicating identical amplification efficiencies. Note that in this example the NTCs are negative, since the probe will not detect primer dimers.

Conclusions

Some technologies arrive at a time that is 'just right', and advances in chemistries and instrumentation make RT- qPCR a technology whose time has come (Walker, 2002). However, it is also clear that appropriate application, quality control and standardization are issues that must

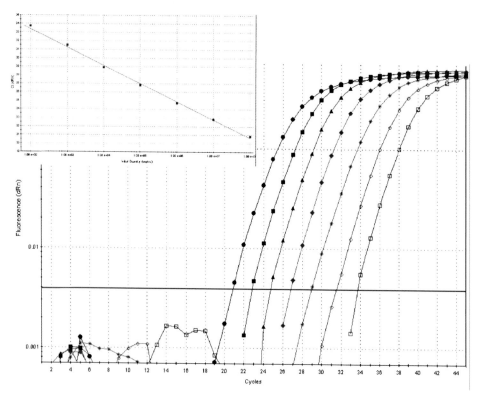

Figure 9.7 The standard curve generated from a plot of C_t against log concentration for a serial dilution of template should be linear, with a slope of −3.323 and an R^2 of greater than 0.98 for samples replicated at least three times. In the example shown, the slope of the SYBR Green assay is −3.357 with an R^2 of 0.999, with a dynamic range of 11 logs, detecting a target present at three copies in eight replicates.

be addressed and it is vital to consider each stage of the experimental protocol, starting with the laboratory set-up, proceeding through sample acquisition and template preparation, and the RT and the PCR steps. Only if every one of these stages is properly validated, is it possible to obtain reliable quantitative data. Of course, choice of chemistries, primers and probes and instruments must be appropriate to whatever is being quantitated. Finally, data must be interpreted correctly, and this remains a real problem.

The problems may be addressed by a next generation of assays that aim to overcome the limitations that are inherent to any RT-PCR reaction. Developments in microfluidics are already allowing the amplification of target nucleic acid from nanolitres of sample (Liu *et al.*, 2002). However, the ideal assay would eliminate the need to amplify a target and it is likely that real-time RT-PCR will be eventually replaced by methods capable of direct analysis of single biological molecules. One such single-molecule fluorescence detection technology is fluorescence correlation spectroscopy, which detects expression by hybridizing two dye-labelled DNA probes to a selected target molecule, which can be either DNA or RNA in solution. The subsequent dual colour cross-correlation analysis allows the quantification of the bio-molecule of interest in absolute numbers down to target concentrations of less than 10^{-12} M (Winter *et al.*, 2004). Combined with single molecule sequencing (Braslavsky *et al.*, 2003) this will eventually permit RNA identification as well as distortion-free quantification. That is the future. For now, qRT-PCR assays, when carried out appropriately, are the method of choice for RNA detection and quantification.

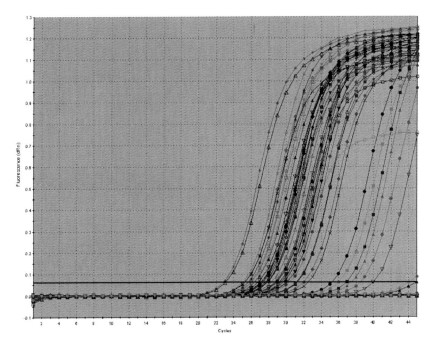

Figure 9.8 Typical amplification plot obtained using a SYBR Green I assay. Note the presence of primer dimers in the NTC.

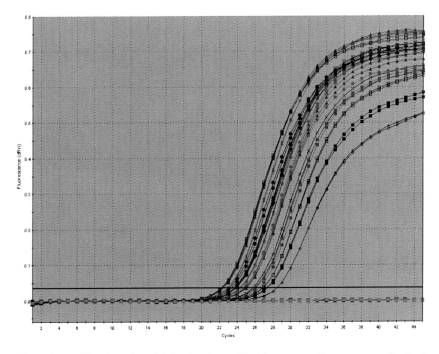

Figure 9.9 Typical amplification plot obtained using a TaqMan assay. There are no C_ts in the NTC, nor should there ever be when using probes.

References

Afzal, M.A., Osterhaus, A.D., Cosby, S.L., Jin, L., Beeler, J., Takeuchi, K. and Kawashima, H. 2003. Comparative evaluation of measles virus-specific RT-PCR methods through an international collaborative study. J. Med. Virol. 70, 171–176.

Akane, A., Matsubara, K., Nakamura, H., Takahashi, S. and Kimura, K. (1994). Identification of the heme compound copurified with deoxyribonucleic acid (DNA) from bloodstains, a major inhibitor of polymerase chain reaction (PCR) amplification. J. Forensic Sci. 39, 362–372.

Al Soud, W.A. and Radstrom, P. (1998). Capacity of nine thermostable DNA polymerases To mediate DNA amplification in the presence of PCR-inhibiting samples. Appl. Environ. Microbiol. 64, 3748–3753.

Al Soud, W.A. and Radstrom, P. 2001. Purification and characterization of PCR-inhibitory components in blood cells. J. Clin. Microbiol. 39, 485–493.

Barnes, W.M. (1994). PCR amplification of up to 35-kb DNA with high fidelity and high yield from lambda bacteriophage templates. Proc. Natl. Acad. Sci. USA 91, 2216–2220.

Bibikova, M., Talantov, D., Chudin, E., Yeakley, J.M., Chen, J., Doucet, D., Wickham, E., Atkins, D., Barker, D., Chee, M., et al. 2004a. Quantitative gene expression profiling in formalin-fixed, paraffin-embedded tissues using universal bead arrays. Am. J. Pathol. 165, 1799–1807.

Bibikova, M., Yeakley, J.M., Chudin, E., Chen, J., Wickham, E., Wang-Rodriguez, J. and Fan, J.B. 2004b. Gene expression profiles in formalin-fixed, paraffin-embedded tissues obtained with a novel assay for microarray analysis. Clin. Chem. 50, 2384–2386.

Braslavsky, I., Hebert, B., Kartalov, E. and Quake, S.R. 2003. Sequence information can be obtained from single DNA molecules. Proc. Natl. Acad. Sci. USA 100, 3960–3964.

Bustin, S.A. 2000. Absolute quantification of mRNA using real-time reverse transcription polymerase chain reaction assays. J. Mol. Endocrinol. 25, 169–193.

Bustin, S.A. 2002. Quantification of mRNA using real-time reverse transcription PCR (RT-PCR): trends and problems. J. Mol. Endocrinol. 29, 23–39.

Bustin, S.A. 2004. A–Z of quantitative PCR (La Jolla, CA: IUL Press).

Bustin, S.A. 2005. Real-time, fluorescence-based quantitative PCR: a snapshot of current procedures and preferences. Expert Rev. Mol. Diagn. 5, 493–498.

Bustin, S.A. and Mueller, R. 2005. Real-time reverse transcription PCR (qRT-PCR) and its potential use in clinical diagnosis. Clin. Sci. (Lond.) 109, 365–379.

Bustin, S.A. and Nolan, T. 2004. Pitfalls of quantitative real-time reverse-transcription polymerase chain reaction. J. Biomol. Tech. 15, 155–166.

Christensen, U.B. and Pedersen, E.B. 2002. Intercalating nucleic acids containing insertions of 1-O-(1-pyrenylmethyl)glycerol: stabilisation of dsDNA and discrimination of DNA over RNA. Nucleic Acids Res. 30, 4918–4925.

Christensen, U.B., Wamberg, M., El-Essawy, F. A., Ismail Ael, H., Nielsen, C.B., Filichev, V.V., Jessen, C.H., Petersen, M. and Pedersen, E.B. 2004. Intercalating nucleic acids: the influence of linker length and intercalator type on their duplex stabilities. Nucleosides Nucleotides Nucleic Acids 23, 207–225.

Cloud, J.L., Hymas, W.C., Turlak, A., Croft, A., Reischl, U., Daly, J.A. and Carroll, K.C. 2003. Description of a multiplex Bordetella pertussis and Bordetella parapertussis LightCycler PCR assay with inhibition control. Diagn. Microbiol. Infect. Dis. 46, 189–195.

Cone, R.W., Hobson, A.C. and Huang, M.L. (1992). Coamplified positive control detects inhibition of polymerase chain reactions. J. Clin. Microbiol. 30, 3185–3189.

Coombs, N.J., Gough, A.C. and Primrose, J.N. (1999). Optimisation of DNA and RNA extraction from archival formalin-fixed tissue. Nucleic Acids Res. 27, e12.

Cusi, M.G., Valassina, M. and Valensin, P.E. (1994). Comparison of M-MLV reverse transcriptase and Tth polymerase activity in RT-PCR of samples with low virus burden. Biotechniques 17, 1034–1036.

de Vries, T.J., Fourkour, A., Punt, C.J., Ruiter, D.J. and van Muijen, G.N. 2000. Analysis of melanoma cells in peripheral blood by reverse transcription-polymerase chain reaction for tyrosinase and MART-1 after mononuclear cell collection with cell preparation tubes: a comparison with the whole blood guanidinium isothiocyanate RNA isolation method. Melanoma Res. 10, 119–126.

Dheda, K., Huggett, J.F., Chang, J. S., Kim, L.U., Bustin, S.A., Johnson, M.A., Rook, G.A. and Zumla, A. 2005. The implications of using an inappropriate reference gene for real-time reverse transcription PCR data normalization. Anal. Biochem. 344, 141–143.

Easton, L. A., Vilcek, S. and Nettleton, P.F. 1994. Evaluation of a 'one tube' reverse transcription-polymerase chain reaction for the detection of ruminant pestiviruses. J. Virol. Methods 50, 343–348.

Filichev, V. V., Christensen, U.B., Pedersen, E.B., Babu, B.R. and Wengel, J. 2004. Locked nucleic acids and intercalating nucleic acids in the design of easily denaturing nucleic acids: thermal stability studies. Chembiochemistry 5, 1673–1679.

Fleige, S. and Pfaffl, M.W. 2006. RNA integrity and the effect on the real-time qRT-PCR performance. Mol. Aspects Med. 27, 126–139.

Fredricks, D. N. and Relman, D. A. (1998). Improved amplification of microbial DNA from blood cultures by removal of the PCR inhibitor sodium polyanetholesulfonate. J. Clin. Microbiol. 36, 2810–2816.

Gabert, J., Beillard, E., van der Velden, V.H., Bi, W., Grimwade, D., Pallisgaard, N., Barbany, G., Cazzaniga, G., Cayuela, J.M., Cave, H., et al. 2003. Standardization and quality control studies of 'real-time' quantitative reverse transcriptase polymerase chain reaction of fusion gene transcripts for residual disease detection in leukemia – a Europe Against Cancer program. Leukemia 17, 2318–2357.

Gerard, C.J., Olsson, K., Ramanathan, R., Reading, C. and Hanania, E.G. 1998. Improved quantitation of minimal residual disease in multiple myeloma using real-time polymerase chain reaction and plasmid-DNA complementarity determining region III standards. Cancer Res. 58, 3957–3964.

Gibson, U.E., Heid, C.A. and Williams, P.M. 1996. A novel method for real time quantitative RT-PCR. Genome Res. 6, 995–1001.

Goossens, K., Van Poucke, M., Van Soom, A., Vandesompele, J., Van Zeveren, A. and Peelman, L.J. 2005. Selection of reference genes for quantitative real-time PCR in bovine preimplantation embryos. BMC Dev. Biol. 5, 27.

Guy, R. A., Payment, P., Krull, U.J. and Horgen, P.A. 2003. Real-time PCR for quantification of Giardia and Cryptosporidium in environmental water samples and sewage. Appl. Environ. Microbiol. 69, 5178–5185.

Halford, W.P., Falco, V.C., Gebhardt, B.M. and Carr, D.J. 1999. The inherent quantitative capacity of the reverse transcription- polymerase chain reaction. Anal. Biochem. 266, 181–191.

Hilscher, C., Vahrson, W. and Dittmer, D.P. 2005. Faster quantitative real-time PCR protocols may lose sensitivity and show increased variability. Nucleic Acids Res. 33, e182.

Huggett, J., Dheda, K., Bustin, S. and Zumla, A. 2005. Real-time RT-PCR normalisation; strategies and considerations. Genes Immun. 6, 279–284.

Jiang, J., Alderisio, K.A., Singh, A. and Xiao, L. 2005. Development of procedures for direct extraction of Cryptosporidium DNA from water concentrates and for relief of PCR inhibitors. Appl. Environ. Microbiol. 71, 1135–1141.

Kanno, J., Aisaki, K., Igarashi, K., Nakatsu, N., Ono, A., Kodama, Y. and Nagao, T. 2006. 'Per cell' normalization method for mRNA measurement by quantitative PCR and microarrays. BMC Genomics 7, 64.

Ke, D., Menard, C., Picard, F.J., Boissinot, M., Ouellette, M., Roy, P.H. and Bergeron, M.G. 2000. Development of Conventional and real-time PCR assays for the rapid detection of group B streptococci. Clin. Chem. 46, 324–331.

Keilholz, U., Willhauck, M., Rimoldi, D., Brasseur, F., Dummer, W., Rass, K., de Vries, T., Blaheta, J., Voit, C., Lethe, B. and Burchill, S. 1998. Reliability of reverse transcription-polymerase chain reaction (RT-PCR)-based assays for the detection of circulating tumour cells: a quality- assurance initiative of the EORTC Melanoma Cooperative Group. Eur. J. Cancer 34, 750–753.

Kwok, S. and Higuchi, R. 1989. Avoiding false positives with PCR. Nature 339, 237–238.

Lacey, H.A., Nolan, T., Greenwood, S.L., Glazier, J.D. and Sibley, C.P. 2005. Gestational profile of Na^+/H^+ exchanger and Cl^-/HCO_3^- anion exchanger mRNA expression in placenta using real-time QPCR. Placenta 26, 93–98.

Ladanyi, A., Soong, R., Tabiti, K., Molnar, B. and Tulassay, Z. 2001. Quantitative reverse transcription-PCR comparison of tumor cell enrichment methods. Clin. Chem. 47, 1860–1863.

Lee, K.H., McKenna, M.J., Sewell, W.F. and Ung, F. 1997. Ribonucleases may limit recovery of ribonucleic acids

from archival human temporal bones. Laryngoscope 107, 1228–1234.

Lefevre, J., Hankins, C., Pourreaux, K., Voyer, H. and Coutlee, F. 2004. Prevalence of selective inhibition of HPV-16 DNA amplification in cervicovaginal lavages. J. Med. Virol. 72, 132–137.

Lekanne Deprez, R.H., Fijnvandraat, A.C., Ruijter, J.M. and Moorman, A.F. 2002. Sensitivity and accuracy of quantitative real-time polymerase chain reaction using SYBR green I depends on cDNA synthesis conditions. Anal. Biochem. 307, 63–69.

Lewis, F. and Maughan, N. J. 2004. Extraction of Total RNA from Formalin-Fixed Paraffin-Embedded Tissue, In A-Z of quantitative PCR, S.A. Bustin, ed. (La Jolla, CA: IUL).

Lewis, F., Maughan, N.J., Smith, V., Hillan, K. and Quirke, P. 2001. Unlocking the archive – gene expression in paraffin-embedded tissue. J. Pathol. 195, 66–71.

Liu, J., Enzelberger, M. and Quake, S. 2002. A nano-liter rotary device for polymerase chain reaction. Electrophoresis 23, 1531–1536.

Liu, W. and Saint, D.A. 2002. Validation of a quantitative method for real time PCR kinetics. Biochem. Biophys. Res. Commun. 294, 347–353.

Madejon, A., Manzano, M. L., Arocena, C., Castillo, I. and Carreno, V. 2000. Effects of delayed freezing of liver biopsies on the detection of hepatitis C virus RNA strands. J. Hepatol. 32, 1019–1025.

Mader, R.M., Schmidt, W.M., Sedivy, R., Rizovski, B., Braun, J., Kalipciyan, M., Exner, M., Steger, G.G. and Mueller, M.W. 2001. Reverse transcriptase template switching during reverse transcriptase-polymerase chain reaction: artificial generation of deletions in ribonucleotide reductase mRNA. J. Lab. Clin. Med. 137, 422–428.

Marubini, E., Verderio, P., Raggi, C.C., Pazzagli, M. and Orlando, C. 2004. Statistical diagnostics emerging from external quality control of real-time PCR. Int. J. Biol. Markers 19, 141–146.

Max, N., Willhauck, M., Wolf, K., Thilo, F., Reinhold, U., Pawlita, M., Thiel, E. and Keilholz, U. 2001. Reliability of PCR-based detection of occult tumour cells: lessons from real-time RT-PCR. Melanoma Res. 11, 371–378.

Mizuno, T., Nagamura, H., Iwamoto, K. S., Ito, T., Fukuhara, T., Tokunaga, M., Tokuoka, S., Mabuchi, K. and Seyama, T. 1998. RNA from decades-old archival tissue blocks for retrospective studies. Diagn. Mol. Pathol. 7, 202–208.

Muller, M.C., Hordt, T., Paschka, P., Merx, K., La Rosee, P., Hehlmann, R. and Hochhaus, A. 2004. Standardization of preanalytical factors for minimal residual disease analysis in chronic myelogenous leukemia. Acta Haematol. 112, 30–33.

Niesters, H.G. 2001. Quantitation of viral load using real-time amplification techniques. Methods 25, 419–429.

Niesters, H.G. 2004. Molecular and diagnostic clinical virology in real time. Clin. Microbiol. Infect. 10, 5–11.

Nolan, T., Hands, R.E., Ogunkolade, B.W. and Bustin, S.A. 2006. SPUD: a qPCR assay for the detection of inhibitors in nucleic acid preparations. Anal. Biochem. *351*, 308–310.

Orlando, C., Pinzani, P. and Pazzagli, M. 1998. Developments in quantitative PCR. Clin. Chem. Lab. Med. *36*, 255–269.

Ozoemena, L.C., Minor, P.D. and Afzal, M.A. 2004. Comparative evaluation of measles virus specific TaqMan PCR and conventional PCR using synthetic and natural RNA templates. J. Med. Virol. *73*, 79–84.

Pasloske, B.L., Walkerpeach, C. R., Obermoeller, R.D., Winkler, M. and DuBois, D.B. (1998). Armored RNA technology for production of ribonuclease-resistant viral RNA controls and standards. J. Clin. Microbiol. *36*, 3590–3594.

Perch-Nielsen, I.R., Bang, D. D., Poulsen, C.R., El-Ali, J. and Wolff, A. 2003. Removal of PCR inhibitors using dielectrophoresis as a selective filter in a microsystem. Lab. Chip *3*, 212–216.

Perez-Novo, C. A., Claeys, C., Speleman, F., Van Cauwenberge, P., Bachert, C. and Vandesompele, J. 2005. Impact of RNA quality on reference gene expression stability. Biotechniques *39*, 52, 54, 56.

Pfaffl, M.W., Tichopad, A., Prgomet, C. and Neuvians, T.P. 2004. Determination of stable housekeeping genes, differentially regulated target genes and sample integrity: BestKeeper – Excel-based tool using pair-wise correlations. Biotechnol. Lett. *26*, 509–515.

Poddar, S.K., Sawyer, M.H. and Connor, J.D. (1998). Effect of inhibitors in clinical specimens on Taq and Tth DNA polymerase-based PCR amplification of influenza A virus. J. Med. Microbiol. *47*, 1131–1135.

Radstrom, P., Knutsson, R., Wolffs, P., Lovenklev, M. and Lofstrom, C. 2004. Pre-PCR processing: strategies to generate PCR-compatible samples. Mol. Biotechnol. *26*, 133–146.

Raggi, C.C., Verderio, P., Pazzagli, M., Marubini, E., Simi, L., Pinzani, P., Paradiso, A. and Orlando, C. 2005. An Italian program of external quality control for quantitative assays based on real-time PCR with Taq-Man probes. Clin. Chem. Lab. Med. *43*, 542–548.

Ramakers, C., Ruijter, J.M., Deprez, R.H. and Moorman, A. F. 2003. Assumption-free analysis of quantitative real-time polymerase chain reaction (PCR) data. Neurosci. Lett. *339*, 62–66.

Schroeder, A., Mueller, O., Stocker, S., Salowsky, R., Leiber, M., Gassmann, M., Lightfoot, S., Menzel, W., Granzow, M. and Ragg, T. 2006. The RIN: an RNA integrity number for assigning integrity values to RNA measurements. BMC Mol. Biol. *7*, 3.

Shanmugam, V., Sell, K. W. and Saha, B.K. 1993. Mistyping ACE heterozygotes. PCR Methods Appl. *3*, 120–121.

Stahlberg, A., Aman, P., Ridell, B., Mostad, P. and Kubista, M. 2003. Quantitative real-time PCR method for detection of B-lymphocyte monoclonality by comparison of kappa and lambda immunoglobulin light chain expression. Clin. Chem. *49*, 51–59.

Stahlberg, A., Hakansson, J., Xian, X., Semb, H. and Kubista, M. 2004a. Properties of the reverse transcription reaction in mRNA quantification. Clin. Chem. *50*, 509–515.

Stahlberg, A., Kubista, M. and Pfaffl, M. 2004b. Comparison of reverse transcriptases in gene expression analysis. Clin. Chem. *50*, 1678–1680.

Stahlberg, A., Zoric, N., Aman, P. and Kubista, M. 2005. Quantitative real-time PCR for cancer detection: the lymphoma case. Expert Rev. Mol. Diagn. *5*, 221–230.

Stangegaard, M., Dufva, I.H. and Dufva, M. 2006. Reverse transcription using random pentadecamer primers increases yield and quality of resulting cDNA. Biotechniques *40*, 649–657.

Stanley, K. K. and Szewczuk, E. 2005. Multiplexed tandem PCR: gene profiling from small amounts of RNA using SYBR Green detection. Nucleic Acids Res. *33*, e180.

Sunen, E., Casas, N., Moreno, B. and Zigorraga, C. 2004. Comparison of two methods for the detection of hepatitis A virus in clam samples (Tapes spp.) by reverse transcription-nested PCR. Int. J. Food Microbiol. *91*, 147–154.

Suslov, O. and Steindler, D.A. 2005. PCR inhibition by reverse transcriptase leads to an overestimation of amplification efficiency. Nucleic Acids Res. *33*, e181.

Tichopad, A., Dilger, M., Schwarz, G. and Pfaffl, M.W. 2003. Standardized determination of real-time PCR efficiency from a single reaction set-up. Nucleic Acids Res. *31*, e122.

Tricarico, C., Pinzani, P., Bianchi, S., Paglierani, M., Distante, V., Pazzagli, M., Bustin, S.A. and Orlando, C. 2002. Quantitative real-time reverse transcription polymerase chain reaction: normalization to rRNA or single housekeeping genes is inappropriate for human tissue biopsies. Anal. Biochem. *309*, 293–300.

Vandesompele, J., De Paepe, A. and Speleman, F. 2002a. Elimination of primer–dimer artifacts and genomic coamplification using a two-step SYBR green I real-time RT-PCR. Anal. Biochem. *303*, 95–98.

Vandesompele, J., De Preter, K., Pattyn, F., Poppe, B., Van Roy, N., De Paepe, A. and Speleman, F. 2002b. Accurate normalization of real-time quantitative RT-PCR data by geometric averaging of multiple internal control genes. Genome Biol. *3*, 0034.0031–0034.0011.

Vlems, F., Soong, R., Diepstra, H., Punt, C., Wobbes, T., Tabiti, K. and van Muijen, G. 2002a. Effect of blood sample handling and reverse transcriptase-polymerase chain reaction assay sensitivity on detection of CK20 expression in healthy donor blood. Diagn. Mol. Pathol. *11*, 90–97.

Vlems, F., Diepstra, J.H., Cornelissen, I.M., Ruers, T.J., Ligtenberg, M.J., Punt, C.J., van Krieken, J.H., Wobbes, T. and van Muijen, G. N. 2002b. Limitations of cytokeratin 20 RT-PCR to detect disseminated tumour cells in blood and bone marrow of patients with colorectal cancer: expression in controls and downregulation in tumour tissue. Mol. Pathol. *55*, 156–163.

Vlems, F.A., Ladanyi, A., Gertler, R., Rosenberg, R., Diepstra, J. H., Roder, C., Nekarda, H., Molnar, B., Tulassay, Z., van Muijen, G. N. and Vogel, I. 2003. Reliability of quantitative reverse-transcriptase-PCR-based detection of tumour cells in the blood between different laboratories using a standardised protocol. Eur. J. Cancer *39*, 388–396.

Walker, N.J. 2002. A technique whose time has come. Science 296, 557–559.

Wall, S.J. and Edwards, D.R. 2002. Quantitative reverse transcription-polymerase chain reaction (RT-PCR): a comparison of primer-dropping, competitive, and real-time RT-PCRs. Anal. Biochem. 300, 269–273.

Weissensteiner, T. and Lanchbury, J.S. (996. Strategy for controlling preferential amplification and avoiding false negatives in PCR typing. Biotechniques 21, 1102–1108.

Winter, H., Korn, K. and Rigler, R. 2004. Direct gene expression analysis. Curr. Pharm. Biotechnol. 5, 191–197.

Zippelius, A., Lutterbuse, R., Riethmuller, G. and Pantel, K. 2000. Analytical variables of reverse transcription-polymerase chain reaction-based detection of disseminated prostate cancer cells. Clin. Cancer Res. 6, 2741–2750.

Validation of Array Data

10

Elisa Wurmbach

Abstract

Microarray techniques allow the parallel assessment of the relative expression of thousands of transcripts in response to different experimental conditions or in different tissues. The ability to correctly identify differentially expressed genes is limited by the signal to noise ratio, the variation in the levels of gene expression, and/or the variability in the measurements due to the assay itself. Therefore, an unequivocal identification of differentially expressed transcripts requires independent confirmation. Quantitative real-time RT-PCR (qPCR) is the method of choice because of its broad range of linearity. Furthermore, it can be easily adapted to systematically study tens to hundreds of different transcripts.

The cDNA microarray technique is introduced as an example, followed by comparisons to different microarray platforms and their characteristics. Data analysis of microarray experiments will show the importance of verification of results. General differences between microarray hybridizations and PCR reactions and, in particular, the performance of different platforms are described and compared. Furthermore, the effects of increasing tissue complexity on detection of differentially expressed transcripts are elucidated with specific examples.

Brief introduction to microarray experiments

Microarray-based genomic techniques enable the simultaneous determination of the expression levels of thousands of genes. This leads to the identification of transcripts with altered expression due to treatment as indicated by relative expression levels of mRNAs in the samples studied.

A number of different types of microarrays are available, but for illustrative purposes, the principles of a cDNA array experiment are described here. Unlike in a Northern blot analysis of gene expression in which experimental RNA samples are fixed to a solid membrane and hybridized with labelled probes, in a cDNA microarray analysis probes are spotted and cross-linked onto glass slides and then hybridized with fluorescently labelled 'targets'. Probe arrays can be made as simple as a few genes of interest or as complex as a whole genome. To make targets, RNA is extracted from cells under both control and experimental conditions or from different tissues for comparison. The mRNAs are reverse transcribed incorporating Cy3- or Cy5-conjugated nucleotides. The labelled cDNA (targets) are pooled and used for the hybridization onto the cDNA array. After hybridization, microarrays are washed to remove excess dye and targets, followed by scanning at appropriate wavelengths (635 nm for Cy5, which appears red, after scanning, and 532 nm for Cy3, which appears green). A visual assessment of the overlaying scans reveals which transcripts are differentially expressed due to the experimental treatment. The colours, which arise from the mixing of colours in the overlay, are shades of yellow, red and green. The determination as to whether red or green indicate up- or down-regulation depends on which fluorophore is used to label the control versus the experimental condition. Yellow spots result from equal amounts of both dyes and illustrate unregulated genes. Ideally, a gene that is not expressed will show in black because no signal is measured. Furthermore it is recommended to do dye reversal experiments to control for bias in the labelling reaction (Wurmbach et al., 2003).

Another method of representing the differential expression of genes is to use a scatterplot, where the signal intensities of each gene from both channels are plotted against each other (Fig. 10.1). Genes that do not differ in expression between the two samples will plot on a diagonal, whereby the signal intensity is in proportion to the abundance of a transcript. With the control samples plotted on the x-axis, up-regulated genes will appear in the upper sector and down-regulated genes in the lower sector. The distance from the diagonal reflects the degree of regulation.

Different microarray platforms: cDNA array, Affymetrix, Codelink, long oligonucleotides

There are four principal types of microarray platforms used in microarray experiments: the cDNA array (mostly two-colour arrays using PCR products as probes), Affymetrix (one-col-

our array using multiple 25-mer oligonucleotide probes for each gene), Codelink (one-colour array using a single 30-mer oligonucleotide probe per gene), and the use of spotted long oligonucleotides as probes, 50 to 80mers.

cDNA microarrays

cDNA arrays (described above) are relatively easy to design, but their fabrication depends on specialized equipment. As a consequence, their quality varies. The quality can be affected by contamination, unverified sequences spotted onto the slide, and heterogeneity in the quantity of DNA in each spot, all of which decrease the reliability of the results. Even if these problems were considered during manufacture, cDNA arrays have inherent disadvantages. For example, the complimentary strand of the spotted probe can compete with the target sample during the hybridization reducing the signal intensity. In addition, the attached DNA strand may not be easily accessible during the hybridization process, resulting in lower signal intensity. Low signal intensities limit the sensitivity with which

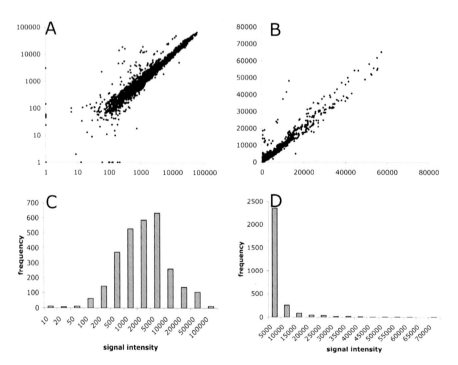

Figure 10.1 Illustration of the same data of a cDNA microarray experiment. A) and B) show the signal intensities from both channels, Cy3 and Cy5 plotted against each other on a log scale (A) or on a linear scale (B). The Cy5-channel data of the hybridization of (A) was analysed for the frequency distribution of the signal intensities. In contrast to a linear representation (D), the logarithmic scale (C) approaches a normal distribution of signal intensities.

differences in expression can be detected. Probe density is also very important for all microarray hybridizations. Increasing probe density can lead to steric hindrance and slow the hybridization (Peterson *et al.*, 2001). Finally, cDNA arrays are easily saturated, rendering the detection of high fold-changes inaccurate.

Affymetrix

Affymetrix microarray chips are composed of short single stranded DNA oligos, (25 mers) synthesized on the chip by photo-lithography (Chee *et al.*, 1996). Each transcript is represented by multiple oligos distributed over the transcript with a bias towards the 3′ end and consequently Affymetrix recommends a 3′ end biased labelling procedure (oligo-dT primers are suggested for target production). Each 25mer is synthesized individually *in situ* and has a sister-spot containing a mismatch in the middle of the oligo. The hybridization signals for the perfect matches will be higher on average than for their mismatched partners due to specific targets. The perfect-match to mismatch signal ratio is used to determine whether a gene was expressed and will result in present or absent calls (Lockhart *et al.*, 1996). To label the target, extracted RNA is reverse transcribed with oligo-dT primers, followed by *in vitro* transcription to obtain biotin labelled cRNA, which is then fragmented to smaller pieces ranging from 50 to 200 nucleotides. In contrast to cDNA arrays, each target is hybridized separately under non-competitive conditions. Therefore, a comparison of treatments requires at least two hybridizations as compared to only a single one for cDNA arrays. The advantage over cDNA microarrays is that there is no spotted complementary strand competing for the hybridization with the cRNA to lower the hybridization signal. In addition, the hybridization procedure is highly standardized, which results, in principle, in reliable and comparable results, independent of the location where experiments are performed.

When comparing the same biological experiment on a cDNA- and Affymetrix-platform, as done by Yuen *et al.* (2002), the results from both platforms correlate (Fig. 10.2). Highly regulated genes appear to be highly regulated on both platforms but the correlation of their actual fold-changes is modest. The correlation for genes with lower fold-changes and non-regulated genes was better. Overall, the rank-order of regulated genes was comparable.

The Codelink platform

The characteristic of Codelink microarrays is that the spotted DNA probes are embedded

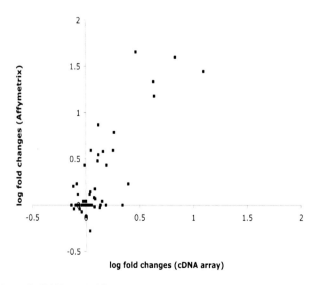

Figure 10.2 Correlation of cDNA and Affymetrix microarray data. The same biological experiment and the same RNA were used as input for both platforms. Scatterplot: log transformed fold-changes from two different microarray platforms are plotted against each other. X-axis: cDNA array data, y-axis: Affymetrix array data. The correlation of these fold-changes is 0.724 (see text for details).

in a 3-D gel matrix, which provides an aqueous probe-environment and holds the probe away from the surface of the slide reducing steric hindrances during the hybridization compared to other platforms (Ramakrishnan *et al.*, 2002). Codelink, similar to Affymetrix, utilizes small oligonucleotides (30mers) positioned onto the slide. But in contrast to Affymetrix, only one oligonucleotide represents each transcript and the synthesis occurs before deposition, allowing purification and validation of the probe. This system also uses a one-colour detection method and provides a hybridization station that partially automates and standardizes the hybridization process. During the hybridization, biotin labelled cRNA targets bind to the spotted DNA single stranded probe. Hybridization conditions (stringency) are calibrated so that most targets with three or more mismatches will not be detected but – due to low signal intensity – will appear within the noise. Thus the stringency of the hybridization allows for the discrimination between highly homologous gene sequences, increasing the specificity of the assay.

Long oligonucleotide arrays

Long oligonucleotides spotted onto glass slides avoid some of the disadvantages arising from cDNA arrays and have the advantages of oligonucleotide arrays. The production and hybridization of long oligo arrays and cDNA arrays is similar. The spots can be generated with equal concentrations of oligos for all genes represented on the slides and with sufficient concentration to prevent a saturation effect, which would result in underestimation of signal intensities. Clontech and Operon sell spotting-ready libraries of 80- and 70-mer probes, respectively. The libraries are designed for specificity to the target genes, are normalized for probe melting-temperatures, and exclude potentially redundant sequences.

Long oligos, like PCR products, can be used to generate focused, custom designed arrays. These arrays are characterized by a careful selection of genes positioned on the array. For example, probes can be selected based on function (kinase, phosphatase, etc.) or by their relationships (signalling pathways). The quality control and robustness can be improved relative to genome-wide arrays, because increasing the

number of genes tested on the array must either increase the false-positive rate (decrease of the specificity) or decrease the sensitivity (Churchill, 2002). In addition, the high expense of global arrays constrains the number of arrays that can be analysed to provide statistically acceptable sensitivity and specificity (Wurmbach *et al.*, 2003).

These different microarray platforms enable many laboratories to choose the method best adjusted to their needs and experimental requirements.

Microarray: normalization, baseline, fold-change

The basic steps for handling and analysing microarray data include normalization, baseline determination and calculation of fold-changes.

Normalization

Normalization corrects for systematic differences across data sets to compare values from corresponding individual genes directly from one array to another. In larger experiments composed of several array hybridizations, normalization is crucial for the comparison of expression levels from various conditions and/or tissues.

Normalization procedures can include multiple steps. One of the first steps is usually the background correction, which accounts for non-specific binding of labelled products and eliminates artefacts or dye residues.

Signal intensities measured in microarray experiments are the raw data used for the normalization and, in a typical microarray experiment, they span up to 4 log orders of magnitude (e.g. they range within 10^1 to 10^5 as shown in Fig. 10.1). In most analysis procedures, the data are first log transformed. The log transformation has several advantages: it reflects the biological significance better, makes it easier to handle the data and makes the distribution of the intensity values nearly symmetrical, whereby the intensity frequencies approximate a normal distribution (Draghici, 2003). (Compare the distribution in log-scale in Fig. 10.1C with the linear-scale in Fig. 10.1D.)

Ordinarily, there are two possibilities to normalize microarray data. One method relies on spiking the labelling reactions with control genes, which should span the whole intensity range to account for genes of different abundances. The other method is based on the overall array intensity: after log transformation of the raw data, the mean signal intensity value is subtracted from each intensity value. Generally, the strategies to normalize the raw data depend on the microarray platform used, e.g. a one- or a two-colour platform.

cDNA arrays need to be corrected for overall differences in the signal intensities of the two wavelengths measured on each slide (cy3 = 532 nm and cy5 = 635 nm) (Quackenbush, 2002) and for colour distortion. This can be accomplished by using the loess function, a locally linear robust scatterplot smoother (Cleveland, 1979; Smyth and Speed, 2003).

Several normalization algorithms are available to normalize Affymetrix data, such as MAS 5.0, RMA and GC-RMA (Affymetrix, 2001; Irizarry et al., 2003; Wu et al., 2004) and most of them are implemented in software-packages to analyse microarray data, e.g. Gene Spring (Agilent Technologies), Gene Traffic (Stratagene) and Multi Experiment Viewer (Saeed et al., 2003).

Baseline and fold-change determination

To compare data from different experimental conditions and to calculate fold-changes, all data from treated samples have to be compared to a baseline or to a reference. Depending on the experimental design, references in microarray studies can be control samples, lacking a specific treatment. As described above, in cDNA arrays, the control is hybridized together with the treated sample. One-colour arrays, like Affymetrix, measure 'absolute' expression levels; therefore, an expression baseline for each gene can only be calculated after a second hybridization with a control sample. In practice, several control hybridizations are performed to reduce the biological variation. The geometric mean of all the control experiments is used as an expression baseline, which makes the calculation more robust against outliers.

Once the data has undergone log transformation, background correction and normalization of signal intensities and colour distortion (two-colour array) or normalization to baseline (one-colour array), the ratio of signal intensi-

ties between experimental and control signals or between multiple experiments and baseline is calculated. This value can be >1, <1 or ~1, reflecting up- and down-regulation or unaffected expression, respectively. This value is referred to as fold-change of expression.

Microarray data analysis

Once the fold change of expression of a set of genes has been determined, statistical analysis of the data must be used to identify significantly differently expressed genes. These candidate targets then need to be verified by repetition and/or other techniques such as quantitative RT-PCR (qPCR).

In many early microarray studies, a fixed fold-change cut-off was used (mostly 2-fold), to identify differentially expressed genes (DeRisi et al., 1996, 1997; Okabe et al., 2001; Smith et al., 2003). These genes were then sorted according to their fold-changes to create a rank order. This arbitrarily chosen threshold may often be inappropriate. For example, at low intensities, where the data are much more variable, genes might be misidentified as being differentially expressed, while at higher intensities, genes that are significantly changed might be missed. Spot replicates on single chips can be used to estimate the noise and calculate confidence levels for gene regulation. Repeating microarray experiments and using significance tests will identify differentially expressed genes with higher accuracy. The budget for the experiment or sample availability may limit the number of possible replicates and may favour other techniques to confirm the regulation of the identified genes. Furthermore, microarrays are very often used to explore a systematically varied set of conditions, e.g. time-course experiments or cancer studies, in which different stages of diseases are compared. Such studies include many biological repeats, which can contribute a certain degree of variation. Therefore, these experimental designs require statistical methods to distinguish between signal and noise.

Properly optimized microarray data are analysed using either unsupervised or supervised methods (Cuperlovic-Culf et al., 2005). Unsupervised methods are used when no non-array data about the sample groups are included.

This comprises for example cluster analysis, where samples are grouped according to their gene expression profile. Hierarchical clustering groups the two most similar samples together. In the following step the third most similar sample is added to this group. This continues until all samples are included and the clustering results in a tree. The lengths of the branches by which the samples are separated reflect the similarity/difference of these samples: the shorter the branches, the more similar the samples. Ideally, all biological repeats should be grouped together. Novel groups or classes can be identified based on such clustering methods (Quackenbush, 2001). Further analysis will then narrow down the expression profile to some characteristically expressed genes, which define these novel groups. Again, to increase the confidence level in the differentially expressed genes, they must be independently validated by other methods such as qPCR.

In supervised methods, samples are assigned to different groups based on the classification of tissue samples, e.g. according to their morphology or similar criteria. Genes with the most distinct expression pattern for the groups are then selected as marker genes. The procedure for selecting markers is essential because the number of marker genes for a particular classification is very small relative to the total number of genes represented on the array. Statistical methods facilitate the identification of marker genes and increase the probability that the selection reflects a biological meaning. Several programs are available that incorporate sophisticated algorithms for clustering, visualization, classification, statistical analysis and biological theme discovery [for example, Gene Traffic (Iobion Informatics, Stratagene), Gene Spring (Agilent Technologies) or Multi Experiment Viewer (MeV)]. Statistical tests include t-tests (Dudoit et al., 2000; Pan, 2002), significant analysis of microarray data (SAM) (Chu et al., 2002; Tusher et al., 2001) and when distinguishing between more than two groups the multiclass test ANOVA (Kerr et al., 2000). Again, it is important to verify the expression of the marker genes by independent methods, such as qPCR, because the subsequent classification of unknown samples relies on them.

Corroboration of gene regulation by qPCR

Microarrays can monitor the expression of thousands of genes simultaneously in a single experiment. But due to biological and technical variations, including the microarray design the precision of microarray results are affected (Churchill, 2002). This can be overcome by using microarrays as a screening tool followed by an independent verification of the changes in gene expression. qPCR has become the routine method, because it is fast to perform, allowing high-throughput RNA quantification, and is accurate over a wide dynamic range, at relatively low costs.

Although, the ranking of regulated genes is often similar, the fold changes obtained from microarray analysis differ from the fold changes measured by qPCR (Czechowski et al., 2004; Rise et al., 2004; Wurmbach et al., 2001). Up-regulated genes found by microarray techniques show often higher levels of regulation when assessed by qPCR. This discrepancy appears to be more pronounced the more strongly a gene is regulated (Fig. 10.3). This is illustrated with an example of a typical biological experiment analysed by microarrays and qPCR. A pituitary derived cell line expressing the receptor of a hormone was treated with the corresponding hormone. Extracted RNA was analysed on a cDNA microarray, on a one-colour array (Affymetrix) and by qPCR (Yuen et al., 2002). Fig. 10.3A shows a typical comparison of the fold changes measured by microarrays (y-axis) and by qPCR (x-axis). Obviously, low fold changes on the microarray relate to low fold changes measured by qPCR and the same is true for more strongly up-regulated genes. Thus the data generated by both methods correlates. The correlation coefficient for the log transformed data of cDNA array vs. qPCR shown in Fig. 10.3A and B was 0.7, and for the one-colour array as shown in Fig. 10.3C, was, 0.84. The correlation coefficient of the one-colour array is higher because it included more not- and low-regulated genes. The representations in Fig. 10.3B and C directly compare the log transformed fold changes from microarrays with qPCR. If the fold-changes from both techniques were in the same range the data would scatter around zero (x-axis). However, both

A

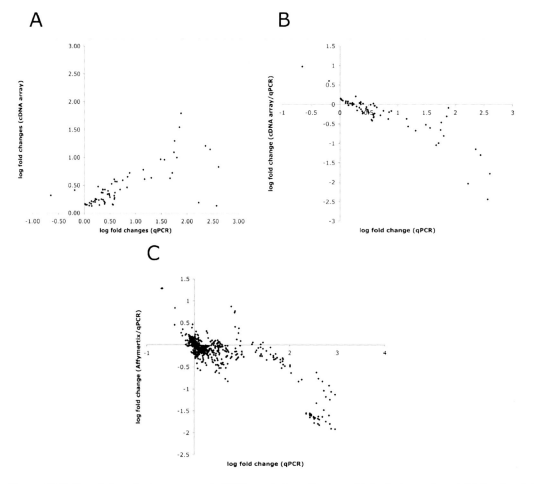

Figure 10.3 Correlation of microarray and qPCR results from the same biological experiment. (A) Scatterplot from log transformed fold-changes from qPCR (x-axis) and cDNA array (y-axis). (B) Comparison of fold-changes between cDNA array and qPCR. (C) Comparison of fold-changes between Affymetrix array and qPCR.

panels clearly reveal that the fold changes generated from microarray measurements (cDNA array and Affymetrix) for differentially expressed genes were in general lower than the fold changes for these genes measured by qPCR. In this experiment, Yuen *et al.* (2002) also showed that the correlation of the cDNA data to the qPCR was roughly linear, which could not be shown for the one-colour array. This enabled them to provide a power law correction to calibrate the cDNA array data. Furthermore, they showed that the cDNA microarray displays a saturation effect, which explains the weak correlation for higher regulated genes.

A strong discrepancy between microarray and qPCR results can be observed for induced

genes. Induced genes are switched on after being at levels below the detection limit. In qPCR experiments, the untreated sample and the non-template control (NTC) result in the same cycle threshold (C_t)-value, while the treated sample reveals a lower C_t-value. Consequently the resulting fold-change can be extremely high, up to thousand fold and is prone to a high standard deviation. Microarrays can hardly reflect this kind of induction of genes, due to their lower sensitivity and higher background. The same genes would show an up-regulation on microarrays, possibly a relatively high up-regulation, which is still significantly lower compared to the qPCR result. Furthermore, the broad range of linearity of qPCR reactions, higher than seven orders of magnitude, results in more accurate measurements, compared to microarrays. The dynamic range for microarrays is only up to four orders of magnitude (Bustin *et al.*, 2005; Wong and Medrano, 2005).

The correlation of microarray data with qPCR data for down-regulated genes is mostly similar as for up-regulated genes. However, Affymetrix flags genes below detection level as 'absent', which makes the calculation of fold-changes for strongly down-regulated genes difficult. This becomes more pronounced for low abundance genes and can result in undetectable changes of the gene expression levels due to the lower sensitivity of microarrays, at levels where qPCR could still measure a down-regulation. In general, qPCR performed on genes identified in microarray experiments provides a more accurate estimation of the fold-change.

Identification of appropriate genes for normalization

The key to comparison of samples is the identification of unaffected genes to which the experiment can be normalized. Normalization is used to correct the qPCR result for differences in RNA quantity and quality, the overall transcriptional activity and variations in the cDNA synthesis and the PCR efficiency. Using microarrays followed by qPCR makes it easy to identify a collection of genes suitable for normalization of the qPCR reactions. In most biological experiments, typical housekeeping genes, like GAPDH, beta-actin or tubulin are used for normalization. However, these 'typical housekeeping genes' may be differentially expressed and affected by the experimental conditions (Hsiao *et al.*, 2001; Warrington *et al.*, 2000). If so, choosing such genes for normalization can lead to misinterpretations or even can make the interpretation of data impossible (Dheda *et al.*, 2005). For example, if a gene used for normalization is differentially affected by the experimental conditions, it will not be possible to gain an accurate estimation of the direction of the regulation of genes with marginally altered levels of transcription. In addition, the degree of regulation of strongly regulated genes will be systematically under- or overestimated. Any gene used for normalization should ideally be a part of the test samples.

Because such internal endogenous controls are handled in the same way as the genes of interest, both measurements underlie the same variations. Thus, although the reference gene can be a housekeeping gene involved in maintaining the cell function it is more important that it should be consistently expressed in the cells or the tissue under investigation and that it does not respond to the experimental treatment.

After performing a microarray experiment it is relatively easy to identify suitable reference genes for use in qPCR. First, they should not be affected by the experimental conditions and, second, they should be expressed in similar abundance (signal intensity) as the regulated genes. When plotting the signal intensities of the different experimental conditions in a scatterplot, all unregulated genes lie on the diagonal (cy3=cy5, or x=y line) as shown in Fig. 10.4. The signal intensity correlates with the abundance of genes. These unchanged genes represent a candidate pool to choose from for normalization. A few candidates should be validated by qPCR prior to their use, in order to confirm that they are not affected by the experimental conditions. Two or three candidates are then used for normalization in qPCR used to corroborate array data. Choosing genes for normalization that belong to different functional classes significantly reduces the chance of possible co-regulation. However, additional changes in experimental conditions may necessitate reassessment of the choice of genes used for normalization.

For best robustness, however, the geometric mean of multiple carefully selected reference genes is taken as a normalization factor (Vandesompele *et al.*, 2002).

In conclusion, analysis by microarray and qPCR makes it relatively straightforward to identify reference genes for use as internal standards that remain constant between experimental conditions.

Complex tissues

Microarray study designs must take account of the problem of distinguishing signal from noise in a massively parallel experiment. Different sources of variation, both technical and biological, directly affect the reproducibility and the accuracy of microarray measurements

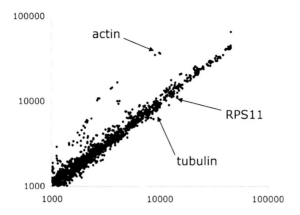

Figure 10.4 Identification of genes for normalization of qPCRs. Scatterplot from a cDNA array: x-axis: signal intensity of Cy5, y-axis: signal intensity of Cy3. Actin is up-regulated in this experiment but tubulin and RPS11 are not. Therefore, the latter can be used for normalization.

(Churchill, 2002). The technical variation arises from the extraction, labelling and hybridization. The precision of microarray experiments can be increased by technical replicates (Lee *et al.*, 2000). The biological variation, intrinsic to all organisms is discussed in more detail below.

In a cell line experiment, biological variation can be tightly controlled. Ideally, all cells react similarly to a change in experimental condition. In contrast when using tissues from animals or clinical samples the biological variation increases due to the heterogeneous cell population in complex tissues. More precisely, a particular experimental treatment might affect only a subset of cells. Therefore the level of change of any transcript in the RNA samples from that tissue is likely to be much lower than in a homogeneous tissue, because the change is masked by the expression of the gene in the unaffected cells. In addition, since any particular gene of interest may only be expressed in a small subset of cells the signal intensity for that gene on the array may be low and difficult to measure reliably. These limitations are referred to as 'dilution effects' (Wurmbach *et al.*, 2002; Wurmbach *et al.*, 2003). Further problems arise when clinical samples are used in microarray experiments, because, in addition to the dilution effects, the individual genotype might cause further variation. In some experimental situations this effect can be limited by assaying RNA samples that are pooled from individual samples and/or by increasing the

number of replicates studied, followed by independent quantification of the transcripts.

Fig. 10.5 shows examples of the effect of samples with increasing complexity on the outcome of microarray data. The same cDNA microarray was utilized for hybridizations with labelled products derived from a cell line, hypothalamus and cortex, respectively. A pituitary derived mouse cell line, L beta T2, expressing the gonadotropin releasing hormone (GnRH) receptor was treated with the hormone (GnRH). Twenty-six regulated genes were identified on the microarray as being differentially expressed by using three replicate experiments. qPCR corroboration confirmed the correct identification of 23 genes (88.5%). Fig. 10.5A shows the strongly regulated transcripts encircled in the scatterplot. The higher the regulation of differentially expressed genes, the farther away they are located from the x=y line, while unregulated transcripts appear on the diagonal.

Using a tissue increases the complexity of the sample and increases the variation of a microarray assay. This effect is illustrated by the following example: hypoglycaemia-associated failure of mice that were fasted for 48 hours and then received an insulin injection was studied in the hypothalamus. Five replicate experiments on microarrays identified 16 candidate genes that changed due to the treatment. qPCR confirmed 12 of these 16 genes (75%). Because only a subset of the cells changed the expression of the

A

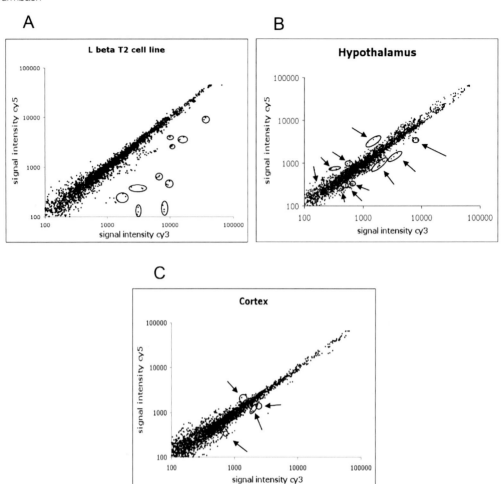

B

C

Figure 10.5 Illustration of effects of sample complexity on microarray experiments ('dilution effects'). A–C show scatterplots of experiments with increasing tissue complexity (A: cell line, B: hypothalamus, and C: cortex). The more complex the tissue, the closer the regulated genes appear to the diagonal (see text for details). Freely adapted from Wurmbach *et al.* (2002).

regulated genes, the fold-changes were smaller compared to the cell line experiment, and thus all regulated genes lay in close proximity to the diagonal (Fig. 10.5B). In addition, the signal intensities for the differentially expressed genes were lower, the number of identified genes was smaller (which could also come from the experimental design), and the percentage of genes confirmed by qPCR was smaller. This reveals the trend of the difficulty of selecting reliably regulated genes using microarrays for complex tissues.

Another example for complex tissues used in a microarray experiment is the comparison from somatosensory cortex samples from

2,5-dimethoxy 4-iodoamphetamine- (DOI) and vehicle-treated mice. An obvious regulation cannot be seen in the scatterplot of a representative experiment shown in Fig. 10.5C. The few confirmed regulated genes were close in proximity to the diagonal with relatively low signal intensities (around 2000) as indicated with arrows in Fig. 10.5C. Seven hybridizations on the array led to 14 candidate genes, of which four (28.6%) were confirmed by qPCR. In the cortex sample two effects have an influence on the measured regulation by the microarray experiment: as above, the first effect is that genes of interest were changed by the treatment only in a subpopulation of cells, leading to lower fold-changes. In addition, a second effect occurred when the treatment affected only a subpopulation of cells expressing the gene of interest, leading to lower signal intensities. Therefore, the differences observed between the cell line and the hypothalamus studies regarding the differential expression detected by microarrays are even more pronounced in the cortex experiment: fewer genes detected by microarrays, lower fold-changes of the differentially expressed genes, smaller percentage of the candidate genes confirmed by qPCR, and lower signal intensity of identified genes, caused by the dilution effects.

There are some approaches available to overcome the dilution effects due to the heterogeneity of cells in complex tissues. The cells of interest can be dissected from the surrounding material, e.g. by finer physical dissection. Tissues that can be disaggregated (e.g. blood tissue), can be tagged, and sorted magnetically or by flow cytometry. Laser microdissection can be useful for isolating specific cells in a non-contact and a contamination-free manner (Burgemeister, 2005). Despite improvements of the technology over the last years (Niyaz *et al.*, 2005) this method increases RNA degradation and provides very little starting material, two issues, which have a direct impact on the outcome of microarray experiments. Integrating an amplification step (Ginsberg, 2005) into the protocol may solve the problem of having too little starting material for performing a hybridization, but might introduce bias in the RNA distribution (Marko *et al.*, 2005).

Conclusions

Microarrays are used to screen many thousands of genes in parallel to detect differentially expressed genes. The objective of a microarray experiment is to correctly identify transcripts that are differentially expressed between groups. Various components that affect this procedure must be optimized and integrated. Measurement variability should be kept as small as possible at all stages of an experiment to improve the outcome. Controllable sources of error in microarray experiments include the biological experiment, RNA quality, labelling method, microarray fabrication, hybridization conditions, and analysis algorithms.

Once identified, regulated genes are verified by qPCR which allows accurate and high throughput RNA quantification over a wide dynamic range with the need for only a small amount of starting material. Ultimately, the combination of microarrays with qPCR enables one to accurately and reliably detect changes in gene expression. A further benefit comes from the fact that the initial microarray analysis also can be used to unambiguously select the proper genes for normalization of the qPCR reactions.

Acknowledgements
I am grateful for the continuous support of Drs Samuel Waxman and Stuart Sealfon. Many thanks also to Drs Andreas Jenny, Justine Rudner, Joseph Delaney and Stuart Sealfon for helpful discussions and careful reading of this manuscript.

References
Affymetrix 2001. Microarray Suite 5.0 User's Guide.

Burgemeister, R. 2005. New aspects of laser microdissection in research and routine. J. Histochem. Cytochem. 53, 409–412.

Bustin, S. A., Benes, V., Nolan, T. and Pfaffl, M.W. 2005. Quantitative real-time RT-PCR – a perspective. J. Mol. Endocrinol. 34, 597–601.

Chee, M., Yang, R., Hubbell, E., Berno, A., Huang, X.C., Stern, D., Winkler, J., Lockhart, D. J., Morris, M.S. and Fodor, S.P. (1996). Accessing genetic information with high-density DNA arrays. Science 274, 610–614.

Chu, G., Narasimhan, B., Tibshirani, R. and Tusher, V. 2002. Significance Analysis of Microarrays, pp. User's Guide and Technical Document.

Churchill, G.A. 2002. Fundamentals of experimental design for cDNA microarrays. Nat. Genet. 32 Suppl., 490–495.

Cleveland, W. S. (1979). Robust locally weighted regression and smoothing scatterplots. J. Am. Stat. Assoc. 74, 829–836.

Cuperlovic-Culf, M., Belacel, N. and Ouellette, R.J. 2005. Determination of tumour marker genes from gene expression data. Drug Discov. Today 10, 429–437.

Czechowski, T., Bari, R.P., Stitt, M., Scheible, W.R. and Udvardi, M.K. 2004. Real-time RT-PCR profiling of over 1400 Arabidopsis transcription factors: unprecedented sensitivity reveals novel root- and shoot-specific genes. Plant J. 38, 366–379.

DeRisi, J., Penland, L., Brown, P.O., Bittner, M. L., Meltzer, P.S., Ray, M., Chen, Y., Su, Y.A. and Trent, J.M. 1996. Use of a cDNA microarray to analyse gene expression patterns in human cancer. Nat. Genet. 14, 457–460.

DeRisi, J.L., Iyer, V.R. and Brown, P.O. 1997. Exploring the metabolic and genetic control of gene expression on a genomic scale. Science 278, 680–686.

Dheda, K., Huggett, J.F., Chang, J.S., Kim, L.U., Bustin, S.A., Johnson, M.A., Rook, G.A. and Zumla, A. 2005. The implications of using an inappropriate reference gene for real-time reverse transcription PCR data normalization. Anal. Biochem. 344, 141–143.

Draghici, S. 2003. Data analysis tools for DNA microarrays (London, Chapman & Hall, CRC Press).

Dudoit, S., Yang, Y.H., Callow, M.J. and Speed, T. 2000. Statistical methods for identifying differentially expressed genes in replicated cDNA microarray experiments. Technical report 2000 Statistics Department, University of California, Berkeley.

Ginsberg, S.D. 2005. RNA amplification strategies for small sample populations. Methods 37, 229–237.

Hsiao, L.L., Dangond, F., Yoshida, T., Hong, R., Jensen, R.V., Misra, J., Dillon, W., Lee, K.F., Clark, K.E., Haverty, P., et al. 2001. A compendium of gene expression in normal human tissues. Physiol. Genomics 7, 97–104.

Irizarry, R.A., Hobbs, B., Collin, F., Beazer-Barclay, Y.D., Antonellis, K. J., Scherf, U. and Speed, T.P. 2003. Exploration, normalization, and summaries of high density oligonucleotide array probe level data. Biostatistics 4, 249–264.

Kerr, M.K., Martin, M. and Churchill, G.A. 2000. Analysis of variance for gene expression microarray data. J. Comput. Biol. 7, 819–837.

Lee, M.L., Kuo, F., Whitmore, G.A. and Sklar, J. 2000. Importance of replication in microarray gene expression studies: statistical methods and evidence from repetitive cDNA hybridizations. Proc. Natl. Acad. Sci. USA 97, 9834–9839.

Lockhart, D.J., Dong, H., Byrne, M.C., Follettie, M.T., Gallo, M.V., Chee, M.S., Mittmann, M., Wang, C., Kobayashi, M., Horton, H. and Brown, E.L. 1996. Expression monitoring by hybridization to high-density oligonucleotide arrays. Nat. Biotechnol. 14, 1675–1680.

Marko, N.F., Frank, B., Quackenbush, J. and Lee, N.H. 2005. A robust method for the amplification of RNA in the sense orientation. BMC Genomics 6, 27.

Niyaz, Y., Stich, M., Sagmuller, B., Burgemeister, R., Friedemann, G., Sauer, U., Gangnus, R. and Schutze, K. 2005. Noncontact laser microdissection and pressure catapulting: sample preparation for genomic, transcriptomic, and proteomic analysis. Methods Mol. Med. 114, 1–24.

Okabe, H., Satoh, S., Kato, T., Kitahara, O., Yanagawa, R., Yamaoka, Y., Tsunoda, T., Furukawa, Y. and Nakamura, Y. 2001. Genome-wide analysis of gene expression in human hepatocellular carcinomas using cDNA microarray: identification of genes involved in viral carcinogenesis and tumour progression. Cancer Res. 61, 2129–2137.

Pan, W. 2002. A comparative review of statistical methods for discovering differentially expressed genes in replicated microarray experiments. Bioinformatics 18, 546–554.

Peterson, A.W., Heaton, R.J. and Georgiadis, R.M. 2001. The effect of surface probe density on DNA hybridization. Nucleic Acids Re. 29, 5163–5168.

Quackenbush, J. 2001. Computational analysis of microarray data. Nat. Rev. Genet. 2, 418–427.

Quackenbush, J. 2002. Microarray data normalization and transformation. Nat. Genet. 32 Suppl., 496–501.

Ramakrishnan, R., Dorris, D., Lublinsky, A., Nguyen, A., Domanus, M., Prokhorova, A., Gieser, L., Touma, E., Lockner, R., Tata, M., et al. 2002. An assessment of Motorola CodeLink microarray performance for gene expression profiling applications. Nucleic Acids Res. 30, e30.

Rise, M.L., Jones, S.R., Brown, G.D., von Schalburg, K.R., Davidson, W.S. and Koop, B.F. 2004. Microarray analyses identify molecular biomarkers of Atlantic salmon macrophage and haematopoietic kidney response to Piscirickettsia salmonis infection. Physiol. Genomics 20, 21–35.

Saeed, A.I., Sharov, V., White, J., Li, J., Liang, W., Bhagabati, N., Braisted, J., Klapa, M., Currier, T., Thiagarajan, M., et al. 2003. TM4, a free, open-source system for microarray data management and analysis. Biotechniques 34, 374–378.

Smith, M.W., Yue, Z.N., Geiss, G. K., Sadovnikova, N.Y., Carter, V.S., Boix, L., Lazaro, C.A., Rosenberg, G.B., Bumgarner, R.E., Fausto, N., et al. 2003. Identification of novel tumour markers in hepatitis C virus-associated hepatocellular carcinoma. Cancer Res. 63, 859–864.

Smyth, G.K. and Speed, T. 2003. Normalization of cDNA microarray data. Methods 31, 265–273.

Tusher, V.G., Tibshirani, R. and Chu, G. 2001. Significance analysis of microarrays applied to the ionizing radiation response. Proc. Natl. Acad. Sci. USA 98, 5116–5121.

Vandesompele, J., De Preter, K., Pattyn, F., Poppe, B., Van Roy, N., De Paepe, A. and Speleman, F. 2002. Accurate normalization of real-time quantitative RT-PCR data by geometric averaging of multiple internal control genes. Genome Biol. 3, research0034.1-research0034.11.

Warrington, J.A., Nair, A., Mahadevappa, M. and Tsyganskaya, M. 2000. Comparison of human adult and foetal expression and identification of 535 housekeeping/maintenance genes. Physiol. Genomics 2, 143–147.

Wong, M.L. and Medrano, J.F. 2005. Real-time PCR for mRNA quantitation. Biotechniques 39, 75–85.

Wu, Z., Irizarry, R.A., Gentleman, R., Martinez-Murillo, F. and Spencer, F. 2004. A model-based background adjustment for oligonucleotide expression arrays. J. Am. Stat. Assoc. 99, 909–917.

Wurmbach, E., Gonzalez-Maeso, J., Yuen, T., Ebersole, B.J., Mastaitis, J.W., Mobbs, C.V. and Sealfon, S.C. 2002. Validated genomic approach to study differentially expressed genes in complex tissues. Neurochem. Res. 27, 1027–1033.

Wurmbach, E., Yuen, T., Ebersole, B. J. and Sealfon, S.C. 2001. Gonadotropin-releasing hormone receptor-coupled gene network organization. J. Biol. Chem. 276, 47195–47201.

Wurmbach, E., Yuen, T. and Sealfon, S.C. 2003. Focused microarray analysis. Methods 31, 306–316.

Yuen, T., Wurmbach, E., Pfeffer, R. L., Ebersole, B. J. and Sealfon, S.C. 2002. Accuracy and calibration of commercial oligonucleotide and custom cDNA microarrays. Nucleic Acids Res 30, e48.

Mutation Detection by Real-Time PCR 11

Elaine Lyon, Rong Mao and Jeffrey Swensen

Abstract

Real-time applications for mutation detection include detecting alterations associated with inherited disease, acquired alterations in oncology, and microbial or viral mutations associated with drug resistance in infectious diseases. Probe chemistries described for these applications include hydrolysis (TaqMan®) and hybridization probes (FRET and Molecular Beacons). Hydrolysis probes detect mutations by allele-specific hybridization at a specific temperature, while hybridization probes allow dynamic detection through a temperature range. Primer chemistries are described for allele-specific amplification and Scorpion primers. Recent progress in scanning amplicons for mutations also includes high resolution melting. The design of each of these methods is described, along with applications.

Introduction

A mutation can be defined as a sequence change in a test sample compared with a sequence of a reference standard. The mutation may cause pathogenic effects which have clinical consequences. It could be a single nucleotide substitution (often referred to as single nucleotide polymorphism, or SNP), involve a small insertion or deletion, or be a large rearrangement, such as a complete gene deletion or duplication. Many techniques for mutation detection currently rely on PCR amplification of a target molecule, followed by a variety of post-amplification analyses, such as restriction enzyme digestion and agarose gel electrophoresis, to detect the specific analyte amplified.

Rapid developments in real-time PCR to detect and quantify DNA have produced a promising array of new platforms for DNA analysis. These platforms offer the potential for higher-throughput, faster, less-expensive, automated genetic analysis in the clinical laboratory. Many chemistries offer single-base discrimination, as well as detection of nucleotide repeats and small deletions and insertions. In addition, they allow PCR monitoring during each cycle of amplification. This in turn offers the potential for improved assay sensitivity and accuracy in mutation detection and quantification because threshold measurements can be made. This chapter will discuss design and application of hydrolysis and hybridization probes, as well as primer-based designs, used in real-time PCR mutation detection.

Hydrolysis TaqMan® probes

TaqMan® probes (Heid, 1996) consist of allele-specific probe sequences labelled at one end with a fluorophore and at the other end with a quencher moiety. In the unhybridized state, fluorescence is quenched because of the close proximity of the fluorophore and quencher. Upon hybridization of the probe to the target sequence during PCR, the 5′–3′ exonuclease activity of *Taq* polymerase cleaves the labelled nucleotide from the 5′ end of the probe (Livak, 1999; McGuigan, 2002). During each cycle of the PCR reaction, a fluorescent signal is generated when the intact probe, which is hybridized to the target allele, is cleaved by the 5′-exonuclease activity of *Taq* polymerase. The PCR primers amplify a specific region of the genomic DNA template, and each fluorescent dye-labelled hybridization probe reports the presence of its associated allele. In each PCR cycle, cleavage of the allele-specific probe produces an exponentially increasing fluorescent signal by freeing the 5′ fluorophore from the 3′ quencher (Fig. 11.1). The use of two probes labelled with different fluorophores, one specific to each allele of the SNP, allows detection of both alleles in a single tube. The quencher dye is technically also a fluorescent dye but usually emits at much longer wavelengths than the reporter dye so that its fluorescence is not detected by a real-time instrument; the subsequent separation of the fluorophore from the quencher produces a concomitant increase in fluorescence (Ginzinger, 2002; De la Vega, 2005).

Probe design

The TaqMan® SNP assay is a single tube PCR which requires a set of locus-specific primers flanking the SNP of interest, and two allele-specific oligonucleotide TaqMan® probes. These probes have a fluorescent reporter dye at the 5′ end, and a fluorescent quencher at the other end. The probe must bind tightly to the template, enabling *Taq* polymerase to cleave nucleotides from the 5′ end of the probes. The probe is designed

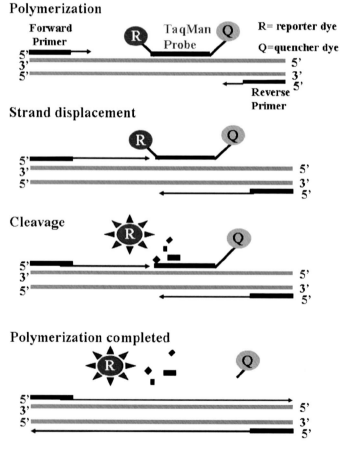

Figure 11.1 Principle of SNP detection with double-dye oligonucleotide TaqMan® probe. The fluorophore linked to inact TaqMan® probe is efficiently quenched by the quencher moiety. Polymerization: TaqMan® probe hybridizes to DNA target during PCR extension. Allelic discrimination is achieved by selective annealing of match probe and template sequence. Strand displacement and Cleavage: The 5′–3′ exonuclease activity of *Taq* polymerase promotes cleavage of the probe between fluorophore and quencher components, separating the two moieties. Once separated from the quencher, the fluorophore emits fluorescent signal. Polymerization completed: Once the TaqMan® probe is cleaved from the targeted DNA, PCR extension continues. A colour version of this figure is located in the plate section at the back of the book.

based on melting temperature (T_m), the temperature at which the binding forces between two strands of DNA become so weak that the duplex becomes separated. Most TaqMan® probes are optimally designed with a T_m of 70°C. However, probe/template duplices with a T_m of 70°C are difficult to design within an A-T rich region. Lower T_m results in *Taq* polymerase pushing the probes from the template strand instead of cleaving the 5' end of the probes. The lower fluorescence yield leads to a higher variation of the real-time PCR data (Livak, 1995 and 1999; McGuigan, 2002).

One technique to resolve regions with high A-T content is the use of minor groove binder (MGB) probes that have been developed for difficult target sequences (Afonina, 1997; Meyer, 2000). In addition to a non-fluorescent 'dark' quencher at the 3' -end, an MBG ligand hyperstabilizes the probe/template duplex by binding to its minor groove. Developed for analysis of SNPs, MGB probes have been used in various applications. Because of their improved binding for templates with low T_m values, MGB probes can potentially be used in both quantification and mutation detection (Kutyavin, 2000; Louis, 2004).

To increase the probability of generating a successful assay design, it is important to mask all known SNPs and other polymorphisms within the region of the target SNP, as well as any ambiguities in the content sequences. This ensures that the primers and probes will be designed to the sequence region most conserved between individuals, and that the oligonucleotides will hybridize efficiently to the intended regions that surround the targeted SNP (Reich, 2003).

Applications

TaqMan® probes have been designed for a variety of applications in genetic studies in mutation detection, pathogen detection, quantification analysis for viral load, oncology, and pharmacogenomics. Since the technique was first described in 1997, more than three hundred publications demonstrate the uses of TaqMan® technology in molecular diagnosis.

Genotyping by TaqMan® PCR for the factor V Leiden mutation, prothrombin (factor II) 20210G>A mutation and the thermolabile methylenetetrahydrofolate reductase (MTHFR) 677C>T mutation has been widely used in the molecular diagnostic laboratory (Happich, 2000; Louis, 2004). An example of genotyping the prothrombin 20210G>A mutation is shown in Fig. 11.2. Real-time detection of amplification products in a closed tube format combines the advantage of a minimized risk of contamination with the advantage of a rapid and automatable method that allows high throughput of samples.

TaqMan® technology has been used for real-time PCR quantification of mutant alleles in mitochondrial (mt) DNA. More than 50 mtDNA mutations associated with human dis-

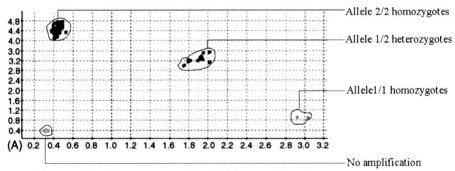

x axis: VIC dye normalized fluorescence, y axis: FAM dye normalized fluorescence

Figure 11.2 TaqMan® assay allelic discrimination scatter graph for SNP detection (using prothrombin mutation 20210G>A as an example). DNA samples with increased FAM or VIC endpoint fluorescence type as homozygous wild type (20210G) or mutant (20210A) respectively; heterozygous appear as an intermediate group with increased fluorescence for both FAM and VIC dye fluorochromes. The no DNA template controls (NTCs) show no amplification.

ease have been described. The majority are single nucleotide changes. A common characteristic of mitochondrial mutations, such as 8344A>G for myoclonus epilepsy and ragged-red fibres (MERRF), is the presence of a mixture of mutant and wild-type mtDNA, a phenomenon referred to as heteroplasmy. Minor allele proportions down to 9% were unambiguously detected and quantified (Bai, 2004).

Infantile spinal muscular atrophy (SMA) is a neuromuscular disease caused by homozygosily dora deletion in the *SMN1* gene in more than 90% of patients. Identification of carriers for the *SMN1* deletion is important for diagnosis and for genetic counselling. However, the duplication of the SMA locus (*SMN1* and *SMN2*) makes the detection of SMA carriers difficult. A quantitative test with TaqMan® technology, using minor groove binder probes, has been developed to distinguish the *SMN1* gene from the *SMN2* gene based on single nucleotide variations between the two genes. In addition, quantitative analysis for the copy number of the *SMN1* gene reliably detects SMA carriers (Anhuf, 2003).

Real-time PCR SNP detection for cytochrome P_{450} CYP2C9 has been used to monitor bone marrow engraftment of donor marrow into a recipient and to detect early disease relapse (Oliver, 2000). It is comparable to restriction fragment length polymorphism (RFLP) and short tandem repeat (STR) methods. However, further investigation of the analytical sensitivity is needed to determine the detection limits of TaqMan® probes in mixed genotypes.

Finally, another important application for SNP genotyping technologies is in pharmacogenetics. Drug metabolism related SNPs are now utilized in pre-clinical and clinical phases of the drug discovery process and are incorporated into clinical practice to predict an individual's response to therapies (Shi, 1999; De la Vega, 2005; Halling, 2005).

Hybridization probes

Molecular beacons
The molecular beacon is a novel amplicon detection technology developed by Tyagi and Kramer in 1996. It is a single-stranded oligonucleotide that possesses a stem-and-loop structure. The central loop portion of a molecular beacon can report the presence of a specific complementary sequence. The stem portions at either end have five to seven base pairs that are complementary. A fluorophore is linked to one end, and a quencher is linked to the other. When the beacon is free in solution (i.e. not hybridized), it forms a stem-loop structure. The stem keeps the fluorophore and quencher in close proximity, causing the fluorescence to be quenched by energy transfer. However, when the oligonucleotide sequence in the hairpin loop hybridizes to its target and forms a rigid double helix, a conformational change occurs that removes the quencher from the vicinity of the fluorophore, thereby permitting fluorescence (Tyagi, 1996, 1998; Schofield, 1997). This feature makes molecular beacon technology ideal for allele discrimination assays, even when the alleles differ by a single nucleotide (Giesendorf, 1998; Marras, 1999). Since amplification and amplicon detection are performed simultaneously in a single closed tube, the risk of laboratory contamination is minimal. Moreover, the addition of molecular beacons to PCR amplification makes real-time monitoring of amplification possible (Marras, 2006; Tan, 2005).

Probe design
For real-time mutation detection, the loop region of molecular beacons is designed to contain 18–25 nucleotides with a melting temperature slightly above the annealing temperature of the PCR primers. This causes the probe–target hybrid to be stable during annealing when the signal is detected (Tyagi, 1996, 1998). The arm sequences (usually 5–7 nucleotides) are then designed to allow the stem to dissociate about 7–10°C above the detection temperature. For example, if the annealing and detection temperature of a reaction is 55°C, then the T_m of the loop region should be approximately 57°C and the T_m of the stem region should be between 62°C and 65°C (Kostrikis, 1998; Sokol, 1998; Ortiz, 1998; Dubertret, 2001).

The melting temperature of the loop sequence of a molecular beacon can be calculated by a variety of primer design programs that use the 'GC' per cent rule (for example: www-genome. wi.mit.edu/cgibin/primer/primer3.cgi). When designing a molecular beacon probe, the SNP position should be near the centre of the loop.

Applications

The fully automated analysis for the 677C>T mutation in the methylenetetrahydrofolate reductase (*MTHFR*) gene was the first described application of molecular beacons for single-nucleotide mutation detection (Giesendorf, 1998). This closed tube method is robust and can test 48 samples within approximately two hours. It is now widely implemented in molecular diagnostic laboratories.

Molecular beacons have also been used to quantitatively detect single base mutations in mitochondrial DNA (mtDNA) in real-time conditions (Szuhai, 2001). In addition to detecting the percent of heteroplasmy, a simultaneous determination of the mtDNA copy number is provided.

The target-specific capability and the availability of different fluorophore-quencher pairs make these probes extremely adaptable for multiplex applications. Molecular beacon technology has advanced to include multicolour systems, surface-immobilization, and a molecular beacon microarray (Hrdlicka, 2005; Wang, 2005; Du, 2005). These advances may significantly contribute to the real-time testing for SNPs and the typing of infectious disease strains (Frutos, 2002; Orru, 2005).

FRET probes (LightCycler™)

Hybridization probes for real-time mutation detection were first described using fluorescence resonance energy transfer (FRET). In FRET, a fluorophore acts as a donor, and is excited at a specific wavelength. When the donor fluorophore is in close proximity to an acceptor fluorophore, the energy is transferred to the acceptor fluorophore, which then emits energy at a higher wavelength. FRET probes were first used with a primer internally labelled with Cy5 (acceptor fluorophore), and a probe spanning the factor V Leiden mutation that was labelled on the 3′ end with the donor fluorophore, fluorescein (Lay, 1997). As the primer extends during PCR, the amplicon incorporates the fluorescein. The probe will hybridize at low temperatures, bringing the fluorophores close together. FRET

then occurs, and the fluorescent signal is detected by an instrument such as the LightCycler™ (Roche Diagnostics). After PCR, amplicons are denatured at 95°C and then cooled to allow probe hybridization. The temperature is raised slowly (0.1–0.3°C/s) and continually monitored to see the probe dissociate from the template dynamically. As the temperature increases, the probe denatures or 'melts' from the amplicon at a characteristic melting temperature (T_m), resulting in a loss of fluorescence. A mutation causes a base mismatch underneath the probe, which will destabilize the probe. Consequently, the probe with a mismatch will melt at a lower temperature. The negative derivative of the melting curve allows easy visualization of both wild-type and mutant alleles. The T_m values of both alleles can be determined. Multiplexed assays are possible either by temperature or by fluorophore. For multiplexed fluorophore assays, a colour compensation file must be created to correct for bleed-through between channels.

To improve the fluorescent signal, an asymmetric PCR can be used. Asymmetric PCR generates less product, because it is no longer exponential and requires more cycles. However, it will preferentially amplify the strand complementary to the probe, and thereby reduce the competition of hybridization with the complementary template strand.

Probe design

The use of dual probes allows end-labelling of both donor and acceptor fluorophores. In this design, one probe of about 20 nucleotides covers the area of the mutation and is referred to as the reporter probe. The other is complementary to approximately 25–30 bp of an adjacent sequence, and is designated the anchor probe because it is designed to melt later than the reporter probe. Probes are blocked at the 3′ end with either a fluorophore or a phosphate to prevent extension from the probe during PCR. The fluorophores are often fluorescein and LC640 or LC705, allowing detection in either of two channels in the LightCycler™ (Bernard, 1998). Later instruments such as the LightCycler 480 (Roche Applied Sciences) expanded the number of channels and the number of samples to allow greater multiplex capabilities and greater throughput. Fig. 11.3

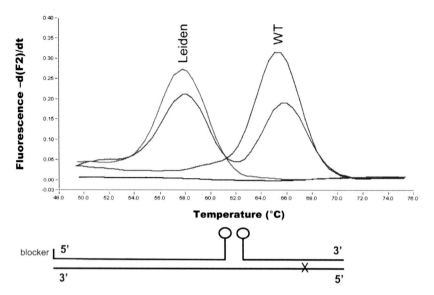

Figure 11.3 Derivative melting curves for factor V Leiden genotyping using FRET labelled dual hybridization probes. After PCR amplification, samples are denatured slowly with continual fluorescent monitoring. The wild-type (WT) and mutant alleles (Leiden, 1691G>A) are shown in the graph. Below is a diagram of dual FRET probes, each with a single fluorophore label. The 'x' indicates the mutation mismatch with the WT probe. The circles indicate fluorophores.

shows a typical melting profile for the factor V Leiden mutation.

As hybridization probes and real-time PCR have evolved, efforts to simplify the chemistry have resulted in the single probe design (Crockett, 2001). This design takes advantage of an inherent quenching ability of guanine bases. The probe is designed with an end-labelled fluorescein that is placed adjacent to a guanine on the opposite strand. When the probe is hybridized, the fluorophore is quenched with a loss of fluorescence. Thus, melting will show a gain of fluorescence and an inverse derivative curve (Fig. 11.4). Alternatively, a single labelled probe able to fluoresce when bound to a target (SimpleProbe™, Roche) can be used. To simplify further, an unlabeled probe can be used with a double-stranded DNA binding dye such as LC® green (Zhou, 2004). The melting curves of unlabelled probes are shown in Fig. 11.5.

Probe stability is dependent on several factors, such as G–C content, probe length, nearest neighbour effects, and number of mismatches (Allawi, 1998a,b; Santa-Lucia, 1996). Since specific base mismatches destabilize to different extents, probes can be designed to maximize T_m differences by the mismatch chosen. For example, a C–T mismatch is less stable than a

G–T mismatch. To optimize the T_m difference between the wild-type and mutant alleles, the probe can be designed to be complementary to either the sense or anti-sense strand. It can be a perfect match to the wild-type, or a perfect match to the mutation. By designing the probe as a mutation-matched probe, false positives due to other mutations with similar T_m values will be avoided. The wild-type will have a single mismatch, and any other variant most likely will have two mismatches with the probe, one at the target mutation position and one at the additional variant position. The optimal T_m difference between the two alleles ranges from 3 to 10°C. To increase the T_m difference, probes can be shortened, or a second mismatch or a base analogue can be incorporated (Chou, 2005b).

Applications
Hybridization probes have been used for detection of single nucleotide variants such as the factor V Leiden or prothrombin 20210G>A mutations. Multiplexing using either the fluorophore or T_m approaches can increase the number of mutations detected. An example of T_m multiplexing is simultaneous detection of the 677C>T and 1298A>C alleles in the *MTHFR* gene (Bernard, 1998). Multiplexing by fluorophores

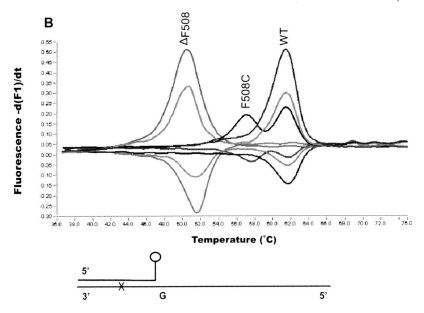

Figure 11.4 Derivative melting curves for F508del and F508C variants in CFTR exon 10, using one probe with a single label. The diagram of this probe design is shown below. The fluorescein dequenching probe (non-inverted curve; SimpleProbe™, Roche) was compared to the fluorescein quenching probe (inverted curve). Courtesy of Dr C.T. Wittwer. The 'x' indicates the mismatch with the WT probe. The circle indicates the fluorophore.

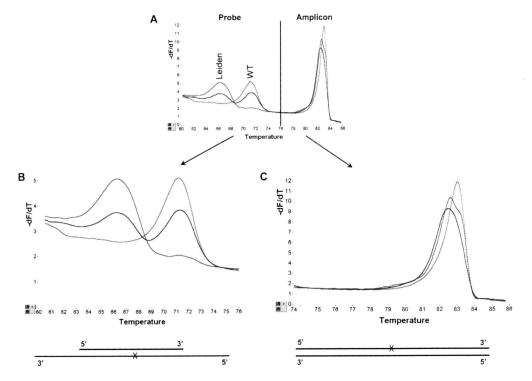

Figure 11.5 High-resolution melting analysis of an unlabelled probe and amplicon using an intercalating dye. (A) The melting range from 60°C to 90°C showing both probe and amplicon melts. (B) An enlargement of the probe melt used for genotyping. (C) An enlargement of the amplicon melt. Courtesy of Dr L.-S. Chou. The diagram below shows probe and amplicon hybridization with 'x' indicating potential mismatches. A colour version of this figure is located in the plate section at the back of the book.

is demonstrated by detection of mutations in the *HFE* gene, with the C282Y mutation detected in the LCRed 640 channel and H63D and S65C in the LCRed 705 channel (Phillips, 2000). Small deletions such as the common F508del mutation in the cystic fibrosis transmembrane regulator (*CFTR*) gene are easily detected. This assay may be used to comply with the American College of Medical Genetics recommendation to confirm that other sequence variants (e.g. F508C, I507V and I506V) do not interfere with the assay (Grody, 2001). A combination of a probe matched to the wild-type sequence and a probe matched to the F508del mutation will differentiate all variants, although either probe alone will not be able to do so (Gundry, 1999; Lyon, 2004).

An advantage of monitoring a temperature range rather than temperature endpoints is the ability to detect and distinguish other sequence variants within the region of the reporter probe. A shift in T_m can indicate a different mutation than the target mutation. At times, the probes may be designed to distinguish two or more variants, such as the previous cystic fibrosis example or the Apolipoprotein B 3500 mutations (von Ahsen, 1999; Aslanidis, 1999). New variants have also been identified through T_m shifts (Lyon, 2005). Examples of these include identifying a factor II C>T mutation at position 20209, one base pair away from the 20210G>A mutation associated with thrombophilia. This T_m shift was obvious visually, but other base alterations may be subtle, such as two described close to the factor V Leiden mutation (Lyon, 1998). Using the T_m difference (ΔT_m) between the wild-type and mutant allele can identify subtle changes better than T_m alone.

Combining mutation detection and quantification is possible by melting curve analysis. In this system, allele areas or heights are used to determine copy number. This has been described for a competitive PCR, using an artificial oligonucleotide with a single base change as the competitor (Lyon, 2001; Millson, 2003). The advantage to this technique is that the competitor and the target have equal PCR efficiencies. This is most useful when a limited dynamic range is required, such as looking for gene deletions or duplications (Millson, 2004; Ruiz-Ponte, 2000).

Primer systems

Allele-specific PCR

An allele-specific PCR approach such as the amplification refractory mutation system (ARMS) (Newton, 1989) may be considered in instances where it proves difficult to achieve specific hybridization with a probe due to sequence composition. With ARMS, the different alleles of a variant are amplified separately by PCR using primers specific for each allele. Genotype is assigned based on the presence or absence of amplification from an allele.

Primer design
In a traditional ARMS assay, each allele of the variant to be tested is placed as the 3′ nucleotide of a separate forward PCR primer. A common reverse primer amplifies a product with each of the allele-specific primers. The reaction for each allele is performed in a separate tube using a DNA polymerase with no proofreading capability, so as not to eliminate a 3′ mismatch. Different nucleotide mismatches have different strengths, so it may be necessary to add additional mismatches in the primers to improve their specificities (Newton, 1989). Separate control PCR products are commonly multiplexed in each reaction to act as amplification controls.

There are difficulties in adapting an ARMS assay to real-time PCR. In addition to requiring two reactions for each variant, real-time PCR does not lend itself to multiplexed amplification controls. In a standard real-time ARMS assay, a generic dsDNA dye such as SYBR® Green I is utilized to follow the reactions. Such dyes are inexpensive, but they lack the ability to differentiate specific PCR products from artefacts. However, a post-amplification melt can determine if a product has the correct T_m. Threshold cycle (C_t) values from the reactions are used to score alleles. Assays can be made quantitative if a particular mutation is expected at variable levels in a sample.

Techniques have been reported that improve the specificity of ARMS assays, including a touchdown thermocycling program (Wu, 2005). Methods that combine both allele-specific reactions in one tube are also very desirable (Rust, 1993; Iwasaki, 2003). In addition to increased simplicity and lower reagent costs, the allele-specific products act as amplification controls for each other, and the competitive PCR improves reaction specificity. However, in a real-time assay, it is difficult to differentiate the mixed products. One reported solution is to incorporate a T_m difference into the allele-specific primers that allows the products to be differentiated by a post-amplification melt (Donohoe, 2000). This is accomplished by making one primer substantially longer than the other, in part through a 5′ GC-rich extension. It is necessary to add additional 5′ primer mismatches to prevent the shorter PCR product from using the longer product as a template. A second reported solution is to utilize allele-specific Scorpion primers with different fluorescent labels (Whitcombe, 1999). This method is discussed in more detail in the following section.

Applications
In addition to testing individual variants in patient samples (Donohoe, 2000), real-time allele-specific PCR has been utilized for genotyping SNPs in pooled DNA samples (Germer, 2000), testing for heteroplasmic mutations in mitochondrial DNA (Bai, 2004), and testing for K-*ras* mutations in tumours (van Heek, 2005).

Scorpion primers
A Scorpion primer system has a fluorophore-labelled probe that is linked to one primer of a PCR amplicon. Upon extension of the PCR product, a target sequence complementary to the probe is created within the amplicon. As the probe binds to this intramolecular target, a quencher group is displaced from the probe's fluorophore with a resultant increase in signal.

Primer design
Scorpion probes are of two basic designs. Both types attach to the 5′ end of a PCR primer through a compound that inhibits PCR extension into the probe. In the first design, the probe consists of a stem-loop structure with a quencher and fluorophore similar to a Molecular Beacon (Whitcombe, 1999). In the second design, the fluorophore-containing probe base-pairs with a complementary oligonucleotide that contains the quencher (Solinas, 2001). In both cases, exten-

sion of the PCR product creates an intramolecular target for the probe. Binding to the target causes the quencher to be displaced from the fluorophore and results in an increased signal.

Applications
Scorpion primers are advantageous because their intramolecular hybridization is specific and rapid, which allows for fast thermocycling. They have been utilized to genotype DNA sequence variants in two basic methods.

First, the intramolecular binding of the probe allows Scorpion primers to be used for real-time allele-specific PCR in which both allele-specific reactions for a variant are multiplexed in a single tube. Each allele-specific Scorpion primer is labelled with a different fluorophore (Whitcombe, 1999). This method is described in more detail in the preceding section.

Second, T_m differences between perfectly matched and mismatched probes can be utilized to obtain the genotype of a variant. The fluorescent signal is monitored at a temperature where the probe dissociates from a target with a mismatch and returns to a quenched state, but remains bound to a matched target. Ideally, probes specific for each allele of a variant are labelled with different fluorophores and multiplexed together. If necessary, FRET excitation can be utilized to take advantage of available channels in a system (Thelwell, 2000; Solinas, 2001).

High-resolution mutation scanning
Melting curve analysis can detect mutations by monitoring with double-stranded DNA dyes rather than with hybridization probes. Heterozygous, homozygous and wild-type samples can be differentiated using small amplicons and high resolution melting (Wittwer, 2003). This technique can be expanded to mutation scanning. In this application, longer amplicons are used to scan the entire coding region of a gene. Increasing the amplicon size can reduce the resolution between a wild-type and a homozygous sample, but it is useful to detect heteroduplexes formed from heterozygous samples. Heteroduplexes are easily detected by melting curves, and have been described for cystic fibrosis (Chou, 2005a) and for the *RET* proto-oncogene (Margraff, 2006).

Combinations of high resolution amplicon melt and unlabelled probes have also been described for high complexity genotyping of human platelet antigen and cystic fibrosis (Liew, 2006; Zhou, 2005). Fig. 11.5a–c show mutation detection by amplicon melt.

Sensitivity and specificity

The sensitivity required for an assay is dependent on the rarity of the target molecule in the sample. In many situations, detection limits approaching the single molecule level are desirable. For example, in the field of oncology a mutation may be present in a single cell in a background of thousands of 'normal' cells. In virology, only one cell out of thousands may be infected. Detection limits for mixed populations vary between SNPs but in one study were found to range between 1.6% and 25% (Tapp, 2000).

For assays requiring very high sensitivity, sensitivity can be compromised by several factors, including the presence of competitors and inhibitors in the sample and interference of the background fluorescence of the system. The major source of background fluorescence arises from the incomplete quenching of the probes in their native state. Background fluorescence also arises from a combination of the physical detection process, the non-specific signals from undesired sample components, and the signal from specific contaminants or competitors.

Assay specificity can be defined as the ability to detect a particular analyte in a complex mixture without interference from other components in the mixture. With real-time PCR assays, specificity can be seriously compromised if suboptimal assay conditions are used.

The ability to discriminate single base differences is an important feature of many assays. The length of the probes, position of mismatch within the probes, fluorescence acquisition temperature, and magnesium chloride concentration of the buffer are all factors that can affect the ability of the assay to detect single base changes.

Conclusion

Mutation detection by real-time PCR includes both amplification and melting techniques. A variety of designs are available, with allele specificity associated with either primers or probes. Specific design parameters have been described in this chapter, along with published applications for these designs. Mutation detection by these methods is established in clinical diagnostic testing. The best model for any laboratory depends on the instrumentation available and compatibility with the fluorophores. Advantages of melting analysis are mutation discovery of novel or unreported mutations, as well as targeting a specific mutation for genotyping.

References

Afonina, I., Zivarts, M., Kutyavin, I., Lukhtanov, E., Gamper, H. and Meyer, R.B. (1997). Efficient priming of PCR with short oligonucleotides conjugated to a minor groove binder. Nucleic Acids Res. 25, 2657–2660.

Allawi, H.T. and SantaLucia, J. Jr, 1998a. Nearest neighbour thermodynamic parameters for internal G.A mismatches in DNA. Biochemistry 8, 2170–2179.

Allawi, H.T. and SantaLucia, J. Jr, 1998b. Thermodynamics of internal C>T mismatches in DNA. Nucleic Acids Res. 11, 2694–2701.

Anhuf, D., Eggermann, T., Rudnik-Schoneborn, S. and Zerres, K. 2003. Determination of SMN1 and SMN2 copy number using TaqMan technology. Hum. Mutat. 22, 74–78.

Aslanidis, C. and Schmitz, G. 1999. High-speed apolipoprotein E genotyping and apolipoprotein B3500 mutation detection using real-time fluorescence PCR and melting curves. Clin. Chem. 45, 1094–1097.

Bai, R.K. and Wong, L.J. 2004. Detection and quantification of heteroplasmic mutant mitochondrial DNA by real-time amplification refractory mutation system quantitative PCR analysis: a single-step approach. Clin. Chem. 50, 996–1001.

Bernard, P.S., Lay, M.J. and Wittwer, C.T. (1998). Integrated amplification and detection of the C677T point mutation in the methylenetetrahydrofolate reductase gene by fluorescence resonance energy transfer and probe melting curves. Anal. Biochem. 255, 101–107.

Chou, L.S., Lyon, E. and Wittwer, C.T. 2005a. A comparison of high-resolution melting analysis with denaturing high-performance liquid chromatography for mutation scanning: Cystic fibrosis transmembrane conductance regulator gene as a model. Am. J. Clin. Pathol. 124, 330–338.

Chou, L.S., Meadows, C., Wittwer, C.T. and Lyon, E. 2005b. Unlabeled oligonucleotide probes modified with locked nucleic acids for improved mismatch discrimination in genotyping by melting analysis. Biotechniques 39, 644, 646, 648 passim.

Crockett, A.O. and Wittwer, C.T. 2001. Fluorescein-labeled oligonucleotides for real-time PCR using the inherent quenching of deoxyguanosine nucleotides. Anal Biochem. 290, 89–97.

De la Vega, F.M., Lazaruk, K.D., Rhodes, M.D. and Wenz, M.H. 2005. Assessment of two flexible and compatible SNP genotyping platforms: TaqMan

SNP genotyping assays and the SNPlex genotyping system. Mutat Res. *573*,111–135.

Donohoe, G.G., Laaksonen, M., Pulkki, K., Ronnemaa, T. and Kairisto, V. 2000. Rapid single-tube screening of the C282Y hemochromatosis mutation by real-time multiplex allele-specific PCR without fluorescent probes. Clin. Chem. *46*, 1540–1547.

Du, H., Strohsahl, C.M., Camera, J., Miller, B.L. and Krauss, T.D. 2005. Sensitivity and specificity of metal surface-immobilized 'molecular beacon' biosensors. J. Am. Chem. Soc. *127*, 7932–7934.

Dubertret, B., Calame, M. and Libchaber, A.J. 2001. Single-mismatch detection using gold-quenched fluorescent oligonucleotides. Nat. Biotechnol. *19*, 365–370.

Frutos, A.G., Pal, S., Quesada, M. and Lahiri, J. 2002. Method for detection of single-base mismatches using bimolecular beacons. J. Am. Chem. Soc. *124*, 2396–2397.

Germer, S., Holland, M.J. and Higuchi, R. 2000. High-throughput SNP allele-frequency determination in pooled DNA samples by kinetic PCR. Genome Res. *10*, 258–266.

Giesendorf, B.A., Vet, J.A., Tyagi, S., Mensink, E.J., Trijbels, F.J. and Blom, H.J. 1998. Molecular beacons: a new approach for semi-automated mutation analysis. Clin. Chem. *44*, 482–486.

Ginzinger, D.G. 2002. Gene quantification using real-time quantitative PCR: an emerging technology hits the mainstream. Exp. Hematol. *30*, 503-r12.

Grody, W.W., Cutting, G.R., Klinger, K.W., Richards, C.S., Watson, M.S., Desnick, R.J; Subcommittee on Cystic Fibrosis Screening, Accreditation of Genetic Services Committee, American College of Medical Genetics, 2001. Laboratory standards and guidelines for population-based cystic fibrosis carrier screening. Genet, Med. *3*, 149–54.

Gundry, CN, Bernard, PS, Herrmann, MG, Reed, GH, Wittwer, CT. 1999. Rapid F508del and F508C assay using fluorescent hybridization probes. Genet. Test. *3*, 365–370.

Halling, J., Petersen, M.S., Damkier, P., Nielsen, F., Grandjean, P., Weihe, P., Lundgren, S., Lundblad, M.S. and Brosen, K. 2005. Polymorphism of CYP2D6, CYP2C19, CYP2C9 and CYP2C8 in the Faroese population. Eur. J. Clin. Pharmacol. *61*, 491–497.

Happich, D., Madlener, K., Schwaab, R., Hanfland, P. and Potzsch, B. 2000. Application of the TaqMan-PCR for genotyping of the prothrombin G20210A mutation and of the thermolabile methylenetetrahydrofolate reductase mutation. Thromb. Haemost. *84*, 144–145.

Heid, C.A., Stevens, J., Livak, K.J. and Williams, P.M. 1996. Real time quantitative PCR Genome Res. *6*, 986–994.

Hrdlicka, P.J., Babu, B.R., Sorensen, M.D., Harrit, N. and Wengel, J. 2005. Multi-labeled pyrene-functionalized 2'-amino-LNA probes for nucleic acid detection in homogeneous fluorescence assays. J. Am. Chem. Soc. *127*, 13293–13299.

Iwasaki, H., Emi, M., Ezura, Y., Ishida, R., Kajita, M., Kodaira, M., Yoshida, H., Suzuki, T., Hosoi, T.,

Inoue, S., Shiraki, M., Swensen, J. and Orimo, H. 2003. Association of a Trp16Ser variation in the gonadotropin releasing hormone signal peptide with bone mineral density, revealed by SNP-dependent PCR typing. Bone *32*, 185–190.

Kostrikis, L.G., Huang, Y., Moore, J.P., Wolinsky, S.M., Zhang, L., Guo, Y., Deutsch, L., Phair, J., Neumann, A.U. and Ho, D.D. (1998). A chemokine receptor CCR2 allele delays HIV-1 disease progression and is associated with a CCR5 promoter mutation. Nat. Med. *4*, 350–353.

Kutyavin, I.V., Afonina, I.A., Mills, A., Gorn, V.V., Lukhtanov, E.A., Belousov, E.S., Singer, M.J., Walburger, D.K., Lokhov, S.G., Gall, A.A., Dempcy, R. and Reed, MW. 2000. 3'-minor groove binder-DNA probes increase sequence specificity at PCR extension temperatures. Nucleic Acids Res. *28*, 655–661.

Lay, M.J. and Wittwer, C.T. 1997. Real-time fluorescence genotyping of Factor V Leiden during rapid-cycle PCR. Clin. Chem. *43*, 2262–2267.

Liew, M., Nelson, L., Margraf, R., Mitchell, S., Erali, M., Mao, R., Lyon, E. and Wittwer, C. 2006. Genotyping of human platelet antigens 1 to 6 and 15 by high-resolution amplicon melting and conventional hybridization probes. J. Mol. Diag. *8*, 97–104.

Livak, K.J. (1999). Allelic discrimination using fluorogenic probes and the 5' nuclease assay. Genet. Anal. *14*,143–149.

Livak, K.J., Flood, S.J., Marmaro, J., Giusti, W. and Deetz, K. 1995. Oligonucleotides with fluorescent dyes at opposite ends provide a quenched probe system useful for detecting PCR product and nucleic acid hybridization. PCR Methods Appl. *4*, 357–362.

Louis, M., Dekairelle, A.F. and Gala, J.L. 2004. Rapid combined genotyping of factor V, prothrombin and methylenetetrahydrofolate reductase single nucleotide polymorphisms using minor groove binding DNA oligonucleotides (MGB probes) and real-time polymerase chain reaction. Clin. Chem. Lab. Med. *42*, 1364–1369.

Lyon, E., Millson, A., Phan, T. and Wittwer, C.T. 1998. Detection and identification of base alterations within the region of Factor V Leiden by fluorescent melting curves. Mol. Diag. *3*, 203–210.

Lyon, E., Millson A., Lowery, M.C., Woods, R. and Wittwer, C.T. 2001. Quantification of HER2/ neu gene amplification by competitive per using fluorescent melting curve analysis. Clin. Chem. *47*, 844–851.

Lyon, E. 2004. Hybridization probes for real-time PCR. Molecular Analysis and Genome Discovery. Ed: Rapley R., Harbron S., Ch 3, 29–42, John Wiley and Sons Ltd, NJ.

Lyon, E. 2005. Discovering rare variants by use of melting temperature shifts seen in melting curve analysis. Clin Chem. *51*, 1331–1332.

Margraf, R.L., Mao, R., Highsmith W.E., Holtegaard L.M. and Wittwer C.T. 2006. Mutation scanning of the RET protooncogene using high-resolution melting analysis. Clin. Chem. *52*, 138–141.

Marras, S.A., Kramer, F.R. and Tyagi, S. 1999. Multiplex detection of single-nucleotide variations using molecular beacons Genet. Anal. *14*, 151–156.

Marras, S.A., Tyagi, S. and Kramer, F.R. 2006. Real-time assays with molecular beacons and other fluorescent nucleic acid hybridization probes. Clin. Chim. Acta *363*, 48–60.

McGuigan, F.E., Ralston, S.H. 2002. Single nucleotide polymorphism detection: allelic discrimination using TaqMan. Psychiatr. Genet. *12*,133–136.

Meyer, R.B. and Hedgpeth, J. 2000. 3'-minor groove binder-DNA probes increase sequence specificity at PCR extension temperatures. Nucleic Acids Res. *28*, 655–61.

Millson, A., Suli, A., Hartung, L., Kunitake, S., Bennett, A., Nordberg, M.C., Hanna, W., Wittwer, C.T., Seth, A. and Lyon, E. 2003. Comparison of two quantitative polymerase chain reaction methods for detecting HER2/neu amplification. J. Mol. Diagn. *5*, 184–190.

Millson, A., Frank, E.L. and Lyon, E. 2004. Cytochrome P450 2D6 deletion genotyping using derivative curve analysis on the LightCycler. In Rapid Cycle Real-time PCR – Methods and Applications, Quantification (Wittwer C., Hahn M., Kaul P., eds.). Springer-Verlag, Heidelberg, pp. 171–178.

Newton, C.R., Graham, A., Heptinstall, L.E., Powell, S.J., Summers, C., Kalsheker, N., Smith, J.C. and Markham, A.F. (1989). Analysis of any point mutation in DNA. The amplification refractory mutation system (ARMS). Nucleic Acids Res. *17*, 2503–2516.

Oliver, D.H., Thompson, R.E., Griffin, C.A. and Eshleman, J.R. 2000. Use of single nucleotide polymorphisms (SNP) and real-time polymerase chain reaction for bone marrow engraftment analysis. J. Mol. Diagn. *2*, 202–208.

Orru, G., Faa, G., Pillai, S., Pilloni, L., Montaldo, C., Pusceddu, G., Piras, V. and Coni, P. 2005. Rapid PCR real-time genotyping of M-Malton alpha1-antitrypsin deficiency alleles by molecular beacons. Diagn. Mol. Pathol. *14*, 237–242.

Ortiz, E., Estrada, G., Lizardi, P.M. (1998). PNA molecular beacons for rapid detection of PCR amplicons. Mol. Cell. Probes *12*, 219–26.

Phillips, M., Meadows, C.A., Huang, M.Y., Millson, A. and Lyon E. 2000. Simultaneous detection of C282Y and H63D hemochromatosis mutations by dual-color probes. Mol. Diag. *5*, 107–116.

Reich, D.E., Gabriel, S.B. and Altshuler, D. 2003. Quality and completeness of SNP databases. Nat. Genet. *33*, 457–458.

Rust, S., Funke, H. and Assmann, G. 1993. Mutagenically separated PCR (MS-PCR): a highly specific one step procedure for easy mutation detection. Nucleic Acids Res. *21*, 3623–3629.

Ruiz-Ponte, C., Loidi, L., Vega, A., Carracedo, A. and Barros, F. 2000. Rapid real-time fluorescent PCR gene dosage test for the diagnosis of DNA duplications and deletions. Clin. Chem. *46*, 1574–1582.

SantaLucia, J. Jr, Allawi, H.T. and Seneviratne, P.A. 1996. Improved nearest-neighbor parameters for predicting DNA duplex stability. Biochemistry *19*, 3555–3562.

Schofield, P., Pell, A.N. and Krause D.O. 1997. Molecular beacons: trial of a fluorescence-based solu-

tion hybridization technique for ecological studies with ruminal bacteria. Appl. Environ. Microbiol. *63*, 1143–1147.

Shi, M.M., Bleavins, M.R., de la Iglesia, F.A. 1999. Technologies for detecting genetic polymorphisms in pharmacogenomics. Mol. Diagn. *4*, 343–351.

Solinas, A., Brown, L.J., McKeen, C., Mellor, J.M., Nicol, J., Thelwell, N. and Brown, T. 2001. Duplex Scorpion primers in SNP analysis and FRET applications. Nucleic Acids Res. *29*, E96.

Sokol, D.L., Zhang, X., Lu, P., and Gewirtz, A.M. 1998. Real time detection of DNA.RNA hybridization in living cells. Proc. Natl. Acad. Sci. USA *95*, 11538–11543.

Szuhai, K., Ouweland, J., Dirks, R., Lemaitre, M., Truffert, J., Janssen, G., Tanke, H., Holme, E., Maassen, J. and Raap, A. 2001. Simultaneous A8344G heteroplasmy and mitochondrial DNA copy number quantification in myoclonus epilepsy and ragged-red fibers (MERRF) syndrome by a multiplex molecular beacon based real-time fluorescence PCR. Nucleic Acids Res. *29*, E13.

Tan, L., Li, Y., Drake, T.J., Moroz, L., Wang, K., Li, J., Munteanu, A., Chaoyong, J.Y., Martinez, K. and Tan, W. 2005. Molecular beacons for bioanalytical applications. Analyst *130*, 1002–1005.

Thelwell, N., Millington, S., Solinas, A., Booth, J. and Brown, T. 2000. Mode of action and application of Scorpion primers to mutation detection. Nucleic Acids Res. *28*, 3752–3761.

Tapp, I., Malmberg, L., Rennel, E., Wik, M. and Syvanen, A.C. 2000. Homogeneous scoring of single-nucleotide polymorphisms: comparison of the 5'-nuclease TaqMan assay and Molecular Beacon probes. Biotechniques *28*, 732–738.

Tyagi, S. and Kramer, F.R. 1996. Molecular beacons: probes that fluoresce upon hybridization. Nat. Biotechnol. *14*, 303–308.

Tyagi, S., Bratu, D.P. and Kramer, F.R. 1998. Multicolor molecular beacons for allele discrimination. Nat. Biotechnol. *16*, 49–53.

van Heek, N.T., Clayton, S.J., Sturm, P.D., Walker, J., Gouma, D.J., Noorduyn, L.A., Offerhaus, G.J. and Fox, J.C. 2005. Comparison of the novel quantitative ARMS assay and an enriched PCR-ASO assay for K-ras mutations with conventional cytology on endobiliary brush cytology from 312 consecutive extrahepatic biliary stenoses. J. Clin. Pathol. *58*, 1315–1320.

von Ahsen, N., Oellerich, M., Armstrong, V.W. and Schutz, E. 1999. Application of a thermodynamic nearest-neighbor model to estimate nucleic acid stability and optimize probe design: prediction of melting points of multiple mutations of apolipoprotein B-3500 and Factor V with hybridization probe genotyping assay on the LightCycler. Clin Chem. *45*, 2094–2101.

Wang, Y., Wang, H., Gao, L., Liu, H. and Lu, Z. 2005. Polyacrylamide gel film immobilized molecular beacon array for single nucleotide mismatch detection. J. Nanosci. Nanotechnol. *5*, 653–658.

Whitcombe, D., Theaker, J., Guy, S.P., Brown, T. and Little, S. 1999. Detection of PCR products using self-probing amplicons and fluorescence. Nat. Biotechnol. *17*, 804–807.

Wittwer, C.T., Reed, G.H., Gundry, C.N., Vandersteen, J.G. and Pryor, R.J. 2003. High-resolution genotyping by amplicon melting analysis using LCGreen. Clin. Chem. *49*, 853–860.

Wu, W.M., Tsai, H.J., Pang, J.H., Wang, H.S., Hong, H.S. and Lee, Y.S. 2005. Touchdown thermocycling program enables a robust single nucleotide polymorphism typing method based on allele-specific real-time polymerase chain reaction. Anal. Biochem. *339*, 290–296.

Zhou, L., Myers, A.N., Vandersteen, J.G., Wang, L. and Wittwer, C.T. 2004. Closed-tube genotyping with unlabeled oligonucleotide probes and a saturating DNA dye. Clin. Chem. *50*, 1296–1298.

Zhou, L., Wang, L., Palais, R., Pryor, R. and Wittwer, C.T. 2005. High-resolution DNA melting analysis for simultaneous mutation scanning and genotyping in solution. Clin. Chem. *51*, 1770–1777.

Real-Time NASBA

12

Julie D. Fox, Catherine Moore and Diana Westmoreland

Abstract

NASBA is an isothermal nucleic acid amplification method which is particularly suited to detection and quantification of genomic, ribosomal or messenger RNA. The product of NASBA is single-stranded RNA of opposite sense to the original target. First developed NASBA methods relied on liquid or gel-based probe-hybridization for post-amplification detection of products. More recently, real-time procedures incorporating amplification and detection in a single step have been reported and applied to a wide range of RNA and some DNA targets. Thus real-time NASBA has proved to be the basis of sensitive and specific assays for detection, quantification and differentiation of RNA and DNA targets. Molecular beacons have most often been utilized in real-time NASBA whether in commercially available kits or as published in-house developed assays. As experience in design of molecular-beacon probes increases and fluorimeters suitable for real-time NASBA become widely available this methodology will be confirmed as a suitable alternative to real-time RT-PCR (and perhaps DNA PCR).

Introduction and background to the methodology

NASBA technology has provided an alternative method to standard procedures with a broad application for the amplification and detection of a range of nucleic acid targets (Compton, 1991). The majority of applications have been developed for detection and analysis of RNA targets including viral genomes, viroids, ribosomal RNA (rRNA) and messenger RNA (mRNA). Advantages of NASBA over methods such as RT-PCR include fast amplification kinetics and selective amplification of RNA in a background of DNA. The amplification is isothermal and thus there is no requirement for thermocycling during the procedure. Single-stranded RNA amplicons are produced by NASBA which can be used directly in subsequent rounds of amplification or probed for detection without the need for denaturation or strand separation.

Sample preparation for NASBA

Amplification inhibitors and maintaining RNA integrity are the main concerns when preparing clinical samples for NASBA. Degradation of associated DNA is not required if the amplification target is non-spliced RNA as standard NASBA protocols do not utilize temperatures high enough to denature DNA. The most widely used method for extraction is based on that described by Boom et al. (1990) with recent modifications of the technique utilizing magnetic silica described. Both traditional and magnetic silica extraction methodologies have been automated and reagents are commercially available. On addition of the sample to lysis buffer containing the chaotropic agent guanidine isothiocyanate (GuSCN), nucleic acid is rapidly released and stabilized. Nucleases are denatured, thus allowing for long-term storage of clinical material. Both manual and automated extraction methods have been validated for a wide range of clinical and environmental samples with the additional advantage that highly dilute samples can be concentrated during extraction.

Primer and probe design

As for other amplification-based procedures, NASBA requires the selection of two target-specific oligonucleotide regions suitable for use as primers in the amplification phase. A region for probe-specific detection of amplified products also needs to be identified. Design of primers and probes, whether for end-point or real-time detection, follows the same general rules and guidelines as those for other assays using nucleic acid amplification. The difference between NASBA and other amplification methods is that in NASBA the T7 RNA polymerase promoter sequence needs to be included on one of the primers. Design criteria for NASBA primers and probes are discussed in detail below.

Amplification of RNA

The amplification methodology is transcription-based with isothermal, homogeneous amplification analogous to the replication of retroviruses (Compton, 1991; Chan and Fox, 1999). The amplification has a non-cyclic and cyclic phase. In the first stage the RNA template is reverse transcribed followed by a second-strand DNA synthesis and transcription of the resulting double-stranded DNA. The anti-sense RNA produced is then amplified in the cyclic phase of the reaction. For this reaction to occur, avian reverse transcriptase (RT), ribonuclease H (RNase H) and T7 RNA polymerase together with two target-specific primers are required. The process proceeds at 41°C typically over a 90 min time period with the major amplification product being anti-sense single-stranded RNA.

NASBA requires a single melt step at 65°C in order to melt out any secondary structure to allow the target directed primers to anneal. The three enzymes involved in the reaction are not thermostable and therefore are added after the melt step. The subsequent single temperature of 41°C required for the amplification process eliminates the need for thermocycling equipment used in standard molecular amplification techniques.

Amplification of DNA

In a modification of the standard procedure, NASBA can be utilized for detection of DNA targets in which thermocycling is required for the first one or two steps (depending if the target is single- or double-stranded). The first published example of a diagnostic NASBA assay for DNA targets was for hepatitis B (Yates *et al.*, 2001) but others have been designed and validated [e.g. for herpes simplex virus (HSV)].

Detection of amplified products

Methods for the detection of NASBA products have been reported using a probe-capture hybridization and electrochemiluminescence (ECL) (Kievits *et al.*, 1991; van Gemen *et al.*, 1994; Chan and Fox, 1999; Deiman *et al.*, 2002). Prior to the development of real-time NASBA, ECL was most often the NASBA detection method of choice. End point detection methods such as ECL have proved useful when setting up real-time NASBA in order to validate procedures (van Gemen *et al.*, 1994; Lanciotti and Kerst, 2001; Deiman *et al.*, 2002; Fox *et al.*, 2002; Hibbitts *et al.*, 2003; Greene *et al.*, 2003; Moore *et al.*, 2004a) and some ECL detection NASBA assays form the basis of diagnostic kits (Moore *et al.*, 2004a; Ginocchio *et al.*, 2005; Landry *et al.*, 2005; Yao *et al.*, 2005).

Post-amplification detection procedures, such as ECL, require additional handling steps and machinery that affects both cost and time required in order to obtain results. The development of fluorescence-based assays combining amplification and detection for PCR and RT-PCR initiated studies to adapt NASBA methodologies into real-time procedures, enabling product detection concurrent with target amplification in closed tubes, and reducing both the handling steps and the risk of contamination in the post-amplification process.

All reported real-time NASBA procedures utilize molecular beacons for detection of amplified products. Molecular beacons are hairpin shaped oligonucleotides with a loop region containing a probe sequence complementary to the target amplicon and a stem with complementary arm sequences located on either end of the probe sequence (Tyagi and Kramer, 1996; Marras *et al.*, 2006). A fluorophore is covalently linked to one arm (5′ end) and a quencher to the other (3′ end). The quencher is a non-fluorescent chromophore that dissipates the energy it receives from the fluorophore as heat. However, when the probe sequence hybridizes to its target it forms a rigid probe–target double helix that is longer and more stable than the stem hybrid. This conformational change separates the quencher from the fluorophore enabling fluorescence to be detected (by opening the hairpin loop described above). Fluorescence increases concurrently with formation of amplified products. An example of results for NASBA with real-time detection using a single fluorophore is given in Fig. 12.1.

Leone *et al.* published the first study in which NASBA was combined with detection using molecular beacons (Leone *et al.*, 1998). Since then methods have been adapted to utilize real-time NASBA in a diagnostic kit format [NucliSens EasyQ® HIV-1, Enterovirus, RSV A+B, human metapneumovirus (hMPV) and Influenza A H5N1 assays from bioMérieux Ltd] with resulting publications on their clinical utility (e.g. Capaul and Gorgievski-Hrisoho, 2005; de Mendoza *et al.*, 2005; Landry *et al.*, 2005; Yao *et al.*, 2005; Moore *et al.*, 2006). 'Home brew' or 'in-house' assays have also been developed using real-time NASBA (see complete reference list and Tables 12.1 and 12.2 for examples). Some real-time NASBA assays incorporate use of an internal control (IC) which is used for interpretation of valid negative results and assessment of extraction efficiency. An example of interpreta-

tion of a qualitative NASBA result using an internal control is given in Fig. 12.2.

Quantification of NASBA products

Quantitative NASBA with end point detection of amplified products is a well-established technique with internal calibrators of known copy number included in the reaction. Calibrators can be included at the beginning of the procedure ensuring they control all steps of the extraction, amplification and detection. They are generally amplified with the same primers as the wild-type target but with different internal sequences for probe detection. The first commercially available HIV-1 quantitative NASBA assay (end-point ECL detection) incorporated three calibrators that produced a calibration curve for the wild type ECL signal against which to calculate viral load (van Gemen *et al.*, 1994). Assays set up in this way have a broad dynamic range (DR) of approximately five logs and proven clinical utility.

With the development of real time NASBA, the kit-based HIV-1 assay was modified for real-time detection. In the new assay, molecular beacons were used for product detection with just a single calibrator incorporated to quantify target input. Weusten *et al.* (2002) described development of the mathematical model for this assay using two molecular beacons. The beacons com-

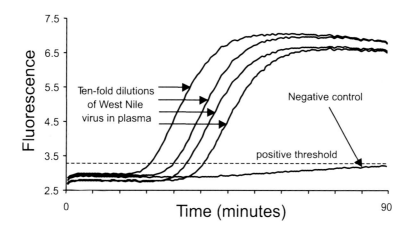

Figure 12.1 Example results for real-time NASBA using a diluted positive control. Results illustrate real-time amplification and detection of diluted West Nile cultured virus spiked into plasma. Raw fluorescence readings (FAM beacon) are given on the Y axis with detection over a 90 minute time period. Primers and molecular beacon are as published previously (Lanciotti and Kerst, 2001; Tilley *et al.*, 2006). Differentiation between positive and negative results is clear and input virus relates closely to time to positivity (TTP).

Table 12.1 Example design characteristics for real-time NASBA

Target (reference)	Primer length (bases)		Molecular beacon (Dabcyl/Dabsyl used as quencher in each case)		
	P1 antisense*	P2 sense**	Target length (bases)	Stem length (bases)	Fluorescent label(s)
HIV-1 (*gag*) (de Baar *et al.*, 2001a)	29	29	18–20	6 (all)	TET, ROX, FAM, TAMRA
HIV-1 (LTR) (de Baar *et al.*, 2001b)	18	20	20	6	FAM
HIV-1 (RT) (de Ronde *et al.*, 2001)	26	20	21	7 (both)	FAM, ROX
CMV IE1 (WT and IC) and *pp67* (Greijer *et al.*, 2001)	17 (*pp67*)	20 (*pp67*)	18 (*pp67*), 19 (IE1 WT and IC)	6 (all)	FAM, ROX, CB
WN virus and SLE virus (envelop gene) (Lanciotti and Kerst, 2001)	23 (both)	22 (both)	22 (WN), 26 (SLE)	6 (WN), 7 (SLE)	FAM
Hepatitis B virus (single-stranded DNA region) (Yates *et al.*, 2001)	28	27	30	6	FAM
HHV8 (multiple targets), U1A control (mRNA for snRNP) (Polstra *et al.*, 2002)	24–28	23,25	22–24	6 (all)	FAM
Bacterial ring rot (16s rRNA) (van Beckhoven *et al.*, 2002)	18	18	20	6	FAM
PVY (coat protein) (Szemes *et al.*, 2002)	21	22	20–22	6 (all)	FAM, Texas red, HEX, TAMRA
PIV1 and PIV3 (HN) (Hibbitts *et al.*, 2003)	21 (both)	21 (both)	21 (PIV1), 20 (PIV3)	7 (both)	FAM
Mycoplasma pneumoniae (16srRNA) (Loens *et al.*, 2003)	22	19	19	6	FAM
Influenza A (N) (Moore *et al.*, 2004b)	20	20	20	6	FAM
Mycobacterium avium subspecies *paratuberculosis* (*dna* A) (Rodriguez-Lazaro *et al.*, 2004)	18 (target), 28 (IC)	18 (target), 36 (IC)	18 (target), 20 (IC)	6 (both)	FAM (target), HEX (IC)
Piscine nodavirus (capsid) (Starkey *et al.*, 2004)	26	23	20	6	FAM

Karenia brevis (rbcL mRNA) (Casper et al., 2004) and IC (Patterson et al., 2005)	25	19	22 (target) 20 (IC)	7 (both)	FAM (target), ROX (IC)
Enterovirus (5' non-coding region) and IC (Casper et al., 2005)	18	20	20 (target), 21 (IC)	6	FAM (target), ROX (IC)
Hepatitis A (5' NCR) (Abd el-Galil et al., 2005)	18	28	20	6	FAM
SARS coronavirus (multiple targets) (Keightley et al., 2005)	21–28	20–23	20–25	6–7	FAM
Plasmodium falciparum (18SrRNA)\(Schneider et al., 2005)	27	20	19	6	FAM
RSVA and B (F) Influenza B (polymerase) Rhinoviruses (5' non-coding region) (Lee et al., 2006)	20–22	21–22	20–23	6	FAM
Chlamydophila pneumoniae (16srRNA) Loens et al., 2006)	25	23	25	7	FAM

Example publications have been selected for this table to reflect the range of samples and target types reported. Only published methodologies are included. Other assay details for unpublished methodologies are available at the Basic Kit Application database, access to which is via bioMérieux Ltd.

*P1 primer has the T7 RNA polymerase promoter sequence added to the target-specific sequence

**In some procedures P2 primer has a generic tail sequence used for end-point detection (e.g. by ECL using the NucliSens® Basic Kit).

LTR = long terminal repeat, HIV = human immunodeficiency virus, CMV = cytomegalovirus, IE = immediate early, IC = internal control, WN = West Nile, SLE = St Louis encephalitis, HN = haemagglutinin neuraminidase, HHV8 = human herpesvirus 8, N = nucleoprotein, ORF = open reading frame, ORF 73 = latency-associated nuclear antigen (LANA), vGCR = viral receptor, vBc1–2 = viral inhibitor of apoptosis, vIL–6 = viral growth factor, PVY = potato virus Y, PIV = parainfluenza virus, FAM = fluorescein, Dabcyl/Dabsyl = Dabcyl-200 or 4 (4'-dimethylaminophenylazo) benzoid acid, ROX= 6-carboxy-X-rhodamine, CB = Cascade blue, TET = tetrachloro-6-carboxyfluorescein, TAMRA = tetramethylrhodamine, ss = single-stranded, snRNP = small nuclear ribonucleoprotein, cfu/ccu = colony forming/counted units, HEX = hexachloro derivative of fluorescein, RSV = respiratory syncytial virus, F = fusion protein.

Examples are in date order with most recent at the end.

Table 12.2 Summary of performance characteristics for real-time NASBA

Reference (for target gene see Table 12.1)	Sensitivity/specificity	Multiplexing and typing	Quantification
de Baar et al. (2001a)	Detection limit 1000 copies input. 92% sensitivity on clinical panel	HIV-1 type-specific assays developed (A, B, C and recombinants)	Qualitative assay but linear relationship between input and TTP reported
de Baar et al. (2001b)	Detection limit 10 copies input	Assay not type specific	TTP used to determine target amounts. DR 10^2–10^7 for M, N and O types
de Ronde et al. (2001)	Not quoted	Assay developed to identify RT variants.	DR good. Mixtures were identified efficiently with variants present at 1–4% detectable
Greijer et al. (2001)	Detection limit 1 –3 × 10^3 molecules/100 μl blood. Specificity equivalent to monoplex (single target) assays.	3 label multiplex undertaken successfully	DR of 10^3–10^6 RNA copies
Lanciotti and Kerst, (2001); further clinical study undertaken (Tilley et al., 2006)	Detection limit 0.10 pfu input for WN virus and 0.15 pfu input for SLE virus. No cross-reaction with range of virus stocks. Clinical sensitivity excellent for WN virus cases	Not applicable	Qualitative assay
Yates et al. (2001)	Detection limit 10 input copies of DNA target	Not applicable	Broad DR from 10^1 to 10^7
Polstra et al. (2002)	Detection limit 50 copies input for each HHV8 target and 1000 copies input copies for U1A	Multiplex or typing assays not performed	TTP used to calculate copy Accurate quantification for HHV8 targets (100 to 1 × 10^7 copies input). Quantification of U1A over 10^3–10^8 copies input
Szemes et al. (2002)	Detection limit 10–100 copies of RNA (100 copies in presence of all four beacons)	Tuber necrotic variants differentiated using four probes	Qualitative assay
Hibbitts and Fox (2002); further clinical study undertaken (Lee et al., 2006)	Detection limit for each assay ≤ 1 $TCID_{50}$ or 100 RNA copies input. No cross-reaction with range of virus stocks.	Multiplex PIV1/PIV3 assay performed with no compromise in sensitivity and specificity.	Qualitative assays but linear relationship between input and TTP reported.
Loens et al. (2003)	Specificity compared with alternative assays excellent. 95% hit-rate of 148 RNA copies in the assay with limit of detection for cultured bacteria of 5 CCU. Clinical sensitivity comparable to NASBA with ECL detection	Multiplexing or typing assays not performed	Qualitative assay
Moore et al. (2004b); further clinical study undertaken (Lee et al., 2006)	Detection limit < 0.1 $TCID_{50}$ virus stock	Multiplexing or typing assays not performed	Qualitative assay

Example publication	Sensitivity/detection limit	Multiplexing	Quantification
Rodriguez-Lazaro et al. (2004)	Detection limit 150–200 cells from culture or 10^3 cells/20mls contaminated drinking water or 10^4 cells/20 ml contaminated milk	Two beacons used successfully	Qualitative assay
Starkey et al., 2004	Detection limit between 1 and 0.1 $TCID_{50}$ or between 10^5 and 10^4 synthetic transcript RNA copies	Multiplexing or typing assays not performed	Qualitative assay
Casper et al. (2004) and Patterson et al. (2005)	Original assay confirmed detection of one K. brevis cell or 1.0 fg of synthetic RNA with no specificity problems. Assay developed to determine precision of detection for unknown low target copies of bacteria	Multiplex used for quantification. Two beacons detected successfully	Negative logarithmic relationship between initial cell number and TTP. Quantification and improved assay precision based on ratio of TTP of target WT and IC products described in follow-up paper
Casper et al. (2005)	Wild-type and IC assays sensitive to 10 viral particles or 10 copies of RNA in the assay. Specificity for human enteroviruses good	Two beacons used successfully	TTP used to generate a standard curves with use of the IC method of analysis increasing precision (see Patterson et al., 2005)
Abd el-Galil et al. (2005)	Detection limit 1 pfu/100 μl cell culture fluid or 10 pfu spiked into concentrated lake water	Multiplexing or typing assays not performed	TTP against pfu used to produce a standard curve. Quantification performed in lake water 1–1000 pfu
Keightley et al. (2005)	Detection limit 10^{-6} to 10^{-7} transcript copies/μl RNA for pol and N gene targets. L and NCR were less sensitive	Multiplexing or typing assays not performed	Qualitative assay
Schneider et al. (2005)	Detection limit 20 parasites/ml blood	Multiplexing or typing assays not performed	TTP against ring stage parasites or gametocytes used to produce standard curves
Lee et al. (2006)	Assays more sensitive than conventional procedures and provided information relevant to patient management in 15% more cases than culture/DFA. Rhinovirus assay picked up enteroviruses when in high copy number. Enterovirus-specific NASBA used to resolve these. No other specificity issues noted.	Two target multiplexing undertaken successfully (PIV1/3, PIV2/4, influenza A/B, RSV A/B)	Qualitative assays
Loens et al. (2006)	Sensitivity 0.1–1 IFU/100 μl of sample or 10 copies of input target. Sensitivity of real-time assay equivalent to conventional NASBA (ECL detection). Assay did not pick up a range of related organisms and so specificity was good	None	Qualitative assay

Example publications have been selected for this table to reflect the range of samples and target types. Examples are in date order with most recent at the end.
$TCID_{50}$ = tissue culture infectious dose 50%, pfu =plaque forming units, DR = dynamic range, TTP = time to positivity, IFU = inclusion forming units, ECL = electrochemiluminescence.
Other abbreviations as for Table 12.1.

Wild type strong positive may compete out internal control

Wild type positive with detectable internal control

Wild type negative has detectable internal control

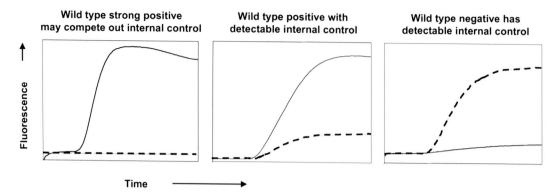

Fluorescence

Time

Figure 12.2 Interpretation of real-time NASBA with an internal control on clinical samples. Typical profiles for a qualitative NASBA incorporating an internal control are given. Results are for the Enterovirus NucliSens® EasyQ assay as described previously (Landry *et al.*, 2005).

prised different loop structures and fluorophores that specifically bound to either sample HIV-1 RNA or calibrator amplicons. The fluorescence curves obtained for wild-type target and calibrator depend upon the NASBA-driven time-dependent growth in RNA levels, and binding of the beacon to this RNA. As the concentration of the calibrator RNA in the reaction is known the levels of the sample HIV-1 RNA can be quantified using the developed mathematical model (Weusten *et al.*, 2002). The performance characteristics for this assay (available in kit format) have been published (de Mendoza *et al.*, 2005; Yao *et al.*, 2005).

The time to positivity (TTP) can also be utilized to estimate levels of nucleic acid target amplified in real-time NASBA. As can be seen in Fig. 12.1 input copy number relates to development of detectable fluorescence. Patterson and colleagues enhanced the precision around RNA quantification by NASBA by comparing the ratio of TTP values for wild-type (WT) and internal control (IC) RNA (Patterson *et al.*, 2005).

Multiplexing of multiple targets in a single reaction

Multiplex real-time NASBA assays have been reported with several molecular beacons present in one reaction. Different fluorophores are measured simultaneously using defined excitation and emission filters. As described above, the commercially available HIV-1 assay utilizes a simple multiplex reaction with molecular beacons specific for wild-type and calibrator amplicons

having different labels (FAM and ROX, respectively). All commercially available qualititative real-time NASBA assays are multiplex in that they incorporate an IC for extraction and amplification based on potato leaf-roll virus. The WT product being detected by a FAM labelled beacon and the IC product detected by a ROX labelled beacon. Other examples of in-house multiplex reactions have been described using both end point (e.g. van Deursen *et al.*, 1999; Fox *et al.*, 2002; Hibbitts *et al.*, 2003; Abd el-Galil *et al.*, 2005; Casper *et al.*, 2005; Ginocchio *et al.*, 2005) and real time (e.g. van Deursen *et al.*, 1999; Lanciotti and Kerst, 2001; de Ronde *et al.*, 2001; Greijer *et al.*, 2001; de Baar *et al.*, 2001a; Szemes *et al.*, 2002; Hibbitts *et al.*, 2003; Casper *et al.*, 2004; Abd el-Galil *et al.*, 2005; Casper *et al.*, 2005; Keightley *et al.*, 2005; Patterson *et al.*, 2005) detection methods. Thus, the feasibility of mixing multiple primer and probe sets together has been shown. As for other multiplexed in-house amplification assays, careful quality control is required to ensure sensitivity and specificity of the resulting assays.

Summary of NASBA results for in-house developed assays

In-house designed assays for RNA targets have proved to be very sensitive and specific (e.g. Chan and Fox, 1999; Lanciotti and Kerst, 2001; Deiman *et al.*, 2002; Fox *et al.*, 2002; Hibbitts and Fox, 2002; Moore *et al.*, 2004b; Tilley *et al.*, 2006; Lee *et al.*, 2006) and in many cases have been used to develop and validate kit-based assays for diagnostic use. Published performance characteristics for in house assays are given in Table 12.2 (selected targets to reflect the range of reported methods).

Methods and protocols for real-time assays

Primer design

A summary of design criteria used in successful applications is given below but further details can be obtained from previous publications (e.g. Deiman et al., 2002) and from a web site dedicated to NASBA applications (access available via bioMérieux Ltd).

The optimum length of amplification sequence has been quoted as approximately 100–250 nucleotides (including target-specific primer regions) although successful published assays have amplified up to 279 bp in real-time NASBA. As for other amplification assays, it is important that primer and probe sequences are checked for any cross-hybridization or internal structure which could compromise the reaction. If possible, it is helpful to avoid stretches of the same nucleotide, to keep the melt-temperature of the primers similar and their GC content close to 50%. The long length of some of the primers utilized in NASBA makes it particularly important that they are purified before use (HPLC or PAGE is usually used). The standard concentration of each primer used in NASBA (end point or real time detection) is 0.2 μM but a few investigators have seen benefit in adjusting the primer concentration, particularly in multiplex reactions (Tables 12.1 and 12.2).

The target-specific region of primer 1 (P1) is usually 20–25 nucleotides although sequences of 17–29 nucleotides have been utilized in real-time NASBA (Table 12.1). The 3′ sequence of P1 complements the target nucleic acid and the 5′ terminal has a T7 RNA polymerase promoter sequence (5′ AATTCTAATACGACTCACTATAGGG). The transcription may be enhanced by an additional purine spacer region of 6–10 nucleotides added between the T7 polymerase promoter tail sequence and the target-specific region (e.g. AGAAGG) as described previously (Deiman et al., 2002).

The second primer (P2) is complementary to the DNA sequence that is produced by extension from primer 1 at the 3′ end. As for P1, the target region for this primer is usually 20–25 nucleotides although variation outside this range is possible (Table 12.1). A generic detection sequence may be used for ECL or other hybridization reactions. This tail sequence is not necessary if only real-time detection is to be used but does not seem to compromise detection using a molecular beacon.

Probe design

The loop sequence of the molecular beacon should be homologous to the region of interest. The stem structure is provided by adding complementary residues to the 5′ and 3′ ends of the loop sequence. If the loop sequence is less than 25 nucleotides a six-nucleotide stem (e.g. 5′-CCAAGC…GCTTGG-3′) is usually used whereas loop sequences longer than 25 nucleotides may have a seven-nucleotide stem (e.g. 5′ CCATGCG….CGCATGG3′). It is important that the probe designed should not form multiple structures. The stability and predicted structure of the beacon can be analysed by using the MFOLD server [http://www.bioinfo.rpi.edu/applications/mfold (Zuker, 2003)]. Details of predicted RNA structures and thermodynamics are available from this web site. In general, a free energy of −3 ± 0.5 kcal/mol for the molecular beacon is recommended (Deiman et al., 2002) although good structures with lower or higher energy values should still be considered. Reported real-time NASBA assays have utilized beacons with a loop varying between 18 and 30 bases. Deliberate mismatches and universal bases (e.g. inosine) can be used to alter probe binding and for discrimination between NASBA products. A wide range of fluorescent labels have been utilized but in each case Dabcyl (or Dabsyl) has been used as a universal quencher. The concentration of molecular beacon in the NASBA assay is generally between 0.5 and 0.05 μM.

Interaction of primers and probe

It has been recognized that different primer sets affected hybridization of the beacon to an RNA amplicon and subsequent amplicon formation. Thus, it is important that probe hybridization does not remove RNA intermediates and therefore reduce further amplification. These factors should be taken into consideration with optimization of primer sets, the length and sequence of the molecular beacon and its final concentration in a reaction.

Nucleic acid extraction

Extraction and preparation of nucleic acid is critical to NASBA assays. Many of the validated procedures in use have benefited from modification of extraction to incorporate a concentrate step which can greatly enhance clinical sensitivity. This is accomplished by extracting from a relatively large volume of sample (up to 2 ml) and then eluting nucleic acid into a small volume of buffer [can be as little as 20–50 µl (Tilley *et al.*, 2006)].

The most widely evaluated procedure for the preparation of nucleic acid for NASBA and other amplifications is that using the NucliSens® extraction kit (bioMérieux Ltd) based on the original method published by Boom *et al.* (1990). More recently, the method has been modified to incorporate the use of magnetic silica in both a semi-automatic system (NucliSens® miniMAG, bioMérieux Ltd) and a fully automated system (easyMAG, bioMérieux Ltd).

As for all nucleic acid amplification procedures, specimen preparation should be carried out in a designated extraction area and gloves should be worn to prevent transfer of nucleases or contaminating nucleic acids (which can lead to false-negative or false-positive results, respectively). For all methods based on 'Boom' extraction the principles and basic steps in sample preparation are essentially the same. The sample must be mixed with the GuSCN containing buffer; for manual extraction the proportion of sample to buffer is 1:9, however, the magnetic silica method utilizes a 2 ml lysis buffer volume into which sample volumes of 0.1–2 ml can be added, thus increasing the potential concentration factor. To the sample/lysis buffer mixture, 50 µl of either type of silica suspension is added. After mixing, the tubes are held for 10 min at room temperature. For the manual methods the tube is microcentrifuged to pellet the silica before the supernatant is carefully withdrawn. The standard wash procedure is undertaken by the

addition of wash buffer followed again by mixing and pelleting by centrifugation. This procedure is repeated four times, once more with wash buffer, twice with 70% ethanol and once with acetone. The tubes are then left open to dry at 56°C before elution in a water based elution buffer.

The use of magnetic silica removes the need for continuous agitation of the silica by the rotation of the tubes in the presence of a magnet (miniMAG) or the changing polarity of the electric magnet (easyMAG). The wash buffers used differ between manual and semi-automated/automated methods in that instead of using ethanol and acetone following the first wash buffer (manual extraction); two borate based buffers of reducing salt concentration are used (automatic extraction). Elution using the miniMAG system is performed at 60°C in a thermo shaker with the tubes on a magnetic rack to remove the eluted nucleic acid. The whole extraction procedure is fully automated on the NucliSens® easyMAG system enabling extraction of nucleic acid from 24 samples in less than 40 minutes.

Real time NASBA using the NucliSens® Amplification Kit

NASBA reactions are carried out according to the general procedure reported previously (Kievits *et al.*, 1991) with some modifications for real-time analysis. The availability of reagents for NASBA in kit format (bioMérieux Ltd) makes the procedure easy to undertake but reagents and enzyme mixes can also be made in-house.

The NucliSens® Basic Kit (bioMérieux Ltd) is used according to the manufacturer's instructions for standard amplification of RNA targets. Some minor changes are needed for amplification from DNA [e.g. (Yates *et al.*, 2001)]. The kit contains the reagent sphere, diluent, stock KCl, nuclease-free water, enzyme mix and enzyme diluent. Optimal KCl concentrations must be determined for each target and are typically between 70–80 mM. NASBA generally works well over a broad KCl concentration which is one parameter that can be adjusted when working up a new procedure. Final concentration of primers is generally 0.05–0.2 µM but can be adjusted. The amplification solution is dispensed into 10 µl/reaction in 1.5 ml microtubes or strips of small tubes/wells depending on the fluorimeter to be used for real-time measurement of amplification. 5 µl of extracted nucleic acid are added to the amplification mix and extracted positive and negative controls are included in each assay. Reaction tubes are incubated at 65°C for 2 min and then held for 2 min at 41°C. Subsequently, 5 µl of NASBA enzyme solution are added to initiate amplification and the NASBA mix immediately transferred to a real-time fluorimeter. A commercially available fluorimeter is available for detection of NASBA products (NucliSens® EasyQ analyser, bioMérieux Ltd) but any fluorimeter capable of holding samples at 41°C may be used for this procedure. Thermocycling is not required for real time NASBA but where real time PCR machines are available these may be adopted for NASBA assays (e.g. use of ABI 7700 as described previously; Lanciotti and Kerst, 2001).

Applications of real-time NASBA

Leone et al. published the first study in which NASBA was combined with detection using molecular beacons (Leone et al., 1998). The target for this assay was potato leaf roll virus with detection sensitivity of 100–1000 synthetic RNA copies. Since then there have been many more reported applications, some published in peer-reviewed journals (Tables 12.1 and 12.2). Although many of the reported methods are set up as qualitative assays the TTP can be used as an indication of RNA load in the sample as shown in Fig. 12.1. Yates et al. (2001) described the quantitative detection of hepatitis B virus DNA by real-time NASBA. Although traditionally NASBA focuses on the amplification of RNA targets, in this study DNA was targeted with the production of RNA amplicons and the assay proved to be both sensitive and have a wide DR. The authors reported that fewer handling steps needed for NASBA resulted in a lower risk of contamination compared with PCR-based assays. A second DNA NASBA assay was more recently described by Deiman et al. (unpublished) for the simultaneous detection of HSV 1 and 2 and an internal control using three fluorophores (FAM, ROX and Cy5). It differed from the Yates method in that an enzyme restriction digest rather than heat denaturation was performed on the DNA prior to amplification.

Other procedures have focussed on development of multiplex reactions or utilization of real-time NASBA for typing and viability testing (Tables 12.1 and 12,2).

Where comparison has been made between different real-time procedures the rapid amplification of RNA by NASBA is apparent. In some cases NASBA has proved to have superior sensitivity and or specificity over real-time RT-PCR (Lanciotti and Kerst, 2001; Moore et al., 2004b; J.D. Fox unpublished;) and should be considered as a suitable alternative amplification procedure.

Conclusions

Real-time NASBA assays are rapid, specific and sensitive with RNA amplification and a target-specific fluorescent signal achieved simultaneously in one tube with measurements obtained through a fluorimeter. Qualitative, quantitative, monoplex and multiplex formats of real-time NASBA have now been described. The methodology seems to be a suitable alternative to other amplification procedures without the need for expensive thermocyclers. Since the original reports of real-time NASBA in 1998 the number of applications, available kits and expansion to include DNA targets is apparent and likely to continue over the next few years.

References

Abd el-Galil, K.H., el Sokkary, M.A., Kheira, S.M., Salazar, A.M., Yates, M.V., Chen, W. and Mulchandani, A. 2005. Real-time nucleic acid sequence-based amplification assay for detection of hepatitis A virus. Appl. Environ. Microbiol. 71, 7113–7116.

Boom, R., Sol, C.J., Salimans, M.M., Jansen, C.L., Wertheim-van Dillen, P.M. and van der, N.J. 1990. Rapid and simple method for purification of nucleic acids. J. Clin. Microbiol. 28, 495–503.

Capaul, S.E. and Gorgievski-Hrisoho, M. 2005. Detection of enterovirus RNA in cerebrospinal fluid (CSF) using NucliSens EasyQ Enterovirus assay. J. Clin. Virol. 32, 236–240.

Casper, E.T., Patterson, S.S., Smith, M.C. and Paul, J.H. 2005. Development and evaluation of a method to detect and quantify enteroviruses using NASBA and internal control RNA (IC-NASBA). J. Virol. Methods 124, 149–155.

Casper, E.T., Paul, J.H., Smith, M.C. and Gray, M. 2004. Detection and quantification of the red tide dinoflagellate Karenia brevis by real-time nucleic acid sequence-based amplification. Appl. Environ. Microbiol. 70, 4727–4732.

Chan, A.B. and Fox, J.D. 1999. NASBA and other transcription-based amplification methods for research and diagnostic microbiology. Rev. Med. Microbiol. 10, 185–196.

Compton, J. 1991. Nucleic acid sequence-based amplification. Nature 350, 91–92.

de Baar, M.P., Timmermans, E.C., Bakker, M., de Rooij, E., van Gemen, B. and Goudsmit, J. 2001a. One-tube real-time isothermal amplification assay to identify and distinguish human immunodeficiency virus type 1 subtypes A, B, and C and circulating recombinant forms AE and AG. J. Clin. Microbiol. 39, 1895–1902.

de Baar, M.P., van Dooren, M.W., de Rooij, E., Bakker, M., van Gemen, B., Goudsmit, J. and de Ronde, A. 2001b. Single rapid real-time monitored isothermal RNA amplification assay for quantification of human immunodeficiency virus type 1 isolates from groups M, N, and O. J. Clin. Microbiol. 39, 1378–1384.

de Mendoza, C., Koppelman, M., Montes, B., Ferre, V., Soriano, V., Cuypers, H., Segondy, M. and Oosterlaken, T. 2005. Multicenter evaluation of the NucliSens EasyQ HIV-1 v1.1 assay for the quantitative detection of HIV-1 RNA in plasma. J. Virol. Methods 127, 54–59.

de Ronde, A., van Dooren, M., van Der, H.L., Bouwhuis, D., de Rooij, E., van Gemen, B., de Boer, R. and Goudsmit, J. 2001. Establishment of new transmissible and drug-sensitive human immunodeficiency virus type 1 wild types due to transmission of nucleoside analogue-resistant virus. J. Virol. 75, 595–602.

Deiman, B., van Aarle, P. and Sillekens, P. 2002. Characteristics and applications of nucleic acid sequence-based amplification (NASBA). Mol. Biotechnol. 20, 163–179.

Fox, J.D., Han, S., Samuelson, A., Zhang, Y., Neale, M.L. and Westmoreland, D. 2002. Development and evaluation of nucleic acid sequence based amplification (NASBA) for diagnosis of enterovirus infections using the NucliSens Basic Kit. J. Clin. Virol. 24, 117–130.

Ginocchio, C.C., Zhang, F., Malhotra, A., Manji, R., Sillekens, P., Foolen, H., Overdyk, M. and Peeters, M. 2005. Development, technical performance, and clinical evaluation of a NucliSens basic kit application for detection of enterovirus RNA in cerebrospinal fluid. J. Clin. Microbiol. 43, 2616–2623.

Greene, S.R., Moe, C.L., Jaykus, L.A., Cronin, M., Grosso, L. and Aarle, P. 2003. Evaluation of the NucliSens Basic Kit assay for detection of Norwalk virus RNA in stool specimens. J. Virol. Methods 108, 123–131.

Greijer, A.E., Verschuuren, E.A., Harmsen, M.C., Dekkers, C.A., Adriaanse, H.M., The, T.H. and Middeldorp, J.M. 2001. Direct quantification of human cytomegalovirus immediate-early and late mRNA levels in blood of lung transplant recipients by competitive nucleic acid sequence-based amplification. J. Clin. Microbiol. 39, 251–259.

Hibbitts, S. and Fox, J.D. 2002. The application of molecular techniques to diagnosis of viral respiratory tract infections. Rev. Med. Microbiol. 13, 177–185.

Hibbitts, S., Rahman, A., John, R., Westmoreland, D. and Fox, J.D. 2003. Development and evaluation of NucliSens basic kit NASBA for diagnosis of parainfluenza virus infection with 'end-point' and 'real-time' detection. J. Virol. Methods 108, 145–155.

Keightley, M.C., Sillekens, P., Schippers, W., Rinaldo, C. and George, K.S. 2005. Real-time NASBA detection of SARS-associated coronavirus and comparison with real-time reverse transcription-PCR. J. Med. Virol. 77, 602–608.

Kievits, T., van Gemen, B., van Strijp, D., Schukkink, R., Dircks, M., Adriaanse, H., Malek, L., Sooknanan, R. and Lens, P. (1991). NASBA isothermal enzymatic in vitro nucleic acid amplification optimized for the diagnosis of HIV-1 infection. J. Virol. Methods 35, 273–286.

Lanciotti, R.S. and Kerst, A.J. 2001. Nucleic acid sequence-based amplification assays for rapid detection of West Nile and St. Louis encephalitis viruses. J. Clin. Microbiol. 39, 4506–4513.

Landry, M.L., Garner, R. and Ferguson, D. 2005. Real-time nucleic acid sequence-based amplification using molecular beacons for detection of enterovirus RNA in clinical specimens. J. Clin. Microbiol. 43, 3136–3139.

Lee, B.E., Robinson, J.L., Khurana, V., Pang, X.L., Preiksaitis, J.K. and Fox, J.D. 2006. Enhanced identification of viral and atypical bacterial pathogens in lower respiratory tract samples with nucleic acid amplification tests. J. Med. Virol. 78, 702–710.

Leone, G., van Schijndel, H., van Gemen, B., Kramer, F.R. and Schoen, C.D. (1998). Molecular beacon probes combined with amplification by NASBA enable homogeneous, real-time detection of RNA. Nucleic Acids Res. 26, 2150–2155.

Loens, K., Beck, T., Goossens, H., Ursi, D., Overdijk, M., Sillekens, P. and Ieven, M. 2006. Development of conventional and real-time nucleic acid sequence-based amplification assays for detection of Chlamydophila pneumoniae in respiratory specimens. J. Clin. Microbiol. 44, 1241–1244.

Loens, K., Ieven, M., Ursi, D., Beck, T., Overdijk, M., Sillekens, P. and Goossens, H. 2003. Detection of Mycoplasma pneumoniae by real-time nucleic acid sequence-based amplification. J. Clin. Microbiol. 41, 4448–4450.

Marras, S.A., Tyagi, S. and Kramer, F.R. 2006. Real-time assays with molecular beacons and other fluorescent nucleic acid hybridization probes. Clin. Chim. Acta 363, 48–60.

Moore, C., Clark, E.M., Gallimore, C.I., Corden, S.A., Gray, J.J. and Westmoreland, D. 2004a. Evaluation of a broadly reactive nucleic acid sequence based amplification assay for the detection of noroviruses in faecal material. J. Clin. Virol. 29, 290–296.

Moore, C., Hibbitts, S., Owen, N., Corden, S.A., Harrison, G., Fox, J., Gelder, C. and Westmoreland, D. 2004b. Development and evaluation of a real-time nucleic acid sequence based amplification assay for rapid detection of influenza A. J. Med. Virol. 74, 619–628.

Moore, C., Valappil, M., Corden, S. and Westmoreland, D. 2006. Enhanced clinical utility of the NucliSens

EasyQ RSV A+B Assay for rapid detection of respiratory syncytial virus in clinical samples. Eur. J. Clin. Microbiol. Infect.. Dis 25, 167–174.

Patterson, S.S., Casper, E.T., Garcia-Rubio, L., Smith, M.C. and Paul, J.H., III 2005. Increased precision of microbial RNA quantification using NASBA with an internal control. J. Microbiol. Methods 60, 343–352.

Polstra, A.M., Goudsmit, J. and Cornelissen, M. 2002. Development of real-time NASBA assays with molecular beacon detection to quantify mRNA coding for HHV-8 lytic and latent genes. BMC. Infect. Dis. 2, 18.

Rodriguez-Lazaro, D., Lloyd, J., Herrewegh, A., Ikonomopoulos, J., D'Agostino, M., Pla, M. and Cook, N. 2004. A molecular beacon-based real-time NASBA assay for detection of *Mycobacterium avium* subsp. *paratuberculosis* in water and milk. FEMS Microbiol. Lett. 237, 119–126.

Schneider, P., Wolters, L., Schoone, G., Schallig, H., Sillekens, P., Hermsen, R. and Sauerwein, R. 2005. Real-time nucleic acid sequence-based amplification is more convenient than real-time PCR for quantification of *Plasmodium falciparum*. J. Clin. Microbiol. 43, 402–405.

Starkey, W.G., Millar, R.M., Jenkins, M.E., Ireland, J.H., Muir, K.F. and Richards, R.H. 2004. Detection of piscine nodaviruses by real-time nucleic acid sequence based amplification (NASBA). Dis. Aquat. Organ 59, 93–100.

Szemes, M., Klerks, M.M., van den Heuvel, J.F. and Schoen, C.D. 2002. Development of a multiplex AmpliDet RNA assay for simultaneous detection and typing of potato virus Y isolates. J. Virol. Methods 100, 83–96.

Tilley, P.A., Fox, J.D., Jayaraman, G.C. and Preiksaitis, J.K. 2006. Nucleic acid testing for West Nile virus RNA in plasma enhances rapid diagnosis of acute infection in symptomatic patients. J. Infect. Dis. 193, 1361–1364.

Tyagi, S. and Kramer, F.R. 1996. Molecular beacons: probes that fluoresce upon hybridization. Nat. Biotechnol. 14, 303–308.

van Beckhoven, J.R., Stead, D.E. and van der Wolf, J.M. 2002. Detection of *Clavibacter michiganensis* subsp. *sepedonicus* by AmpliDet RNA, a new technology based on real time monitoring of NASBA amplicons with a molecular beacon. J. Appl. Microbiol. 93, 840–849.

van Deursen, P.B., Gunther, A.W., van Riel, C.C., van der Eijnden, M.M., Vos, H.L., van Gemen, B., van Strijp, D.A., Tackent, N.M. and Bertina, R.M. 1999. A novel quantitative multiplex NASBA method: application to measuring tissue factor and CD14 mRNA levels in human monocytes. Nucleic Acids Res. 27, e15.

van Gemen, B., van Beuningen, R., Nabbe, A., van Strijp, D., Jurriaans, S., Lens, P. and Kievits, T. 1994. A one-tube quantitative HIV-1 RNA NASBA nucleic acid amplification assay using electrochemiluminescent (ECL) labelled probes. J. Virol. Methods 49, 157–167.

Weusten, J.J., Carpay, W.M., Oosterlaken, T.A., van Zuijlen, M.C. and van de Wiel, P.A. 2002. Principles of quantitation of viral loads using nucleic acid sequence-based amplification in combination with homogeneous detection using molecular beacons. Nucleic Acids Res. 30, e26.

Yao, J., Liu, Z., Ko, L.S., Pan, G. and Jiang, Y. 2005. Quantitative detection of HIV-1 RNA using NucliSens EasyQ HIV-1 assay. J. Virol. Methods 129, 40–46.

Yates, S., Penning, M., Goudsmit, J., Frantzen, I., van de, W.B., van Strijp, D. and van Gemen, B. 2001. Quantitative detection of hepatitis B virus DNA by real-time nucleic acid sequence-based amplification with molecular beacon detection. J. Clin. Microbiol. 39, 3656–3665.

Zuker, M. 2003. Mfold web server for nucleic acid folding and hybridization prediction. Nucleic Acids Res. 31, 3406–3415.

Applications in Clinical Microbiology

13

Andrew David Sails

Abstract

The introduction of real-time PCR assays to the clinical microbiology laboratory has led to significant improvements in the diagnosis of infectious disease. There has been an explosion of interest in this technique since its introduction and several hundred reports have been published describing applications in clinical bacteriology, parasitology and virology. There are few areas of clinical microbiology which remain unaffected by this new method. It has been particularly useful to detect slow growing or difficult to grow infectious agents. However, its greatest impact is probably its use for the quantitation of target organisms in samples. The ability to monitor the PCR reaction in real-time allows accurate quantitation of target sequence over at least six orders of magnitude. The closed-tube format which removes the need for post-amplification manipulation of the PCR products also reduces the likelihood of amplicon carryover to subsequent reactions reducing the risk of false-positives. As more laboratories begin to utilize these methods standardization of assay protocols for use in diagnostic clinical microbiology is needed, plus participation in external quality control schemes is required to ensure quality of testing.

Introduction

The first PCR methods to be described for clinical microbiology utilized gel electrophoresis for the detection of PCR amplification products. Although these assays proved useful, their specificity and sensitivity was compromised by this rather cumbersome end-point detection method.

Specificity of detection could be improved by incorporating a solid phase hybridization such as Southern blotting; however, this was labour intensive and time consuming requiring further manipulation of the PCR product. Detection of PCR products by solid phase hybridization also limited the numbers of samples that could be processed, and the methods used were difficult to standardize between laboratories. The overall time taken to produce a result from a PCR assay could be two or three days and the test required a significant level of technical skill limiting the use of PCR to specialized laboratories only. The introduction of enzyme-linked hybridization probe formats (PCR-ELISA) for the detection of amplification products did improve the detection process; however, they still required manipulation of the amplification products following PCR. Manipulation of the amplified product increases the likelihood of contaminating subsequent PCR reactions leading to false positives a phenomenon known as amplicon carryover. PCR-ELISA facilitated the introduction of quantitative PCR (QPCR) assays; however, the range and accuracy of quantitation was limited. The more recent introduction of real-time platforms for PCR has revolutionized molecular diagnostic detection methods in clinical microbiology. These closed tube systems virtually eliminate the risk of amplicon carryover because the samples are not opened following thermal cycling. Many of these new platforms process samples more rapidly than conventional block-based thermal cyclers making pathogen testing much more rapid. In addition, the ability to monitor the reaction in real-time provides results immediately after cycling and facilitates quantitation of the original target sequence over many orders of magnitude. Real-time platforms can differentiate between several closely related sequences within the same reaction therefore assays can be multiplexed to detect a range of pathogens within the same tube. Many of the assays described to date have utilized the Idaho LightCycler or the Roche LightCycler instrument, both of which I will refer to as LC for the purposes of this review. Some of the other commonly used platforms for real-time PCR are the Applied Biosystems ABI Prism 7000, 7500, and 7900 Sequence Detection Systems, and the Cepheid Smart Cycler.

The real-time PCR method has been applied in virtually all areas of clinical microbiology and has proven useful in a wide range of applications. Since the first edition of this text there have been many more papers published in this rapidly expanding field. In this chapter I have tried to provide an updated overview of some of the areas and applications where real-time PCR has made a significant impact in clinical microbiology and the diagnosis of infectious disease. This review is not intended to be exhaustive and does not attempt to describe every real-time PCR study which has been published therefore I would like apologize in advance to authors of papers which I have overlooked in the preparation of this chapter. For a very comprehensive review of all of the real-time PCR assays currently described in the literature I would direct the reader to the review by Espy and colleagues (Espy *et al.*, 2006).

Real-time PCR applications in clinical bacteriology

Real-time PCR assays have been described for a number of bacterial pathogens, some of which have been presented in Table 13.1. Some of the areas in which real-time PCR methods have made an impact on clinical bacteriology are described below.

Detection of bacterial respiratory pathogens by real-time PCR

Bordetella pertussis causes whooping cough, a respiratory disease occurring mainly in adolescents. Laboratory diagnosis has traditionally relied on isolation of the organism by culture which is highly specific but sensitivity varies between 6 and 95% depending on the time of sampling (Kösters *et al.*, 2002). The results of culture are also dependent on the quality of specimen provided. *B. pertussis* is a slow growing, fastidious organism and therefore isolation may take from 3 to 12 days. Reliable and rapid diagnosis facilitates the administration of appropriate treatment and prophylaxis of contacts. This has led to the development of a number of conventional PCR assays for the detection of *B. pertussis*. More recently a number of real-time PCR assays have been developed for the detection of *B. pertussis* in clinical samples. Reischl *et al.* (2001) described a LC assay for the detection of *B. pertussis* targeting the IS481 sequence. However, specificity studies demonstrated that the assay was also positive with strains of the closely related species *Bordetella holmseii* and sequencing of the PCR products revealed sequence homology between the two species. The authors concluded that the specificity and predictive value of IS481-based PCR assays might be compromised. A real-time 5′ nuclease assay was described by Kösters *et al.* (2001) that targeted the IS481 sequence of *B. pertussis* and the IS1001 sequence of the closely related species *Bordetella parapertussis*. The assays demonstrated high sensitivity; however, similarly to the previous study, the species *B. holmseii* gave a positive signal in the IS481 assay. The assays were applied to 182 samples [nasopharyngeal swabs, nasopharyngeal aspirates (NPAs), pharyngeal swabs, and tracheal swabs] from patients with and without symptoms of pertussis and the results compared to conventional culture. The real-time PCR assay demonstrated an increased sensitivity over culture with 28 patients being PCR positive/culture negative. Although 24 of these patients did meet the CDC clinical case definition for pertussis the results must be interpreted with caution because *B. holmesii* also produced a positive signal in the assay in the specificity studies. Overall the PCR assay appeared to have a much greater sensitivity than culture with a detection rate which was nearly doubled compared to culture which was similar to the previously reported studies using conventional PCR methods. A single-tube multiplex real-time

Table 13.1 Application of real-time PCR to the detection of clinically significant bacterial pathogens

Organism	Gene targeted	Detection/ quantitation	Sensitivity	Comments	Reference
Bartonella species	ribC	D	Not reported		Zeaiter et al. (2003)
Bordetella pertussis, B. holmesii	IS481	D	Not reported		Reischl et al. (2001)
B. pertussis, B. parapertussis	IS481, IS1001	D	0.1 CFU and 10 CFU/PCR		Kösters et al. (2001)
B. pertussis, B. parapertussis	IS481, IS1001, ptg	D	0.75 CFU/PCR	Multiplex	Sloan et al. (2002)
B. pertussis, B. parapertussis	IS481, IS1001	D	0.1 CFU and 10 CFU/PCR		Kösters et al. (2002)
B. pertussis, B. parapertussis	IS481, IS1001	D	1 CFU and 5 CFU/PCR	Multiplex	Cloud et al. (2003)
Borrelia burgdorferi	flagellin	Q	1–3 spirochaetes/PCR		Pahl et al. (1999)
B. burgdorferi	recA	Q	Not reported		Morrison et al. (1999)
Borrelia garnii, Borrelia afzelii, B. burgdorferi	recA	D	Not reported		Pietila et al. (2000)
Borrelia species	ospA	Q	1 to 10 spirochaetes/PCR		Rauter et al. (2002)
B. burgdorferi	flagellin	Q	1 to 10 spirochaetes/PCR		Piesman et al. (2001)
B. burgdorferi	flagellin	Q	1 to 10 spirocahetes/PCR		Zeidner et al. (2001)
B. burgdorferi	recA, p66	D	Not reported		Mommert et al. (2001)
Campylobacter species	16S rRNA	D	Not reported		Logan et al. (2001)
Campylobacter jejuni	Novel ORF[a]	Q	12 GE[b]		Sails et al. (2003)
C. jejuni	Novel sequence[c]	Q	1 CFU/PCR		Nogva et al. (2000)
Chlamydophila pneumoniae	ompA	D	0.001 IFU[d]/PCR		Tondella et al. (2002)
C. pneumoniae	ompA	Q	Not reported		Kuoppa et al. (2002)
C. pneumoniae	ompA	Q	10^{-6} IFU/PCR		Apfalter et al. (2003)
C. pneumoniae	16S rRNA	D	0.02 IFU/PCR		Reischl et al. (2003b)
C. pneumoniae	Pst1 gene fragment	D	10 gene copies	Multiplex[e]	Welti et al. (2003)
Chlamydia trachomatis	Cryptic plasmid, MOMP	Q	1–3 copies/PCR	Multiplex	Jalal et al. (2006)
C. trachomatis	Cryptic plasmid, omp1 (L–2 specific)	Q	25–50 copies/PCR	Multiplex	Halse et al. (2006)
Clostridium difficile	tcdA, tcdB	D	10 gene copies	Multiplex	Bélanger et al. (2003)
Escherichia coli (VTEC)	stx_1, stx_2	D	1 CFU/PCR	Multiplex	Bellin et al. (2001)
E. coli (VTEC)	stx_1, stx_2	D	10 gene copies	Multiplex	Bélanger et al. (2002)

Table 13.1 *continued*

Organism	Gene targeted	Detection/ quantitation	Sensitivity	Comments	Reference
E. coli (VTEC)	stx_1, stx_2, *eae*	D	5.8–580 CFU/beef, 1.2–1200CFU/faeces	Multiplex	Sharma *et al.* (1999)
E. coli (VTEC)	stx_1, stx_2, *eae*, *hlyA*	D	Not reported	Multiplex	Reischl *et al.* (2002a)
E. coli (VTEC)	stx_1, stx_2, *eae*[f]	D	Not reported	22 assays	Nielson and Anderson (2003)
E. coli (VTEC)	stx_1, stx_2, *eae*	D	10^4 CFU/g faeces		Sharma and Dean-Nystrom (2003)
Haemophilus influenzae	*bexA*	D	Not reported	Multiplex[g]	Corless *et al.* (2001)
H. influenzae	Region 2 capsulation locus	Q	1 CFU/PCR		Marty *et al.* (2004)
Helicobacter pylori	*ureC*	Q	10 gene copies		He *et al.* (2002)
Legionella pnuemophila	*mip*	D	2.5 CFU/PCR		Ballard *et al.* (2000)
Legionella species, L. pnuemophila	5S rRNA, *mip*	D	<10 CFU/PCR		Hayden *et al.* (2001)
Legionella species, L. pnuemophila	16S rRNA, *mip*	D	1 fg DNA/PCR		Wellinghausen *et al.* (2001)
Legionella species, L. pnuemophila	16S rRNA	D	3 GE/PCR	Multiplex	Reischl *et al.* (2002b)
Legionella species	16S rRNA	D	2 CFU/PCR		Rantakokko-Jalava and Jalava (2001)
L. pnuemophila	*mip*	D	10 CFU/PCR		Wilson *et al.* (2003)
L. pnuemophila	*mip*	D	10 GE/PCR	Multiplex[e]	Welti *et al.* (2003)
Moraxella catarrhalis	*copB*	D	1 CFU/PCR		Greiner *et al.* (2003)
Mycobacterium tuberculosis	IS6110	Q	Not reported		Desjardin *et al.* (1998)
M. tuberculosis complex	ITS	D	800 gene copies/PCR		Kraus *et al.* (2001)
M. tuberculosis complex	ITS	D	Not reported		Miller *et al.* (2002)
M. tuberculosis complex	16S rRNA gene	D	Not reported		Shrestha *et al.* 2003)
M. tuberculosis complex	16S rRNA gene	D	50 genomes/PCR		Burggraf *et al.* (2005a)
Mycoplasma genitalium	16S rRNA	Q	10 gene copies/PCR		Deguchi *et al.* (2002)
M. genitalium	16S rRNA	Q	10 gene copies/PCR		Yoshida *et al.* (2002)
Mycoplasma pneumoniae	P1 gene	D	10 gene copies	Multiplex[e]	Welti *et al.* (2003)
M. pneumoniae	P1 gene	D	20–200 cells/PCR		Ursi *et al.* (2003)

Organism	Target	Assay	Sensitivity	Format	Reference
M. pneumoniae	P1 gene	D	Not reported		Hardegger et al. (2000)
Neisseria gonorrhoeae	*cppB* gene	D	Not reported		Whiley et al. (2002)
N. gonorrhoeae	*porA* psuedogene	D	Not reported		Whiley et al. (2004)
N. gonorrhoeae	*cppB*, 16S rRNA	D	Not reported		Boel et al. (2005)
N. gonorrhoeae	*opa*	D	1 GE/PCR		Geraats-Peters et al. (2005)
Neisseria meningitidis	*ctrA*	D	Not reported	Multiplex[g]	Corless et al. (2001)
N. meningitidis	16S rRNA, *sacC*, *siaD*, *porA*	D	Not reported	Genogrouping PCR	Molling et al. (2002)
Staphylococcus aureus	*nucA*	D	2 CFU/PCR		Palomares et al. (2003)
CNS[h]	16S rRNA	D	Not reported	Multiplex	Edwards et al. (2001a)
S. aureus, MRSA[i]	Novel sequence[j], *mecA*	D	25 GE/PCR	Multiplex	Reischl et al. (2000)
S. aureus, MRSA	Sa442[k], *mecA*	D	Not reported		Tan et al. (2001)
MRSA	Sa442, *mecA*	D	Not reported	Multiplex	Grisold et al. (2002)
MRSA	*nucA*, *mecA*	D	Not reported	Multiplex	Fang et al. (2003)
MRSA	SCC*mec*, *orfX*	D	25 CFU/sample		Huletsky et al. (2004)
Group A Streptococci	not stated	D	Not reported		Uhl et al. (2003)
Streptococcus pneumoniae	*ply*	Q	1 CFU/PCR		Greiner et al. (2001)
S. pneumoniae	*lytA*	D	4 GE/PCR		McAvin et al. (2001)
S. pneumoniae	*ply*	D	Not reported	Multiplex[g]	Corless et al. (2001)
Tropheryma whipplei	16S 23S rRNA ITS, *rpoB*	Q	Not reported		Fenollar et al. (2002)
Yersinia enterocolitica	*bipA* and novel target sequence[l]	D	Not reported	Multiplex	Aarts et al. (2001)

[a] 256-bp region of an open reading frame adjacent to and downstream from a novel two-component regulatory gene specific for *C. jejuni*.
[b] GE, genome equivalents.
[c] 86-bp fragment including positions 381121 to 381206 of the published *C. jejuni* strain NCTC 11168 genome.
[d] IFU, inclusion-forming unit.
[e] Multiplex assay including *C. pneumoniae*, *M. pneumoniae*, and *L. pnuemophila*.
[f] Multiplex PCR for *stx*$_1$, *stx*$_2$, *ehlyA*, *katP*, *espP*, *etpD*, *eae*, *tir*, *espD* (including all variants), and *saa*.
[g] Multiplex assay for *N. meningitidis*, *H. influenzae*, and *S. pneumoniae*.
[h] CNS, coagulase-negative staphylococcus.
[i] MRSA, methicillin-resistant *S. aureus*.
[j] Novel *S. aureus*-specific fragment.
[k] Sa442, Novel *S. aureus*-specific fragment.
[l] Novel 129-bp gene fragment specific for pathogenic strains of *Y. enterocolitica*.

LC assay was reported by Sloan *et al.* (2002) that also targeted IS481 and IS1001 in *B. pertussis* and *B. parapertussis*. Again the IS481 assay cross-reacted with four *B. holmseii* strains, similarly to previously reported assays. The authors considered that the sensitivity of using the IS481 as a target outweighed the limitations of specificity and reported that this assay has been adopted at their institution as the primary diagnostic test for this organism. The real-time molecular beacon assay of Templeton *et al.* (2003) also targeted the IS481 and IS1001 regions and incorporated a phocine herpesvirus (PhHV) internal control to monitor nucleic acid extraction and PCR amplification (Niesters, 2001). Duplexing the assays with the internal control PCR and spiking the clinical samples with PhHV did not adversely affect the clinical sensitivity of the two assays. The use of such a control which is spiked into the clinical sample and then co-extracted with the sample should be advocated for all diagnostic PCR assays. The detection of *B. pertussis* and *B. parapertussis* by real-time PCR is a good example of where improvements in patient diagnosis have arisen from the introduction of PCR-based tests. In this case PCR-based diagnostic tests have replaced conventional methods because of their increased levels of sensitivity and the rapid timeframe in which results can be obtained. In the future real-time PCR assays for other slow growing pathogens and those, which are difficult to recover from clinical samples by culture, will be developed.

The slow growing respiratory pathogen *Mycobacterium tuberculosis* was also one of the first bacterial pathogens to be investigated by real-time PCR. Desjardin *et al.* (1998) developed a 5' nuclease QPCR assay targeting the IS6110 element specific for *M. tuberculosis*. They used this assay to determine if changes in the amount of *M. tuberculosis* DNA present in sputum correlated with the numbers of viable bacilli therefore allowing the therapeutic response of patients to be monitored. They compared the results to acid-fast bacilli (AFB) microscopy and culture. Prior to the start of therapy levels of AFB, *M. tuberculosis* DNA and cultivable bacilli were similar indicating that the assay may be useful for measuring the pathogen load prior to treatment. However, following the initiation of therapy levels of AFB and *M. tuberculosis* DNA did not correlate with cultivable bacilli indicating that the assay was not suitable for monitoring treatment efficacy.

Real-time PCR has also been applied to the direct detection of *M. tuberculosis* in clinical samples as a method to aid diagnosis. Kraus *et al.* (2001), developed a LC assay which targeted 220-bp fragment of the ITS region of *M. tuberculosis*. Hybridization probes were designed and the assay was demonstrated to be specific for members of the *M. tuberculosis* complex (*M. tuberculosis*, *Mycobacterium bovis* and *Mycobacterium africanum*). However, the assay was not validated for the detection of these organisms directly in clinical samples. Miller *et al.* (2002) extended this work by applying the LC assay of Kraus *et al.* (2001) to the detection of *M. tuberculosis* in AFB smear-positive respiratory specimens and BacT/ALERT MP culture bottles. The LC assay demonstrated a sensitivity of 98.1% and a 100% specificity for the AFB smear-positive samples and 100% sensitivity and specificity for 232 BacT/ALERT samples with 114 samples being positive in both culture and the LC assay. Shrestha *et al.* (2003) described a LC hybridization probe assay which targeted a region of the 16S rRNA gene of Mycobacteria which included the hypervariable region A. The hybridization probes were designed to have 100% sequence homology with the *M. tuberculosis* sequence but had ≥ 1 bp mismatches with other clinically significant mycobacteria. Melting curve analysis of the amplicon following cycling allowed differentiation between species based on differences in binding of the hybridization probes to the region A. Application of the assay to 18 reference strains and 168 clinical mycobacterial isolates demonstrated that the assay consistently detected and differentiated *M. tuberculosis* from non-tuberculous mycobacteria (NTM). The assay was applied to 50 clinical specimens which had previously been shown to be culture positive for *M. tuberculosis*. In addition three samples which were acid-fast bacilli (AFB) smear positive but which yielded non-tuberculous mycobacteria on culture were also included. The results of the LC assay were compared to the COBAS Amplicor *M. tuberculosis* assay and the previous culture results. The PCR assays detected *M.*

tuberculosis in 48 of the 50 samples, with two samples (one urine and a biopsy) being negative in both of the assays. Interestingly five samples which were smear negative (but culture positive for *M. tuberculosis*) were found to be PCR positive for *M. tuberculosis* in both PCR assays indicating that PCR-based detection methods may be more sensitive than microscopy for some samples. Melting curve analysis following thermal cycling of the three AFB smear positive samples correctly identified them as containing the non-tuberculous mycobacteria *M. kansasii*, and *M. avium/M. intracellulare* complex. Burggraf *et al.* (2005a) further developed this real-time assay by incorporating an internal control (IC) and by transferring it to the LightCycler 2.0 instrument with its greater sample volume of 100 μl compared to 20 μl. The IC control monitors the function of the reaction components and can detect the presence of potential PCR inhibitors within each individual reaction. The IC was designed to contain the primer and probe sequences; however, the probe binding site incorporated three mismatches to facilitate discrimination between the IC and positive clinical samples (Burggraf *et al.*, 2005b). Application of this modified assay to 146 clinical samples (respiratory and non-respiratory) demonstrated very good agreement (100% sensitivity, 98.6% specificity) with the COBAS Amplicor *M. tuberculosis* assay.

Real-time PCR assays may prove useful for the rapid confirmation of AFB smear-positive samples in the clinical laboratory; however, further studies are require to validate their sensitivity and specificity.

Legionella pneumophila is the causative agent of Legionnaires' disease; a nosocomial or community-acquired pulmonary infection, which can be severe and life threatening. Infection occurs in immunocompromised patients and is acquired through inhalation of *L. pneumophila* from a contaminated environmental source such as the water system of large buildings, e.g. hospitals. Rapid identification of the source of infection is important during outbreaks to limit further cases of infection. Occasionally other *Legionella* species such as *Legionella micdadei*, *Legionella bozemanii*, *Legionella dumoffii*, and *Legionella longbeachae* also cause opportunistic pneumonia. The isolation of *Legionella* from water samples is the current gold

standard; however, it is limited by the organisms fastidious growth requirements, prolonged incubation periods (up to 2 weeks), the overgrowth of other bacteria and the presence of viable but non-culturable *Legionella*. Ballard *et al.* (2000) developed a LC bi-probe assay specific for the macrophage infectivity potentiator (*mip*) gene of *L. pneumophila*. The limit of detection (LOD) of the assay was 2.5 CFU/reaction equivalent to 1000 CFU/l of water. The assay was applied to the detection of *L. pneumophila* in 14 natural water samples and 10 laboratory microcosms of which 11 water samples and all 10 microcosms were culture positive for *L. pneumophila*. All 10 of the microcosms were positive in the LC assay; however, only 6 of the 11 culture-positive water samples were positive in the assay. Three of the samples contained <200 CFU/l and were below the detection limit of the assay and the other two samples contained PCR inhibitors. Overall the LC assay was a rapid method which has the potential for screening significant numbers of samples in short time period. This may prove useful in outbreak situations were the reservoir of infection normally contains >1000 CFU/l.

A quantitative genus-specific LC assay was developed by Wellinghausen *et al.* (2001) for the detection and quantitation of legionellae in hospital water samples. The assay targeted the 16S rRNA gene and included an internal inhibitor control based on a cloned fragment of lambda phage. A dual-colour hybridization probe assay design was used and detected products were quantified with external quantitative standards (QS) composed of *L. pneumophila* DNA. The LOD of the assay was demonstrated to be approximately 1 fg of DNA (equivalent to one *Legionella* organism) and it detected all 44 *Legionella* species and serogroups. The assay was applied to 77 water samples from three hospitals, with 54 of the samples being shown to be positive by culture. The LC assay detected *Legionella* in 76 of the 77 samples, with the quantitative results being generally 25-fold higher than those recovered by quantitative culture. These higher results may have been due to the assay detecting DNA from dead and non-culturable but viable cells. The authors developed a second LC assay based on the *mip* gene that was specific for *L. pneumophila* and re-tested the samples in the assay. All 76 samples were positive in the second assay correlating with the results of the first assay. The degree of contamination of hospital water supplies has been shown to correlate with the incidence of nosocomial infection; however, the exact levels of contamination associated with infection have not been clearly established. The application of this quantitative assay to the monitoring of *Legionella* contamination levels in hospital water supplies may provide further data on the association between the level of contamination and risk of infection. This may facilitate the identification of critical levels of contamination useful for the monitoring of hospital water supplies thereby reducing the risk to patients.

The current gold standard for the diagnosis of *Legionella* infection is laboratory culture of the organism from clinical samples. Diagnosis can be made from a variety of clinical specimen types; however, the bacterial culture of bronchoscopy or lung biopsy specimens is the most sensitive method of detection. Rapid diagnosis of infection in immunocompromised patients, in whom the fatality rate can be as high as 50%, is important because the infection often responds to antimicrobial therapy. The isolation of *Legionella* from patient clinical samples is hampered by the fastidious requirements of the organism and prolonged incubation periods of up to 2 weeks to ensure maximum recovery.

Alternative, more rapid tests include direct fluorescent antibody assays (DFA) on respiratory secretions or urine antigen detection tests have been developed; however, these too have some limitations. DFA can cross-react with other species, leading to false-positive results, and DFA has a low sensitivity of detection. Antigen testing of urine has a relatively high sensitivity (85%) but can detect only a limited range of pathogenic *Legionella* species. Serological tests have been developed and are highly sensitive; however, their use is limited to epidemiological studies due to the significant time taken for patients to seroconvert.

PCR-based diagnostic assays for the detection of *Legionella* species in clinical samples have been developed based on the detection of target regions in the 16S rRNA, 5S rRNA or the *mip* gene. Hayden *et al.* (2001) described a *Legionella* genus-specific LC assay targeting the 5S rRNA gene and a *L. pneumophila*-specific LC assay targeting the *mip* gene for the detection of *Legionella* in clinical samples. The assays were applied to the detection of *Legionella* in bronchoalveolar lavage (BAL) and open lung biopsy samples from 35 patients with the samples being previously archived following conventional tests. The results of the LC assays were compared to culture, DFA, and *in situ* hybridization (ISH) the gold standard being culture. For the BAL specimens both the genus-specific and the *L. pneumophila*-specific LC assays demonstrated 100% sensitivity and specificity with nine of the samples being culture positive for *L. pneumophila*. The assays both demonstrated a reduced sensitivity for detection within the tissue samples with the genus-specific PCR having a sensitivity of 68.8% and the *L. pneumophila*-specific assay having a sensitivity of 17%. This reduced sensitivity may have been due in part to the samples previously being formalin-fixed prior to recovery and testing. In addition, the *mip* gene is present as a single copy in the *Legionella* genome unlike the 5S rRNA which is present in multiple copies.

These assays require further validation such as a prospective study to establish their utility for the detection of *Legionella* in clinical samples, especially tissue samples. Rantakokko-Jalava and Jalava (2001) described a LC SYBR Green I assay for the detection of *Legionella* species, which targeted the 16S rRNA gene. This assay was applied to the detection of *Legionella* in 71 clinical samples from hospital patients with acute pneumonia; however, only two samples were culture positive for *L. pneumophila* but they were positive in the PCR assay. No culture-negative samples produced a positive result in the assay; however, the assay requires further application to clinical samples to validate its specificity and sensitivity for the detection of *L. pneumophila* in patient samples.

Reischl *et al.* (2002b) extended the previous study of Wellinghausen *et al.* (2001) by using the *Legionella*-specific primers and probes of the previous study in a LC assay for the detection of *Legionella* organisms in 26 culture-positive and 42 culture-negative BAL specimens. The study also utilized a second LC assay that targeted a *L. pneumophila*-specific fragment of the 16S rRNA gene. Specificity studies revealed that some of the non-*L. pneumophila* species cross-reacted with the *L. pneumophila* probes and produced amplification plots therefore melting curve analysis was required for unambiguous identification. The assay demonstrated 100% sensitivity and specificity for the BAL specimens tested; however, its usefulness requires further investigation through prospective studies of samples from patients with atypical pneumonia.

The *L. pneumophila mip* gene was also targeted in the LC hybridization assay of Wilson *et al.* (2003) which demonstrated a LOD of 10 fg DNA equivalent to approximately ten organisms. The specificity of the assay was validated with a range of *L. pneumophila* strains, other *Legionella* species and species from unrelated genera. However, the authors did note that the assay might produce a positive result with *Legionella worsleinsis* and *Legionella fairfieldensis* due to the homology of the target sequence within these species and the target species *L. pneumophila*. The assay was applied to seven culture positive and 41 *L. pneumophila* culture negative clinical specimens with the assay results correlating with the culture results. Overall these real-time PCR assays may be a useful alternative to culture for the detection of *L. pneumophila* in clinical samples from patients with acute pneumonia; however, further clinical trials are needed to validate this approach to the diagnosis of *Legionella* infection.

Chlamydophila pneumoniae is an obligate intracellular pathogen that has been implicated as a cause of upper and lower respiratory tract infection in humans. It is also thought to be responsible for approximately 10% cases of community-acquired pneumonia. Diagnosis is often based on serology or cell culture; however, these methods often produce inconclusive results. PCR detection methods may provide more rapid and reliable diagnosis of *C. pneumoniae* infection. Several real-time PCR assays have been described for *C. pneumoniae* many of which have targeted the major outer membrane protein gene (*ompA*) (Tondella *et al.*, 2002; Kuoppa *et al.*, 2002; Apfalter *et al.*, 2003), which is highly conserved within the species. Alternative gene targets have included the 16S rRNA gene (Reischl *et al.*, 2003b) and the *Pst*I gene fragment (Welti *et al.*, 2003). The assay of Kuoppa *et al.* (2002)

was applied to the detection of *C. pneumoniae* in respiratory specimens, with the results being compared with a conventional PCR assay. The real-time assay was demonstrated to be at least as sensitive for the detection of *C. pneumoniae* as an 'in-house' nested touchdown PCR; however, the real-time assay was more rapid and less labour intensive. The assay of Reischl *et al.* (2003b) targeting the 16S rRNA gene was demonstrated to have a LOD of 0.02 inclusion-forming units (IFU) per PCR reaction equivalent to 1 IFU per ml of BAL. The assay was applied to 90 clinical samples from patients with pneumonia with 12 samples previously shown to be positive for *C. pneumoniae* also being positive in the real-time PCR assay. Although these results are promising further prospective studies are required to confirm the utility of this assay for the diagnosis of *C. pneumoniae* respiratory infections.

A TaqMan quantitative real-time assay was described for *Streptococcus pneumoniae* which targeted the pneumolysin gene (Greiner *et al.*, 2001). The assay was applied to the detection of *S. pneumoniae* in nasopharyngeal secretions (NPSs) from patients with respiratory tract infections with the results being compared to conventional culture. The assay performed with a sensitivity of 100% and a specificity of 96% compared to the culture results. In addition the numbers of *S. pneumoniae* organisms detected by the real-time PCR assay correlated with the numbers detected by semiquantitative culture. The same assay was utilized by Yang *et al.* (2005) in a study investigating the clinical utility of using real-time PCR testing for *S. pneumoniae* in adult patients with community-acquired pneumonia (CAP). A prospective study of emergency department patients with CAP was performed with sputum samples being tested for the presence of *S. pneumoniae* by real-time PCR, the results being compared with those of a composite reference standard comprising Gram staining of sputum samples and sputum/blood cultures. Sputum samples from 129 patients were tested in the real-time PCR assay and the data summarized in a receiver operating characteristic (ROC) curve to determine the diagnostic efficacy of the assay. Sensitivity and specificity were 90.0% and 80.0%, respectively; and positive and negative predictive values were 58.7% and 96.2%, respectively. The authors concluded that the real-time PCR assay had a favourable accuracy for diagnosis of pneumococcal pneumonia in adult patients with CAP and it may be a useful supplementary diagnostic test for clinicians, particularly those practicing in the acute-care setting, where rapid pathogen detection may assist selection of appropriate antibiotic treatment.

Real-time PCR assays have been described for a number of other respiratory pathogens including *Mycoplasma pneumoniae* (Hardegger *et al.*, 2000; Welti *et al.*, 2003; Ursi *et al.*, 2003), *Haemophilus influenzae* (Marty *et al.*, 2004), and *Moraxella catarrhalis* (Greiner *et al.*, 2003).

Detection of bacterial meningitis by real-time PCR

Bacterial meningitis is a serious disease that affects the central nervous system (CNS) causing significant morbidity and mortality. Three pathogens are responsible for the majority of bacterial meningitis infections, these are *Neisseria meningitidis*, *Streptococcus pneumoniae* and *Haemophilus influenzae*. The traditional method for the diagnosis of bacterial meningitis is the culture of the causative organism from cerebrospinal fluid (CSF) taken by lumbar puncture. However, culture can take up to three days to produce a result and following the practice of taking the clinical sample after the initiation of antimicrobial therapy the ability to recover the causative organism by culture has become more difficult. PCR-based diagnostic tests offer an alternative method to detect these organisms even after initiation of antimicrobial therapy with bacterial DNA remaining detectable after the organism can no longer be recovered by culture. The first diagnostic PCR assays described were based on gel electrophoresis or PCR-ELISA endpoint detection methods. Guiver *et al.* (2000) described the first real-time PCR assay for the detection of *N. meningitidis* in whole blood, CSF, plasma, and serum samples. Assays were developed and evaluated targeting the meningococcal capsular transfer gene (*ctrA*), the insertion sequence IS1106, and two assays targeting the sialytransferase gene (*siaD*) for serogroup B and C determination. The specificity of the assays was investigated with a diverse range of meningococci serotypes and sero-subtypes plus species from other genera. The *ctrA* assay was demonstrated to be specific for *N. meningitidis*; however, the IS1106 assay produced false-positive results with non-meningococcal isolates. The *siaD* B and C assays exclusively detected serogroup B and serogroup C meningococci respectively with all other serogroups being negative in the assay. Application of the assays to tenfold dilutions of meningococci demonstrated that the LOD of the assays was less than one viable cell. The authors found that the adoption of the TaqMan platform increased the maximum number of samples that could be processed in a single day from 50 to 200 when compared to their current PCR-ELISA format. The authors adopted the *ctrA* assay as their primary screening test followed by serogroup determination using the *siaD* B and C assays. Following introduction of this assay as a routine test there was a 56% increase in the number of laboratory confirmed cases of meningococcal disease compared to culture only confirmed cases. This study is an excellent example of how the adoption of real-time PCR methods has significantly improved the recognition of the prevalence of meningococci in meningitis and is therefore an important model for other infectious diseases.

The *ctrA* assay was used in another study to determine if bacterial loads in meningococcal disease correlate with disease severity (Hackett *et al.*, 2002). The authors demonstrated that patients with meningococcal disease had higher bacterial loads than previously determined by quantitative culture. On hospital admission bacterial load was significantly higher in patients with severe disease with the maximum load being seen in patients who died. Corless *et al.* (2001) developed a real-time multiplex assay for the simultaneous detection of *N. meningitidis*, *H. influenzae*, and *S. pneumoniae* in suspected cases of meningitis and septicaemia. The assay targeted the *ctrA* gene of *N. meningitidis*, the capsulation (*bexA*) gene of *H. influenzae* and the pneumolysin gene (*ply*) of *S. pneumoniae*. The meningococcal *ctrA* PCR demonstrated a sensitivity of 88.4% when tested against samples from culture-confirmed cases of meningococcal disease and the *H. influenzae* *bexA* assay demonstrated 100% sensitivity when tested against nine culture-confirmed cases of *H. influenzae* disease. The *S. pneumoniae* *ply* PCR assay demonstrated a sensitivity of 91.8% when applied to 36 samples from culture-confirmed *S. pneumoniae* disease. Co-amplification of two gene targets without a loss in sensitivity was demonstrated and the three primers and probe sets were combined into a multiplex assay. The multiplex PCR was then applied to 4113 culture-negative clinical samples and cases of meningococcal, *H. influenzae* and pneumococcal disease that had not previously been confirmed by culture were identified. The application of this multiplex PCR in a routine screening service would provide improved non-culture diagnosis and case ascertainment of bacterial meningitis and septicaemia.

Detection of bacterial genitourinary pathogens by real-time PCR

Neisseria gonorrhoeae is one of the most prevalent sexually transmitted pathogens in the world with a significant proportion of infections, especially in women being asymptomatic. Unfortunately these asymptomatic infections can lead to serious long-term consequences including pelvic inflammatory disease, ectopic pregnancy, neonatal conjunctivitis and infertility. The increasing incidence rate of *N. gonorrhoeae* infection, and the high prevalence of asymptomatic infection highlights the need for sensitive and specific tests for the diagnosis of both symptomatic and asymptomatic infection. The current gold standard for the diagnosis of *N. gonorrhoeae* infection is the isolation of the organism from clinical samples on selective culture media. Unfortunately although the sensitivity of culture is usually between 80 and 95% for acute infections, this falls to approximately 50% in women with chronic infections. The sensitivity of culture can also be compromised by poor specimen collection, and inadequate transport and storage prior to testing.

Nucleic acid amplification methods such as PCR and the ligase chain reaction have proven useful for the detection of *N. gonorrhoeae* and a range of commercial assays are available; however, many of these assays have issues with sensitivity and specificity of detection. The COBAS AMPLICOR *Chlamydia trachomatis/Neisseria gonorrhoeae* (CA CT/NG) assay has been proven to be a useful diagnostic test for the detection of *C. trachomatis* in urine and cervical

swab samples (Morre *et al.*, 1999). However, the *N. gonorrhoeae* component (CA NG) of the assay lacks specificity with some non-gonorrhoeae *Neisseria* species and even some lactobacilli strains being reported to give false-positive results (van Doornum *et al.*, 2001). Therefore all CA NG-positive samples must be confirmed in a second more specific assay to ensure confidence in the reported result. Originally many of these confirmatory tests were based on conventional PCR methodologies; however, they have begun to be replaced with real-time PCR-based tests. Whiley *et al.* (2002) described a LC hybridization probe assay targeting the *cppB* gene of *N. gonorrhoeae* which they applied to 152 urine specimens submitted for routine examination for gonorrhoea infection, the results being compared to an in-house nested PCR-ELISA assay also targeting the *cppB* gene. The PCR-ELISA assay detected *N. gonorrhoeae* in 31 of the samples and the LC assay detected *N. gonorrhoeae* in 29 of the samples, with 121 samples being negative in both methods. The two samples negative in the LC assay were from asymptomatic patients who were culture negative and probably had low levels of target organisms in their samples.

In a later study, Whiley *et al.* (2004) noted the presence of the *cppB* gene in strains of some commensal *Neisseria* species and its absence in some *N. gonorrhoeae* strains which led them to speculate that there may be a risk of false-positive and false-negative results in assays targeting this gene. In this second study the authors developed another LC-based assay (NGpapLC) targeting the *porA* pseudogene in *N. gonorrhoeae*. They applied this new assay to 282 clinical samples with the results being compared to the results of their 'in house' testing algorithm which combined the CA CT/NG assay and the LC *cppB* gene assay (Whiley *et al.*, 2002). The authors concluded that the NGpapLC assay was suitable for the confirmation of *N. gonorrhoeae* positive results obtained using the CA CT/NG assay. This evaluation lacked the gold standard for *N. gonorrhoeae* detection i.e. and isolation of the causative organism from clinical samples by selective culture. Therefore the authors carried out a further study to investigate the clinical specificity of the NGpapLC assay using 636 swab specimens with the results being compared to conventional selective culture (Whiley *et al.*, 2005). Samples included urethral, cervical, throat and anal swabs submitted from patients being screened for sexually transmitted disease. Of the 636 samples, 19 were positive by culture and LC PCR with 613 samples being negative for *N. gonorrhoeae* by both methods. For cervical and urethral specimens the NGpapLC assay demonstrated 100% sensitivity and 100% specificity compared to selective culture. However, four samples (one rectal and three throat swabs) were positive in the NGpapLC assay but negative by culture. These potential false-positive samples were tested in three additional PCR assays targeting the *cppB* gene (Whiley *et al.*, 2002), the *ORF1* and *gyrA* genes. All four samples were positive in the additional assays indicating the NGpapLC assay results were unlikely to be false positives and that the culture results were more likely to be falsely negative. To further validate the specificity of the assay the authors applied it to a panel of 182 strains of *Neisseria* species including 147 isolates of *N. gonorrhoeae* and demonstrated that only the *N. gonorrhoeae* strains were positive in the assay. Overall the authors demonstrated that the new NGpapLC assay was suitable for the sensitive and specific detection of *N. gonorrhoeae* in a range of clinical sample types.

Boel *et al.* (2005) compared the sensitivity and specificity of two PCR-ELISA and two LC real-time in-house assays targeting the *cppB* and 16S rRNA genes as targets for the confirmation of *N. gonorrhoeae* positive results obtained using the CA CT/NG assay. The 16S rRNA gene real-time PCR was unique to this study; however, the *cppB* gene real-time assay had been previously described by Whiley *et al.* (2002). The assays were applied to 765 male and female genital and nasopharyngeal samples all positive for *N. gonorrhoaeae* using the CA CT/ NG assay with 229 samples being confirmed as positive and 534 not being confirmed as positive for *N. gonorrhoeae* using the additional assays. Both the PCR-ELISA and real-time 16S rRNA gene assays had high positive predictive values of 99.1% and 100% respectively. However, 13 of the confirmed samples lacked the *cppB* gene which compromised sensitivity of the PCR-ELISA and real-time PCR assays targeting the *cppB* gene resulting in negative predictive values of 96.8%

and 97.6% respectively. This study demonstrated that the 16S rRNA gene is preferable to the *cppB* gene as a target for the confirmation of CA NG-positive samples. This 16S rRNA gene real-time LC assay was evaluated against a TaqMan assay targeting the 5′ un-translated region of the *opa* genes for the detection of *N. gonorrhoeae* in the study of Geraats-Peters *et al.* (2005). The specificity of the new *opa* gene assay was investigated with a panel of 448 well-defined *N. gonorrhoeae* isolates and a wide range of non-target organisms including 10 other different *Neisseriaceae* isolates. The assay was demonstrated to be specific with all 448 *N. gonorrhoeae* strains giving a positive signal in the assay and no signal was observed for any of the non-target species. The authors then compared the clinical sensitivity of the LC and TaqMan real-time assays by applying them to 122 clinical samples previously shown to be positive for *N. gonorrhoeae* in the CA CT/NG assay. Of the 122 CA NG-positive samples 68% were shown to be negative in both the LC and TaqMan assays with 30% samples being positive in both, again illustrating the large numbers of false positives detected in the CA CT/NG assay. Three samples were positive for *N. gonorrhoeae* in the *opa* TaqMan assay but were negative in the LC 16S rRNA assay indicating that the *opa* assay may be slightly more sensitive for the detection of the target organism in clinical samples.

Chlamydia trachomatis is the most common sexually transmitted bacterial pathogen. There are 15 servovars within the species which are associated with a variety of disease manifestations, serovars A–C cause trachoma, serovars D–K cause genital tract infections, and serovars L-1 to L-3 cause lymphogranuloma venereum. The enhanced sensitivity of nucleic acid amplification tests (NAATs) in comparison with cell culture or antigen tests has led to them becoming the gold standard for the detection of *C. trachomatis* in clinical samples. A large number of in-house and commercial NAATs have been developed for the detection of *C. trachomatis*; however, these are beginning to be replaced by real-time PCR-based assays. Koenig *et al.* (2004) adapted a previously described conventional assay (Mahony *et al.*, 1997) into a LC SYBR green assay and compared it to the BD ProbeTec ET (BDPT) assay for the detection of *C. trachomatis* in clinical samples. Samples previously shown to be positive in the BDPT assay (n=114) were re-tested in the LC assay and 91.2% were found to be positive. The authors concluded that the assay was a cost-effective confirmatory test for BDPT positive samples.

A duplex TaqMan format real-time PCR was described by Jalal *et al.* (2006) which included two *C. trachomatis*-specific targets plus an internal control to facilitate the detection, identification and quantification of *C. trachomatis* in clinical samples. The *C. trachomatis*-specific targets were the cryptic plasmid (CP) and the major outer membrane protein (MOMP), with the internal control (IC) being an artificial single-stranded DNA of a random sequence previously checked against a nucleotide-nucleotide BLAST in GenBank. The analytical sensitivity of the assay was as low as three gene copies of the MOMP and one gene copy for CP with the MOMP having a linear dynamic range of detection of between 25 and 250,000 gene copies per reaction. To validate the assay it was applied to 146 Amplicor PCR-positive and 122 Amplicor PCR-negative samples obtained from 56 males and 212 females. There was a 100% correlation between the results of the TaqMan multiplex assay and the Amplicor assay; however, six Amplicor PCR-positive samples only generated a signal in the CP component of the assay. Repeat testing of these samples confirmed the presence of CP DNA in the absence of *C. trachomatis* genomic DNA for these six samples. The strategy of targeting both the CP and MOMP genes eliminates the possibility of false-negative results and reduces the need for confirmatory testing in almost all samples reducing turnaround times and workload. In addition some *C. trachomatis* strains lack the CP therefore this dual target approach should also detect such strains in clinical samples. Although quantitation of *C. trachomatis* in clinical samples is not currently considered useful for the clinical management of patients it may prove useful in investigating the pathogenesis of infection and the response to therapy.

Lymphogranuloma venereum (LGV) is an invasive disease caused by *C. trachomatis* serovars L1, L2 and L3. It has a variety of manifestations often starting with a small painless blister or sore where the infection entered the body. The infection can then spread to the lymph glands in the groin causing swelling and inflammation (inguinal syndrome) and/or acute haemorrhagic proctitis (anorectal syndrome). If left untreated the infection can lead to complications including genital elephantitis and destruction of the rectum, forming rectal strictures and fistulae. Diagnosis of LGV can be difficult because the symptoms mimic those of other sexually transmitted infections such as chancroid, herpes, syphilis and granuloma inguinale. Diagnostic tests include culture, immunofluoresence and serology; however, NAATs are preferable due to their enhanced sensitivity. Halse *et al.* (2006) recently described a multiplexed TaqMan assay for the rapid detection of *C. trachomatis* and identification of serovar L-2, the serovar most often associated with LGV in the USA. The assay targeted the ORF2 region of the cryptic plasmid (Pickett *et al.*, 2005), specific for *C. trachomatis* and a conserved region of the *omp1* gene specific for serovar L-2. The analytical limit of detection was 25 genome equivalents for the *C. trachomatis* assay and 50 genome equivalents for the L-2-specific assay. The assays were validated using a panel of 30 spiked rectal swab samples and then were applied to 70 rectal swab samples collected from symptomatic patients. Forty-five of the samples were negative in both assays, 25 were positive in the *C. trachomatis*-specific assay and 13 were positive in the serovar L-2-specific assay. Unfortunately the results of the assay were not compared to other established tests such as culture or other NAATs therefore the sensitivity and specificity of the new assays could not be determined. Application of these assays to further clinical samples with the results being compared to established tests is required to validation these assays for the detection and identification of *C. trachomatis* serovar L-2.

Real-time PCR applications in clinical parasitology

Real-time PCR methods have been utilized in several areas of clinical parasitology (Table 13.2). Examples include real-time PCR methods for the detection of faecal parasites which are more sensitive and specific than the conventional methods of microscopy or serology. Real-time PCR assays also facilitate sensitive and accurate

Table 13.2 Application of real-time PCR to the detection of clinically significant parasites

Organism	Gene targeted	Detection/ quantitation	Sensitivity	Reference
Cryptosporidium species	16S rRNA	D	5 oocysts/PCR	Limor *et al.* (2002)
Cryptosporidium parvum	Cp11, 18S rRNA	D	Not reported	Higgins *et al.* (2001)
C. parvum	β-Tubulin, GP900/poly(T)	D	1 oocyst/PCR	Tanriverdi *et al.* (2002)
Cyclospora cayetanensis	18S rRNA	Q	1 oocyst/PCR	Varma *et al.* (2003)
Dientamoeba fragilis	16S rRNA	D	1 plasmid copy	Stark *et al.* (2006)
Entamoeba histolytica, Entamoeba dispar	rDNA	D	0.1 parasite/g faeces	Blessmann *et al.* (2002)
Plasmodium spp.	18S rRNA	D	0.2 GE[a]/PCR	Mangold *et al.* (2005)
Plasmodium falciparum	18S rRNA	Q	20 parasites/ml blood	Hermsen *et al.* (2001)
P. falciparum, P. vivax, P. malariae, P. ovale	16S rRNA	D	0.1 parasite/PCR	Lee *et al.* (2002)
Plasmodium yoelii	18S rRNA	Q	0.5 parasites/PCR	Witney *et al.* (2001)
Encephalitozoon intestinalis	16S rRNA	Q	10^2–10^4 spores/ml faeces	Wolk *et al.* (2002)
Enterocytozoon bienusi	16S rRNA	Q	10 GE/μl stool	Menotti *et al.* (2003)
Giardia lamblia	16S rRNA	D	0.5 *Giardia* cysts	Verweij *et al.* (2003)
Giardia lamblia	β-Giardin	D	Not reported	Guy *et al.* (2004)
Giardia lamblia	18S rRNA	D	1.25 trophozoites	Ng *et al.* (2005)
Giardia lamblia, Entamoeba histolytica, Cryptosporidian parvum	16S rRNA	D	Not reported	Verweij *et al.* (2004)
Leishmania donovani and *L. brasiliensis* complexes	18S rRNA	Q	94.1 parasites/ml blood	Schulz *et al.* (2003)
Leishmania infantum	DNA polymerase	Q	5 GE/PCR	Bretagne *et al.* (2001)
Leishmania species	kDNA (kinetoplast)	Q	0.1 parasites/PCR	Nicolas *et al.* (2002)
Toxoplasma gondii	B1 gene	Q	0.05 tachyzoite/PCR	Lin *et al.* (2000)
T. gondii	B1 gene	Q	0.75 parasites/PCR	Costa *et al.* (2000)
T. gondii	ITS1 of 18S rRNA	Q	1 bradyzoite	Jauregui *et al.* (2001)
T. gondii	B1 gene and a multicopy genomic fragment	Q	2.5 GE/PCR	Reischl *et al.* (2003a)
Trichomonas vaginalis	β-Tubulin gene	D	1.06 organisms	Hardick *et al.* (2003)
Trypanosoma cruzi	195-bp repeat[b]	Q	0.01 parasites/PCR	Cummings and Tarleton (2003)

[a]GE:, genome equivalents.
[b]195-bp repeat unique to *Trypanosoma cruzi*.

quantitation of parasitic burden during infection even at very low levels and therefore may be useful for monitoring the efficacy of vaccines at pre-symptomatic levels. This ability to quantify very low-levels of parasitaemia may also be useful to assess the efficacy of new treatments during drug-trials and for diagnosing low level infections or carrier states. In the future real-time PCR assays will be developed for other endemic and emerging parasitic diseases leading to improvements in the prevention, diagnosis, and treatment of these important human pathogens. Some of the areas of clinical parasitology in which real-time PCR assays have been introduced are reviewed below.

Detection of enteric parasites by real-time PCR

Cryptosporidia are coccidian protozoan parasites that infect a range of vertebrate hosts. More than ten species of *Cryptosporidium* have been described and within the species group *Cryptosporidium parvum* there are also different genotypes. A number of species and *C. parvum* genotypes have been demonstrated to cause disease in humans, *C. parvum* human genotype being the most prevalent, with *C. parvum* bovine genotype, *Cryptosporidium meleagridis*, *Cryptosporidium felis* and *Cryptosporidium canis* (in order of prevalence) also causing disease. These organisms cause a self-limited diarrhoeal disease in immunocompetent individuals; however, infection in immunocompromised patients can cause a severe life-threatening disease with prolonged diarrhoeal illness. Conventional diagnosis is based on identification of oocysts in stained faecal smears; however, this lacks sensitivity and specificity and subsequent species or genotype identification following detection is not possible. A real-time 5′ nuclease QPCR assay was described by Higgins *et al.* (2001), for the Cp11 and 18S rRNA genes of *C. parvum*. The assay was applied to the detection of oocysts in both human and cow and calf faecal samples; however, the probe hybridized to both human-infective and non-infective species limiting its usefulness. A LC hybridization probe assay was described by Limor *et al.* (2002) for the detection and differentiation of most *Cryptosporidium* species and C. *parvum* genotypes. The assay had a high sensitiv-

ity (five oocyst detection level) and differentiated between species and genotypes by melting curve analysis. Although this assay appears useful it requires further validation through the application of it to human faecal samples from patients with *Cryptosporidium* infection. A 157-bp fragment of the β-tubulin gene of *C. parvum* was targeted by the real-time LC assay of Tanriverdi *et al.* (2002). The assay had a LOD of a single purified oocyst per reaction and was the first assay that could differentiate between *C. parvum* genotypes 1 and 2 on the basis of melting curve analysis.

Cyclospora cayetanensis is a coccidian parasite that has recently become recognized as an emerging pathogen of humans. *C. cayetanesis* causes prolonged diarrhoea, nausea, abdominal cramps, and anorexia and weight loss. Transmission of infection occurs through environmentally resistant oocysts that are shed in the faeces of infected individuals. Similarly to other enteric parasites diagnosis is dependent on faecal microscopy which lacks sensitivity and specificity. A 5′ nuclease assay was described which targeted the 18S rRNA sequence of *C. cayetanesis*, and was demonstrated to be sensitive and specific for the detection of *C. cayetanesis* oocysts (Varma *et al.*, 2003). The LOD of the assay was one oocyst per 5μl reaction; however, the application of the assay to faecal samples from patients has not been reported.

Entamoeba histolytica is an intestinal protozoan parasite which is endemic in many parts of the world and is responsible for millions of cases of dysentery and liver disease per year (Blessmann *et al.*, 2002). Laboratory diagnosis of *E. histolytica* infection has been based on faecal microscopy of fresh or fixed stool samples. This method is hampered by the existence of the closely related but non-pathogenic species *Entamoeba dispar* that is morphologically indistinguishable from *E. histolytica*. Conventional PCR methods have been described for the detection and differentiation of these two species; however, these are time consuming and have limited throughput. Blessmann *et al.* (2002) developed a LC assay which targeted a 310-bp fragment of the high copy number ribosomal DNA. The assay utilized two sets of primers, each specific for one of the two species although the rDNA of the two species has a 98.4% sequence homology. The rDNA resides on an episomal plasmid that has a high copy number per cell (approximately 200 copies) which gave the assay a high sensitivity and a LOD of 0.1 parasite/g faeces. This represents a significant increase in sensitivity of detection when compared to microscopy, which has a sensitivity of approximately 70% if a single faecal smear is examined.

The flagellate protozoan pathogen *Giardia lamblia* is the most commonly diagnosed intestinal parasite in the world. Conventional diagnosis is reliant on microscopic examination of stool samples which is time-consuming and the sensitivity of microscopy is dependant on the number of samples examined, the use of concentration techniques and ultimately on the experience of the technician. Alternative methods including antigen detection by ELISA have been developed and provide rapid detection within 1–2 hours; however, these methods are not sensitive enough to detect low levels of cysts in stools. PCR-based methods may provide enhanced sensitivity of detection particularly for stool samples containing low numbers of parasites. Verweij *et al* (2003) developed a TaqMan real-time PCR targeting the small subunit (SSU) rRNA gene of G. *lamblia* which was applied to stool samples from 104 patients previously shown to be either microscopy and/or antigen positive and stool samples from 24 patients microscopy and/or antigen negative. The real-time assay detected G. *lamblia* in 102 of the 104 samples and all 24 microscopy and/or antigen negative samples were negative in the real-time assay. Ten of the 104 samples were negative by microscopy but positive by PCR and antigen detection, indicating that these methods are more sensitive than microscopy for the detection of G. *lamblia* in stool samples.

Phylogenetic sequence analyses have identified that there are two major groups of G. *lamblia* which cause human infection, assemblage A and assemblage B. The clinical or epidemiological significance of the two assemblages is poorly understood with studies reporting differing prevalences for the two groups within symptomatic and asymptomatic patients. Current detection methods do not differentiate between these two genotype groups; however, more complete data on the prevalence of them may enhance our understanding of epidemiology of

infection. Ng *et al* (2005a) recently described a G. *lamblia* multiplex TaqMan assay which can detect and differentiate between assemblage A and assemblage B strains. The assay targets the 18S rRNA with individual forward primers, one specific for A and one specific for B, plus a single reverse primer and individual scorpion probes specific for individual genotype groups. The assemblage A-specific component of the assay was demonstrated to be approximately ten fold more sensitive than the assemblage B assay with a LOD of 20 trophozoites per PCR for A and 200 per PCR for B respectively. Sensitivity of detection was increased at least 16-fold using a sequence-specific oligonucleotide to capture the DNA target giving an improved LOD of 1.25 trophozoites and 12.5 trophozoites per PCR for assemblages A and B respectively. To validate the assay the authors applied it to the detection of G. *lamblia* in 97 human stool samples the results being compared to microscopy using saline wet-mount staining with Lugol's iodine. All PCR positive specimens were also tested in a commercial ELISA test for *Giardia* antigen. Microscopy detected *Giardia* in 22 samples, of which 21 were positive in the PCR assay. The PCR detected *Giardia* in 35 samples, 13 of which were negative by microscopy; however, 9 of these 13 were confirmed as trued positives in the *Giardia*-specific ELISA. The majority of the PCR-positive samples were genogroup B ($n = 32$) with only three samples being genogroup A; however, determining the significance of this type distribution will require larger prospective studies.

Microsporidia are obligate intracellular parasites that cause chronic diarrhoea in human immunodeficiency virus (HIV)-infected patients and in other immunocompromised populations. *Enterocytozoon bieneusi* is the most commonly encountered species in humans (Menotti et al., 2003); however, other species have also been associated with infection including *Encephalitozoon intestinalis* (Wolk et al., 2002). Diagnosis of intestinal infection is usually performed by faecal microscopy with stains such as the modified trichrome stain or Uvitex 2B; however, such methods lack sensitivity and cannot distinguish between species. A number of conventional PCR assays have been developed for the detection of these pathogens but they are not quantitative and are laborious. A LC PCR assay based on hybridization probes for the detection and quantification of *Encephalitozoon* species in faecal samples was described by Wolk *et al.* (2002). Melting temperature analysis differentiated the three *Encephalitozoon* species (*E. intestinalis, E. cuniculi,* and *E. hellem*). The assay targeted a 268-bp region of the 16s rRNA gene and demonstrated a sensitivity of 10^2 to 10^4 spores/ml faeces compared to the sensitivity of trichrome blue stain of $\geq 1.0 \times 10^6$ spores/ml faeces. A real-time QPCR assay for the detection of *Enterocytozoon bieneusi* DNA in stool specimens from immunocompromised patients with intestinal microsporidiosis was described by Menotti *et al.* (2003). Similarly to the assay of Wolk *et al.* (2002) this 5′ nuclease assay targeted the 16s rRNA gene of *E. bieneusi* and provided a quantitative range of detection of 10^1 to 10^7 copies of *E. bieneusi* DNA with a LOD of 10 copies. This assay was applied to the detection and quantitation of *E. bieneusi* DNA in the stools of patients in a randomized comparative trial of the efficacy of fumagillin for the treatment of microsporidial intestinal disease in the immunocompromised. The parasitic burden (*E. bieneusi* DNA in stool) remained stable in the placebo group ($n = 6$); however, the parasitic burden in patients treated with fumagillin ($n = 6$) dropped to below the limit of detection. The real-time PCR assay performed better than semi-quantitative assessments by microscopy to measure parasitic burden.

Dientamoeba fragilis is a protozoan parasite which infects the mucosa of the large intestine. It has been associated with gastrointestinal disease although infection can be either symptomatic or asymptomatic. Symptoms that have been associated with infection include diarrhoea, abdominal pain, anorexia, nausea, vomiting, fatigue, and weight loss. Diagnosis is based on detection of trophozoites in stained faecal smears; however, *D. fragilis* trophozoites can be easily overlooked because they are pale-staining and their nuclei may resemble those of *Endolimax nana* or *Entamoeba hartmanni*. PCR-based assay may provide more sensitive and specific detection to aid in the diagnosis of *D. fragilis* infection. Stark *et al.* (2006) described a TaqMan assay targeting the SSU rRNA of *D. fragilis* which was applied to the detection of *D. fragilis* in 200 stool samples from patients with gastrointestinal disease. The

results of the TaqMan assay were compared to a conventional PCR assay also targeting the SSU rRNA gene and microscopic examination of stained faecal smears using modified iron-haematoxylin. The TaqMan assay detected *D. fragilis* in 51 of the samples, 49 of which were positive by microscopy, and 48 of these were positive in the conventional PCR assay. The study indicated that the TaqMan assay was slightly more sensitive than microscopy for the detection of *D. fragilis* in faecal samples; however, a larger prospective study is required to establish the clinical sensitivity of the assay for the routine diagnosis of *D. fragilis*-associated disease.

Detection and quantitation of malaria parasites by real-time PCR

Malaria is a major global health problem causing up to 500 million clinical cases and 2.7 million deaths annually. Malaria is caused by a parasitic protozoan of the genus *Plasmodium*. The gold standard for malaria diagnosis is the microscopic examination of a blood smear; however, this may lack sensitivity when patients have low level parasitaemias. In addition blood smear analysis cannot provide a quantitative assessment of the level of parasitaemia in the patient which may be relevant for monitoring the effectiveness of anti-malarial therapy by measuring reductions in parasitaemia. Hermsen *et al.* (2001) described a real-time 5′ nuclease QPCR assay targeting the 18s rRNA gene of *Plasmodium falciparum*. The assay was used to measure parasitaemia in five experimentally infected volunteers and the results compared to microscopic analysis of blood smears. The numbers of parasites detected by blood smear showed a good correlation with the real-time PCR assay. Similarly, Witney *et al.* (2001) developed a 5′ nuclease QPCR assay for detection of 18S rRNA gene of *Plasmodium yoelii*. They used the assay to detect and quantify liver-stage parasites in experimentally infected mice as a means of evaluating the efficacy of pre-erythrocytic malaria vaccines. The authors demonstrated the assay to be a rapid and reproducible way of accurately measuring liver stage parasitic burden and vaccine efficacy in rodent malarial models. Real-time QPCR assays for malarial parasites may also be useful for monitoring the effectiveness of anti-malarial therapy during anti-malarial drug trials (Lee *et al.*, 2002).

Mangold *et al.* (2005) described a LC SYBR green assay targeting a conserved region of the 18S rRNA gene of *Plasmodium* spp. which differentiated between individual species by melting curve analysis post amplification. The assay could detect and identify *P. falciparum*, *P. vivax*, *P. ovale*, and *P. malariae* directly from blood samples spotted onto filter paper. The assay was validated by applying it to 358 patient blood samples in a blinded study. Seventy-four of the blood samples were positive in the LC assay, with the species *P. falciparum*, *P. vivax*, and *P. ovale* being detected and identified. The PCR results of all 74 samples correlated 100% with the previous microscopy results; however, the LC assay detected three mixed infections of *P. falciparum* and *P. ovale*, only one of which was detected by microscopy. Two samples which were positive by microscopy were negative in the PCR assay; however, the parasitaemia levels were considered to be below the LOD of the assay. The assay appears to be a useful tool to assist in the diagnosis of malaria either as a complementary or confirmatory test to routine microscopy.

Real-time PCR applications in clinical virology

Real-time PCR methods have proven to be useful tools in the diagnosis and management of a wide range of viral diseases (Niesters, 2002). Table 13.3 lists some of the real-time PCR assays that have been described which target a wide range of viral disease agents. The speed and specificity, plus their ability to directly quantify the target organism without post-PCR manipulations have made them the method of choice for a wide range of applications. Some of the areas in which real-time PCR methods have been applied in clinical virology will be reviewed below.

Detection of respiratory viruses by real-time PCR

Viruses that commonly cause respiratory disease include influenza virus, parainfluenza virus, respiratory syncytial virus, adenovirus, and the more recently recognized human metapneumovirus. The recent emergence of the corona virus-associated severe acute respiratory syndrome (SARS) and influenza A H5N1 has highlighted the need for rapid and specific diagnostic tests to differentiate between patients suffering from

more common respiratory pathogens and those infected with these highly pathogenic viruses.

Influenza virus is a common respiratory pathogen that is highly contagious and responsible for considerable morbidity and mortality especially during the winter months. The elderly, the immunocompromised and patients with respiratory problems such as emphysema and asthma are particularly vulnerable to developing severe illness. Rapid detection of the virus may also be useful for timely initiation of antiviral therapy using new anti-influenza virus compounds such as neuraminidase inhibitors. Conventional detection methods for influenza virus include cell culture, shell vial culturing, antigen detection and serological tests. However, these tests are either too slow to allow timely diagnosis or they lack sensitivity and specificity. The first real-time PCR assay for the detection of influenza virus was described by van Elden *et al.* (2001). The assay simultaneously detects influenza virus A and B using a multiplex 5′ nuclease format. Primers were designed to target the matrix protein of influenza A virus and the haemagglutinin gene segment of influenza B virus. The assays were demonstrated to have a sensitivity of 13 viral RNA copies and 11 viral RNA copies for influenza A virus and influenza B virus respectively. The assays were applied to the detection of influenza virus in 98 clinical samples from patients with upper or lower respiratory tract symptoms with the results being compared to conventional viral culture and shell vial culturing. The real-time PCR assay was found to be more sensitive than the conventional methods and follow-up of six symptomatic patients demonstrated that influenza virus could be detected for up to seven days after infection using the assay. During this time the patients were still symptomatic although the virus could no longer be isolated by culture after day one or two for the majority of the patients. Using serial dilutions of electron microscope counted stocks of influenza virus the authors demonstrated that the assay could be used in a quantitative format. The assay may be useful for determining viral load during antiviral therapy; however, this needs to be confirmed by future studies.

Respiratory syncytial virus (RSV) is the most common etiological agent of lower respiratory tract disease in children. The virus has

Table 13.3 Application of real-time PCR to the detection of clinically significant viruses

Organism	Gene(s) targeted	Detection/ quantitation	Sensitivity	Reference
Adenovirus	Hexon gene	Q	1.5×10^1 to 1.5×10^8 copies/PCR	Heim et al. (2003)
BK virus	T-ag gene	Q	10 copies/PCR	McNees et al. (2005)
BK virus	LT-ag gene	Q	10 copies/ml	Si-Mohamed et al. (2006)
Human cytomegalovirus (HCMV)	major immediate-early gene	Q	10^1 to 10^7 copies/PCR	Nitsche et al. (2000)
HCMV	US17 gene	Q	20 to 10^7 copies/PCR	Machida et al. (2000)
HCMV	UL83 region	Q	10 to 10^6 copies/PCR	Gault et al. (2001)
HCMV	UL83 gene	Q	10 to 10^4 copies/PCR	Griscelli et al. (2001)
HCMV	glycoprotein B	Q	2×10^3 to 5×10^8 CMV DNA copies/ml blood	Kearns et al. (2001b)
HCMV	DNA polymerase gene	Q	10^2 to 10^6 copies/PCR	Sanchez et al. (2002)
HCMV	US17 gene	Q	500 to 50,000 CMV DNA copies/ml plasma	Stocher and Berg. (2002)
HCMV	UL123 gene	Q	250 copies/ml plasma	Leruez-Ville et al. (2003)
HCMV	Immediate-early antigen gene	Q	Not reported	Li et al. (2003)
Coronavirus (SARS associated)	N gene	D	Not reported	Hui et al. (2004)
Dengue virus	Conserved core regions	Q	10 to 10^7 PFU/ml	Shu et al. (2003)
Dengue virus	N5S, capsid (C), UTR[a]	D	0.1 to 1.1 PFU detection limit	Callahan et al. (2001)
Ebola, Marburg, Lassa, Crimean-Congo haemorrhagic fever, Rift Valley fever, Dengue, yellow fever virus	L, GPC, NP, G2, 5' non-coding region, 3' non-coding region	Q	8.6 to 16 RNA copies/PCR	Drosten et al. (2002)
Epstein–Barr virus (EBV)	BALF5 gene	Q	2 to 10^7 copies EBV DNA/PCR	Kimura et al. (1999)
EBV	BNRF1 p143 gene	Q	100 to 10^7 copies/ml plasma or serum	Niesters et al. (2000)
EBV	EBNA1 gene	Q	10 copies/PCR	Stevens et al. (2002)
EBV	BZLF1 gene	Q	10 to 10^9 copies/PCR	Patel et al. (2003)
Enterovirus	5'-Non-coding region	D	11.8 enterovirus GE/PCR	Verstrepen et al. (2001)
Enterovirus	5'-Non-coding region	D	50 enterovirus GE/PCR	Monpoeho et al. (2002)
Enterovirus	5'-Non-coding region	D	Not reported	Nijhuis et al. (2002)
Enterovirus	5'-Non-coding region	D	0.1 $TCID_{50}$[b]	Rabenau et al. (2002)

Table 13.3 *continued*

Organism	Gene(s) targeted	Detection/ quantitation	Sensitivity	Reference
Enterovirus	5'-Non-coding region	D	510 copies/ml CSF	Lai et al. (2003)
Enterovirus	5'-Non-coding region	Q	<10 copies/PCR	Mohamed et al. (2004)
Hepatitis B virus (HBV)	S and X genes	Q	10^1 to 10^8 DNA copies/PCR	Abe et al. (1999)
HBV	Pre S gene	Q	373 to 10^8 copies/PCR	Pas et al. (2000)
Hepatitis C virus (HCV)	5'-Non-coding region	Q	10^3 to 10^7 copies/PCR	Martell et al. (1999)
HCV	5'-Non-coding region	Q	64 to 4,180,000 IU/ml serum	Kleiber et al. (2000)
HCV	Not reported	Q	10^3 to 10^7 RNA copies/PCR	Yang et al. (2002)
Herpes simplex virus (HSV)	DNA polymerase	D	Not reported	Espy et al. (2000b)
HSV	DNA polymerase	D	12.5 copies/PCR	Kessler et al. (2000)
HSV, varicella zoster virus (VZV), CMV	gD, gene, gG gene, DNA polymerase, gene 38	D	580 copies/ml (HSV-1), 430 copies (HSV-2)	van Doornum et al. (2003)
HSV	glycoprotein gene	Q	<10 to 10^8 copies/PCR	Ryncarz et al. (1999)
HSV	DNA polymerase	D	335 copies/ml cervical lavage fluid	Aryee et al. (2005)
HSV, VZV	gD, gene, gG gene, DNA polymerase	D	<10 plasmid copies/PCR	Weidmann et al. (2003)
HSV (genital herpes infection)	glycoprotein B gene	D	Not reported	Scoular et al. (2002)
HSV	gB region	D	Not reported	Namvar et al. (2005)
Human Herpesvirus 6 (HHV-6)	U22 gene	Q	10 GE/PCR	Collot et al. (2002)
HHV-6	U65-U66 genes	Q	10 GE/PCR	Gautheret-Dejean et al. (2002)
HHV-6	U67 gene	Q	1 GE/PCR (1 to 10^6 GE/PCR)	Locatelli et al. (2000)
HHV-6	large tegument protein gene	Q	10 GE/PCR	Kearns et al. (2001a)
HHV-6, human herpesvirus 7 (HHV-7)	DNA polymerase gene	Q	10 GE/PCR	Safrontez et al. (2003)
Human HHV-7	U100 gene	Q	14 GE/PCR	Fernandez et al. (2002)
Human herpesvirus (HHV-8)	Major capsid protein gene	Q	10 GE/PCR	Broccolo et al. (2002)
HHV-8	K5, ORF[c]25, ORF37, ORF47, ORF56	Q	5–50 GE/PCR	Stamey et al. (2001)

Virus	Gene target	Type	Detection limit	Reference
Human metapneumovirus (hMPV)	N gene	D	Not reported	Mackay et al. (2003)
hMPV	N, M, F, P, and L genes	D	100 copies/PCR	Cote et al. (2003)
hMPV	fusion protein gene	Q	10 copies/PCR	Kuypers et al. (2005)
hMPV	Nucleoprotein gene	D	5–10 copies/PCR	Maertzdorf et al. (2004)
hMPV	Nucleocapsid gene	D	Not reported	Sumino et al. (2005)
hMPV	L gene	D	Not reported	Scheltinga et al. (2005)
Influenza viruses A and B	matrix protein (A) and haemagglutinin (B)	Q	13 copies/PCR (A) and 11 copies/PCR (B)	van Elden et al. (2001)
Influenza viruses A and B	Matrix protein and fusion gene	D	50 copies/PCR	Boivin et al. (2004)
Influenza viruses A and B	Matrix protein	D	Not reported	Smith et al. (2003)
Influenza A virus	Matrix gene	D	120–350 copies/PCR	Stone et al. (2004)
Influenza A (H5)	Haemagglutinin	Q	Not reported	Ng et al. (2005)
Influenza A (H5N1)	H5, N1 and M genes	D	10^2–10^3 copies/μl	Payungporn et al. (2006)
JK virus	T-ag gene	Q	10 copies/PCR	McNees et al. (2005)
Norovirus	ORF 1-ORF 2	Q	10 copies/PCR	Kageyama et al. (2003)
Norovirus	Polymerase, capA	Q	5 copies/PCR	Pang et al. (2004)
Respiratory syncytial virus (RSV)	N gene	Q	2.3×10^{-3} to 2.3×10^2 PFU	Hu et al. (2003)
RSV	N gene	Q	Not reported	Gueudin et al. (2003)
RSV	Matrix protein gene/polymerase	Q	10 copies/PCR	Kuypers et al. (2004)
Human rhinovirus	5'-non-coding region	D	Not reported	Scheltinga et al. (2005)
Smallpox virus	Haemagglutinin	D	5 to 10 GE/PCR	Espy et al. (2002)
Smallpox virus	Haemagglutinin	D	25 GE	Sofi Ibrahim et al. (2003)
SV-40 virus	T-ag gene	Q	10 copies/PCR	McNees et al. (2005)
VZV	Gene 28 and gene 29	D	Not reported	Espy et al. (2000a)
West Nile virus (WNV)	Novel sequence	D	<1 PFU virus/PCR	Lanciotti et al. (2000)

[a] UTR, untranslated 3' region.
[b] TCID$_{50}$, 50% tissue culture infective dose.
[c] ORF, open reading frame.

been classified into two subgroups, A and B, based on antigenic and genetic variations in the structural proteins. Detection of RSV is based on virus culture, which is slow and relatively insensitive, or ELISA or immunofluoresence (IF) which also have limited sensitivity and specificity. Conventional RT-PCR assays have been described for the detection of RSV and have been demonstrated to be sensitive and specific diagnostic methods; however, they do require post-PCR analysis. Hu et al. (2003) described a real-time QPCR assay for the detection, subgrouping and quantitation of RSV. The assays utilized two pairs of primers targeting the conserved nucleocapsid (N) gene specific for detection and subgrouping of the A and B virus subtypes. The assays were demonstrated to be sensitive, 0.023 PFU or two copies viral RNA, and 0.018 PFU or nine copies of the viral RNA for the RSV A and B subtypes respectively. The assays were applied to the detection of RSV in 175 NPA from children with respiratory disease with the results being compared to culture and IF. Overall, the new assay was demonstrated to be about 40% more sensitive than culture and 10% more sensitive than IF for the detection of RSV in the samples in the study. In addition the new assay was able to subgroup the 36 RSV-positive samples directly with ten being identified as RSV A and 26 identified as RSV B. The assay also demonstrated a linear range of quantitation of between 2.3×10^{-3} to 2.3×10^{2} PFU per reaction when applied to a virus stock sample. This QPCR assay may prove useful in studies to determine the relationship between viral load and subgroup with disease severity.

Gueudin et al. (2003) described a LC RT-PCR assay for the quantitative detection of RSV in nasal aspirates of children with respiratory disease. The assay was developed to investigate the relationship between disease severity and the amount of RSV in nasal aspirates. Similarly to Hu et al. (2003) they targeted the N gene but they also included a second control QPCR assay in the study specific for the GAPDH housekeeping gene. RSV particles are mainly cell-associated and nasal aspirate samples are non-homogeneous therefore the quantitative GAPDH RT-PCR assay was used to standardize the quantitative results of the RSV assay. RSV and GADPH were quantified in nasal aspirates from 75 children hospitalized for acute respiratory tract disease with the results being compared to culture and IF. The QPCR assay was more sensitive than the other two methods with 42 samples positive in the assay, IF detected RSV in 31 samples and culture detected RSV in 34 samples. The RSV RNA quantitation results for the RSV positive samples were then compared to the severity of disease in the patients. The mean number of RSV RNA copies was slightly higher in the severe disease group (4.05×10^{7} copies) compared to the non-severe group (9.1×10^{6} copies) with the amounts of RNA in individual samples being roughly related to severity of disease. However, when the GADPH RNA quantitation results were taken into account as in indication of total cell number, the differences were no longer significant. The authors concluded that although the assay may not prove useful as a quantitative assay it is suitable as a qualitative diagnostic test for RSV infection.

Human metapneumovirus (hMPV) is a newly identified member of the *Paramyxoviridae* that has been isolated from NPA from patients with respiratory disease. The virus has been shown to cause a disease similar to RSV with patients displaying varying symptoms ranging from a wheeze to bronchiolitis. The virus is difficult to grow in cell culture therefore a rapid and sensitive diagnostic test is required to determine its role in respiratory disease in humans. Mackay et al. (2003) and Cote et al. (2003) both recently described real-time RT-PCR assays for the detection of hMPV. Mackay et al. (2003) developed a LC 5′ nuclease assay targeting the N gene sequence of hMPV that they compared to a PCR-ELISA assay they had also developed. The analytical sensitivity of the two assays was the same; however, the assay proved more sensitive when applied to the detection of hMPV in clinical samples. In the study of Cote et al. (2003) five sets of primers targeting the viral nucleoprotein (N), matrix (M), fusion (F), phosphoprotein (P), and polymerase (L) genes were compared in LC SYBR green quantitative RT-PCR assays. The assays were applied to 20 viral cultures with characteristics of hMPV cytopathic effect and the PCR positivity rates were 100%, 90%, 75%, 60% and 55% using the N, L,

M, P and F primers. The five sets of primers were also evaluated for their ability to detect hMPV in 10 NPA samples. All of the 10 samples were positive for hMPV using the N primers and the other primers detected hMPV in six, eight, three and eight of the samples using the M, F, P, and L gene primers respectively. The authors concluded that the N primers were the most suitable for a diagnostic assay due to their superior sensitivity of approximately 100 copies per reaction. The hMPV assays described in these two studies may prove useful in the detection of hMPV in samples from patients with respiratory tract infections; however, they require further validation through application to larger numbers of clinical samples. These studies will also provide more accurate assessment of the prevalence of hMPV as a cause of human respiratory disease.

Recent studies have demonstrated that hMPV can be grouped into at least four genetic lineages based on phylogenetic analysis of sequences obtained for part of the fusion protein and the complete attachment protein open reading frames (van den Hoogen et al., 2004). It is important that real-time assays can detect all four genetic lineages of hMPV with an equal sensitivity. Maertzdorf et al. (2004) developed a TaqMan assay targeting a conserved region of the nucleoprotein gene using primers and a probe designed to detect virus strains from all known genetic lineages of hMPV. The assay was validated by applying it to the detection of hMPV in clinical samples previously found to be either positive or negative in a conventional RT-PCR assay specific for the L gene of hMPV. The results of the real-time assay correlated 100% with those of the conventional assay indicating that this assay may be useful as a diagnostic test for hMPV. However, only small numbers of samples were investigated in the validation study therefore larger clinical trials are required to validate the use of the assay in the diagnosis of hMPV infection.

Detection of herpes viruses by real-time PCR

The human herpes virus family includes the clinically important pathogen herpes simplex viruses 1 and 2 (HSV-1 and HSV-2), varicella-zoster virus (VZV), and the human herpes viruses 6, 7, and 8 (HHV-6, HHV-7, HHV-8). HSV causes a wide spectrum of clinical disease with a variety of anatomical sites being infected including the skin, lips, oral cavity, eyes, genital tract and the CNS. Disseminated HSV infection is a serious condition that can occur in immunocompromised patients and in neonatal infection acquired by transmission of the virus through the infected birth canal of the mother. Although CNS infection by HSV can be fatal, effective therapeutic management of infections is possible; however, this requires rapid diagnosis to enable timely administration of antiviral therapy. Conventional methods for the detection of HSV are based on cell culture, which is the current gold standard for all specimens except CSF samples from patients with CNS infections. PCR has become the gold standard for diagnosing CNS infections; however, conventional PCR methods were slow and lacked specificity. This has led to the development of a number of real-time PCR assays for the rapid detection of HSV in clinical samples. Kessler et al. (2000) described a LC 5' nuclease assay targeting the DNA polymerase gene of HSV, which was compared to an 'in-house' conventional PCR for the detection of HSV in 59 CSF specimens. The real-time assay and the conventional PCR assay were demonstrated to have very similar levels of sensitivity. A real-time QPCR assay for the detection of HSV was described by Ryncarz et al. (1999). The assay was demonstrated to have a linear range of detection of <10 to 10^8 copies of HSV DNA per reaction. The assay was applied to the detection of HSV in 335 genital tract specimens from HSV-2 sero-positive patients and 380 CSF samples with the results being compared to tissue culture and an 'in-house' gel-based liquid hybridization system. Overall the new assay was slightly less sensitive than the gel-based liquid hybridization system.

Scoular et al. (2002) compared a LC-based real-time PCR assay targeting the glycoprotein B gene with viral culture for the diagnosis of genital herpes in a genitourinary medicine (GUM) clinic. Two swabs were taken from 236 patients with clinical symptoms suggestive of genital herpes, one swab being tested using the LC assay and one tested using routine viral culture. One hundred and nine patients had a positive result for HSV by either culture and/

or PCR, with 88 patients being both culture and PCR positive and twenty-one patients positive only by PCR. Overall the use of PCR increased the sensitivity of detection by 24% leading the authors to advocate the use of real-time PCR as an efficient, sensitive and rapid test to aid the diagnosis of genital herpes infection. A similar study by Ramaswamy et al. (2004) compared the HSV LC assay of Espy et al., (2000b) with viral culture for the diagnosis of genital herpes in routine clinical practice. Samples were collected from 233 GUM clinic attendees with symptoms of genital herpes infection. HSV was detected in 79 samples by culture; however, PCR detected HSV in 132 samples again demonstrating the increased sensitivity of real-time PCR in comparison with routine culture. The authors noted that PCR significantly increased HSV detection in both early (< 5 days) and late (> 5 days) presenters and in both first and recurrent episodes. HSV detection and typing by LC PCR is very rapid with results available within four hours leading to a significant reduction in labour compared to culture. Although the sensitivity of virus culture is suboptimal compared to PCR for the detection of HSV in genital samples a recent survey in the UK found that only 20% of laboratories routinely tested genital samples by PCR with the remaining 80% using viral culture as the method of choice (Geretti and Brown, 2005). However, 68% of the laboratories routinely used HSV PCR for DNA detection in CSF therefore the authors recommended that clinicians and virologists should discuss ways of implementing PCR-based testing for genital swabs to enable greater diagnostic accuracy.

VZV is the causative agent of chicken pox and shingles; however, in immunocompromised patients it can cause severe disease of the CNS or respiratory tract. Disseminated CNS infections can be fatal; however, rapid diagnosis allows timely initiation of antiviral therapy. Weidmann et al. (2003) developed a panel of three 5′ nuclease assays on the LC for the detection of HSV-1, HSV-2, and VZV in clinical samples. All three real-time assays demonstrated a high analytical LOD (<10 plasmid copies per assay) similar to the authors 'in-house' nested conventional PCR assays. The authors applied the new assays to the detection of HSV and VZV in 106 clinical

samples including CSF, skin swabs, biopsies, vitreous body samples, blood samples and BAL. These were previously collected from immuno-compromised patients, tested in the 'in-house' assays and then stored at −20°C for up to five years before re-testing in the real-time assays. Results of the 'in-house' assay previously determined that 46 samples were positive for HSV-1, four were positive for HSV-2, and 56 were positive for VZV. In comparison to the 'in-house' assays, the sensitivity of the real-time assays for the detection of HSV-1, HSV-2, and VZV in the samples was 95%, 100%, and 96% respectively. A few samples, which had previously shown to be positive in the 'in-house' assays, were negative in the real-time assays; however, this may be due to template degradation during storage of the sample for up to five years prior to re-testing.

In a similar study, van Doornum et al. (2003) described a set of 5′ nuclease assays for the detection of HSV-1, HSV-2, VZV, and human cytomegalovirus (HCMV). A universal control DNA, seal herpesvirus type 1 (PhHV-1) was added to each clinical sample prior to extraction and testing to monitor the DNA extraction and subsequent DNA amplification. Each of the viral targets was detected in separate assays because multiplexing the primers reduced the sensitivity of detection. The PhHV-1 was added at a low concentration to each clinical sample and was then extracted and detected to provide an internal control for inter-assay precision and reproducibility. This novel internal control system would be useful in other real-time assays for microbial pathogens and is a model system for other pathogens.

Detection of enterovirus infections by real-time PCR

Enteroviruses (EV) are common causes of CNS infections including viral meningitis and encephalitis. It has been estimated that they are responsible for 70–80% of the cases of viral meningitis. The current gold standard for the diagnosis of enteroviral infections of the CNS is growth of the virus in cell culture. However, this test is slow requiring several days to produce results plus there can be difficulties propagating some EV types in cell culture. Rapid and reliable diagnosis of EV in CNS infections would provide a rational basis for antimicrobial therapy and limit the need for unnecessary procedures and irrelevant treatment. This has led to the development of alternative molecular based methods for the detection of EV in clinical specimens. The first assays to be described were conventional RT-PCR assays often using a nested format but more recently real-time assays have been developed for the rapid diagnosis of EV infections. Verstrepen et al. (2001) developed a 5′ nuclease real-time QPCR assay utilizing a single-tube RT and amplification format. The assay targeted the 5′ non-coding region, which was also targeted by several other real-time assays, and was demonstrated to have a LOD of 11.8 EV genome equivalents (GE) per PCR reaction. The assay was applied to the detection of EV in 70 CSF specimens from patients with suspected viral meningitis, with 19 of the samples being positive in the assay and 17 being positive by culture. The assay of Monpoeho et al. (2002) which included an internal inhibition control was applied to the detection of EV in 104 CSF samples collected during an outbreak of EV meningitis with the results being compared to viral culture. The assay detected EV in 61 of the CSF samples; however, 41 of these were culture negative for EV. Using a 'consensus positive' for EV infection, defined as a culture-positive or RT-PCR positive result and clinical evidence of EV meningitis the sensitivity and negative predictive of the assay was 96.8% and 95.3% respectively. Further application of real-time PCR assays for EV, and other agents of viral meningitis, to samples from patients with viral meningitis will validate their use as the primary diagnostic test for viral CNS infections. Introduction of rapid tests will hopefully improve the diagnosis of viral CNS infections and lead to improvements in the management and treatment of patients.

Detection of viral haemorrhagic fever viruses by real-time PCR

Viral haemorrhagic fever (VHF) is a clinical syndrome caused by a number of different viruses including Marburg virus (MBGV), Ebola virus (EBOV), Lassa virus (LASV), Junin, Machupo, Sabia, and Guanarito viruses, Rift Valley fever virus (RVFV), Hanta viruses, yellow fever virus (YFV) and Dengue virus (DENV). Clinical manifestations of VHF infections include diarrhoea, myalgia, cough, headache, pneumonia, encephalopathy, and hepatitis and VHF can prove to be fatal. The most characteristic manifestation is haemorrhage although non-haemorrhagic infections do occur. Definitive diagnosis is usually based on laboratory tests with rapid definitive diagnosis being important to identify the causative virus so specific treatment can be initiated and perhaps more importantly to exclude non-VHF infections. Rapid diagnosis is also important for case management such as the use of isolation procedures and the tracing of patient contacts. Conventional PCR assays have been described for all of the agents of VHF; however, many of them are relatively slow and require a separate RT step prior to amplification and detection of the amplification product. Drosten et al. (2002) recently described a set of six one-step real-time PCR assays for the detection of MBGV/EBOV, LASV, CCHFV, RVFV, DENV, and YFV. The assays can be performed in two separate runs on the LC with a pair of different universal cycling conditions. The RVFV, DENV, and YFV assays were 5′ nuclease assays and the MGBV/EBOV, LASV and CCHFV assays were SYBR Green intercalation assays. The six assays were demonstrated to be specific for the individual agents and their analytical sensitivity using spiked negative human plasma was determined to be between 1,545 to 2,835 viral GE/ml serum (8.6 to 16 RNA copies per assay). The assays were applied to the detection of VHF agents in 30 samples from suspected cases of VHF with RNA extraction and PCR detection taking less than 6 h. The amounts of VHF RNA in the samples were quantified using a QS run in parallel with the patient samples in the assay. The patients were found to have high viral DNA loads in their serum with virus DNA concentrations in acute phase patients being orders of magnitude above the detection limit of the assays indicating that the assays were sufficiently sensitive to diagnose VHF during acute phase infection.

Additional real-time PCR assays targeting dengue virus (DENV) the causative agent of dengue fever, one of the VHF agents have also been described. Shu et al. (2003) developed group and serotype-specific on-step SYBR green LC assays targeting conserved sequences in the core region of the DENV genome. The assays were demonstrated to detect, differentiate and quantify all four different DENV serotypes in acute phase serum. The assays had a dynamic range of detection of 10 to 10^7 PFU/ml for cell-culture derived DENV and a detection limit of between 4.1 and 10 PFU/ml depending on the serotype. The assays were applied to the detection of DENV in 193 acute phase serum samples from confirmed dengue fever patients with the real-time assays appearing to be more sensitive than the conventional cell culture method. In a similar study, Callahan et al. (2001) developed 5′ nuclease assays for the detection and differentiation of serotypes 1 to 4 and a group-specific assay for the detection of DENV. The assays were applied to 67 dengue viraemic human sera and demonstrated high sensitivity and specificity.

The suspicion of a VHF infection in a patient immediately raises a number of questions which require answers as soon as possible. Most importantly, the infectious agent must be determined so its potential for further spread can be assessed and appropriate control measures and contact tracing can be instigated. Real-time PCR methods may prove very useful to determine the infectious agent in such circumstances.

Viral genome quantitation by real-time PCR

Viral genome quantitation has become increasingly useful in the diagnosis and management of viral disease. For example, viral quantitation or viral load (VL) testing by real-time PCR has proven useful as a diagnostic marker in human cytomegalovirus (HCMV) infection with such testing aiding the diagnosis of active disease. It is also a useful prognostic marker in infections such as HIV for assessing disease progression and as a therapeutic marker for monitoring the efficacy of antiviral chemotherapy in patients with chronic HIV, HBV or HCV infection. The measurement of VL has proven useful in assessing an individuals' infectivity, for example to estimate the risk of vertical transmission of HIV between mother and child. The role of VL testing by PCR in clinical virology has been reviewed by Niesters (2001), Berger and Preiser (2002) and Mackay et al. 2002. The first generation of QPCR methods to be described used end-point

detection utilizing internal or external controls of known concentrations, which were amplified in parallel with the samples of interest. Following PCR cycling the unknown test samples were compared to the control values and a quantitative value assigned to the test samples. Because the final amount of accumulated product at the end of the PCR process is very susceptible to minor variations in reagents and sample matrices, there are limitations on the accuracy of QPCR methods based on end-point detection. Real-time PCR methods provide much more accurate quantitation because reactions are quantified by the cycle at which amplification is detected rather than the amount of PCR product accumulated after a fixed number of cycles. Using fluorescent monitoring of the PCR reaction, the increase in specific amplification product can be monitored in real-time, which allows the accurate quantification over six orders of magnitude of the DNA or RNA target sequence.

Quantitation of HCMV by real-time QPCR has become a very useful diagnostic test for the management of immunocompromised patients following solid organ or bone marrow transplantation (BMT), in individuals with HIV infection and in other patients on immunosuppressive therapy. HCMV can cause a severe disease in such individuals and accurate quantitation methods have demonstrated that patients with high viral load are at greater risk of developing disease. In addition VL testing is useful for monitoring the efficacy of antiviral therapy. Various genes have been targeted in these assays. Nitsche et al. (2000) compared ten different primer and probe combinations targeting the major immediate-early gene locus in a 5′ nuclease assay for HCMV quantitation, the results being compared to the pp65 antigen detection assay. Sensitive detection over six orders of magnitude was demonstrated with the HCMV DNA quantitation results correlating with the pp65 antigen assay in plasma samples from bone marrow transplant (BMT) patients. The US17 gene was targeted in the 5′ nuclease QPCR assay of Machida et al. (2000) in a pilot study to detect HCMV in BMT patients; however, only one patient was demonstrated to have definite HCMV disease. Kearns et al. (2001b) compared a LC QPCR assay targeting the glycoprotein

B gene with a 5′ nuclease quantitative HCMV assay previously validated in a comparison with the pp65 antigenaemia assay for the detection of HCMV in whole blood extracts from immunocompromised patients. The viral load ranges correlated between the two assays with a LOD of approximately 10 copies and a linear range of quantitation of between 2×10^3 to 5×10^8 copies. The authors also successfully applied the LC QPCR assay to the detection of HCMV in urine and respiratory samples (Kearns et al., 2001c, 2002b).

Griscelli et al. (2001) described a 5′ nuclease QPCR assay that targeted the UL83 gene of HCMV and facilitated normalization of the quantitative data through the detection and quantitation of a human housekeeping gene. The UL83 gene encoding the pp65 antigen copy number was quantified and the glyceraldehyde-3-phosphate dehydrogenase (GAPDH) gene copy number was used to normalize the data. A QS containing the both the UL83 and GAPDH genes was used to generate a standard curve with the assay demonstrating a linear range of detection of 10 to 10^4 copies per PCR. The QPCR assay was shown to be more sensitive than the pp65 antigen assay in BMT patients and has the potential to be multiplexed into a single-tube assay. A multiplex 5′ nuclease QPCR for the detection of the HCMV DNA polymerase gene with normalization of the data using a human apoprotein B (HAPB) gene assay was also described (Sanchez et al., 2002). The assay was applied to the detection and quantitation of HCMV in plasma from adult lung transplant patients. The second PCR also acts as an internal control (IC) against PCR inhibition, which can lead to false-negatives. All of these assays provided quantitation data, based on the assumption that the amplification efficiencies of the QS and samples were equal which is not always the case. The use of ICs in a LC QPCR assay for HCMV was investigated by Stocher and Berg (2002). In their LC assays they spiked patient samples with a known amount of heterologous competitor DNA and used an algorithm to normalize the data due to the varying amplification efficiencies between the QS and samples. The IC was a plasmid construct that could be co-amplified with the same HCMV primers but contained the

neomycin phosphotransferase gene. This method successfully controlled for PCR inhibition in the patient samples and is a model for the design of other real-time QPCR assays for the detection of other microbial targets. In these studies HCMV VL in various blood compartments such as peripheral blood leukocytes (PBL) and plasma were investigated; however, the usefulness of VL detection in whole blood (WB) in comparison with PBL was not reported. Using a LC assay, Mengelle et al. (2003) demonstrated that automated DNA extraction and quantitation of HCMV DNA from WB samples provided acceptable results when compared to manual extraction and detection. The authors also demonstrated good correlation between viral loads in WB and PBL, leading them to conclude that the use of WB instead of separated, counted PBL can be used to monitor HCMV-infected patients. Li et al. (2003) determined HCMV loads in WB samples from transplant patients and demonstrated that the loads correlated with those determined using the antigenaemia assay. They demonstrated that antigenaemia values of 1 to 2, 10 and 50 positive cells per 2×10^5 leucocytes correlated with HCMV viral loads of 1000, 4000, and 10,000 copies/ml respectively, and proposed these as cut-off points for initiating antiviral therapy in patient groups with high, intermediate, and low-risk of HCMV diseases. An example of the results of monitoring HCMV viral load in a paediatric severe combined immunodeficiency patient using real-time PCR is presented in Fig. 13.1.

Epstein–Barr virus (EBV) infects more than 90% of the population worldwide and in immunocompetent individuals the virus establishes a life-long asymptomatic infection. In a minority of immunocompetent individuals the virus causes infectious mononucleosis and EBV-related malignancies such as Burkitt's lymphoma and nasopharyngeal carcinoma. In immunocompromised patients, active EBV infection is a strong risk factor for the development of post-transplant lymphoproliferative disease (PTLD), AIDS-related lymphoma, and X-linked proliferative syndrome. The monitoring of EBV load in transplant patients has been shown to be a useful tool in the diagnosis and management of PTLD. A number of real-time

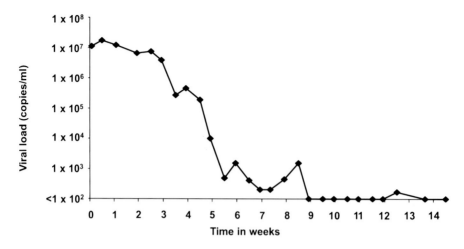

Figure 13.1 Monitoring HCMV viral load in a paediatric severe combined immunodeficiency patient. HCMV viral load was measured in EDTA blood specimens from the patient using a LC real-time PCR assay targeting the glycoprotein B gene. The patient was commenced on ganciclovir, cidofovir and foscarnet on weeks 0, 1 and 2 respectively. This demonstrates the usefulness of viral load testing to monitor response to antiviral therapy.

QPCR assays for EBV VL testing have been described. Kimura et al. (1999) developed a 5' nuclease QPCR assay targeting the BALF5 gene encoding the viral DNA polymerase that had a linear range of 2 to 10^7 copies EBV DNA. The assay was used to measure the EBV VL in peripheral blood mononuclear cells (PBMNC) in patients with symptomatic EBV infection. The authors established the usefulness of the assay for diagnosing symptomatic EBV infection and proposed a diagnostic cut-off value of $10^{2.5}$ copies/μg PBMNC DNA (for symptomatic EBV disease). In this study all patients with PTLD and chronic active EBV infection had EBV viral loads much greater than this cut-off value. Niesters et al. (2000) applied a 5' nuclease QPCR assay targeting the BNRF1 p143 gene encoding the non-glycosylated membrane protein, to the quantitation of EBV in transplant patients. Plasma and serum samples were screened over a range of between 100 and 10^7 copies of DNA per ml using two sample preparation methods. EBV DNA was detected in 19.2% of immunosuppressed solid-organ transplant patients without symptoms of EBV disease with a mean VL of 440 copies/ml. EBV DNA was also detected in all transplant patients with PTLD with a mean VL of 544,570 copies/ml and no EBV DNA was detected in healthy individuals or in non-immunocompromised control groups and a mean

VL of 6400 copies/ml was detectable in patients with infectious mononucleosis. The detection of latent EBV in the immunosuppressed patients without symptoms of EBV disease is in keeping with previous reports suggesting the presence of EBV DNA alone does not always indicate active EBV disease. Two LC QPCR assays have also been described for EBV VL testing, one targeting the EBNA1 gene (Stevens et al., 2002) and one targeting the BZLF1 gene (Patel et al., 2003). These assays were also demonstrated to be useful for the diagnosis of active EBV disease. Although these QPCR assays have proven useful in the diagnosis and management of patients with active EBV disease, inter-laboratory standardization of these methods has not yet been achieved. Future prospective studies investigating the EBV VL in immunosuppressed patients using these assays will help to establish clinically relevant cut-off values, preferred sample types and extraction methods, and standardized QS and ICs.

Approximately 300 million people worldwide are chronically infected with Hepatitis B virus (HBV). The measurement of HBV DNA in patient serum has become a very useful tool in the management of HBV infection. HBV VL testing is useful to assess treatment response in patients undergoing antiviral therapy and can also be a marker for the emergence of resistant viral strains in patients undergoing lamivudine or famciclovir treatment. HBV VL testing is also useful to assess the infectivity of HBV carriers, for example to estimate the risk of vertical transmission from female HBV carriers to their infants. Conventional QPCR assays were developed for the quantitation of HBV in patients including several commercially produced tests. Although these commercially produced assays have proven to be useful many have a limited linear range of quantitation. The development of real-time PCR platforms has led to the development of a number of more rapid, real-time QPCR assays for HBV VL testing. Abe et al. (1999) described three 5′ nuclease assays, two targeting the HBV surface gene (S gene) and one targeting the X gene. The assays were sensitive (LOD= 10 copies/PCR) and had a linear range of quantitation of 10^1 to 10^8 DNA copies per PCR. The assays were applied to 46 serum samples from patients with chronic HBV disease and the results compared to a commercially produced signal amplification assay with enzymatically labelled branched DNA (bDNA). The results between the two assays correlated; however, the real-time assay was more sensitive with eight of the 46 samples only being positive in this assay and not in the signal amplification assay. These eight samples were demonstrated to have HBV DNA levels below the detection limit of the bDNA assay with the real-time QPCR assay being approximately 10^4 to 10^5 times more sensitive. The real-time QPCR assay is a sensitive assay with a wide linear range of quantitation however; the usefulness of this assay as a tool for the management of chronically infected HBV patients has yet to be established. The real-time QPCR assay of Pas et al. (2000) targeted the pre-S gene of HBV and was demonstrated to have a dynamic range of detection of 373 to 10^{10} genome copies per ml. Application of the assay to the monitoring of four patients during antiviral treatment demonstrated the results of the new assay correlated closely with those from two commercial assays, the HBV Digene Hybrid Capture II Microplate assay and the Roche HBV MONITOR assay. All of the assays detected a rapid decrease in HBV VL in the four patients following initiation of lamivudine treatment. To obtain the same dynamic range of detection demonstrated by the real-time QPCR assay, some of the samples had to be re-tested

in the commercial assays following a dilution or concentration step. This wide range of quantitation is a clear advantage of the real-time assay plus the likelihood of false-positives through amplicon carryover is also eliminated.

The RNA virus, Hepatitis C (HCV) is the most common cause of non-A and non-B viral hepatitis. Most infection is asymptomatic; however, in approximately 70–80% of patients, the infection becomes chronic. Chronic HCV infection can lead to a range of clinical outcomes from mild non-progressive liver damage, to severe chronic hepatitis that can progress to cirrhosis, end-stage liver disease, and hepatocellular carcinoma. Chronic infection is characterized by persistent viraemia, which is the standard diagnostic marker for chronic infection. HCV VL testing has only a limited value for determining disease prognosis; however, it is a valuable tool for monitoring the success of antiviral therapy. HCV VL testing is also useful for estimating infectivity and determining the risk of mother-to-infant transmission.

Because HCV is an RNA virus HCV VL testing requires a reverse transcriptase step prior to PCR amplification and detection. Martell et al. (1999) described a 5′ nuclease single-tube QPCR assay targeting the 5′ non-coding region of the HCV genome. The assay demonstrated a linear 5-log range of detection with very low intra-assay and inter-assay coefficients of variation. Application of the assay to 79 RNA samples from the sera of infected patients revealed a correlation between the real-time assay and the Quantiplex 2.0 bDNA assay and the Superquant assay. A second real-time assay was described by Kleiber et al. (2000) which also targeted the 5′ non-coding region of the HCV genome with the assay including an IC for inhibition. All HCV genotypes were amplified with equal efficiency and the assay had a dynamic range of quantitation of 64 to 4,180,000 IU/ml and a LOD of 40 IU/ml. The application of the assay to interferon-treated patients distinguished responders from non-responders and responder-relapsers. Both of these assays may prove useful in monitoring the efficacy and response to antiviral treatment in patients with chronic HCV infection.

There are six species of human adenovirus (HAdV) with 51 types associated with a variety of diseases affecting all organ systems. Recently, studies have shown that HAdV can cause serious life-threatening infections in immunocompromised hosts such as haematopoietic stem cell transplant patients. These HAdV infections are frequently associated with a viraemia with one or more organ systems being involved producing symptoms including pneumonia, meningitis, hepatitis, rash, diarrhoea, or cystitis. The gold standard for the detection of HAdV is virus isolation; however, that is slow and can require up to 3 weeks for cytopathic effects to develop. This led to the development of PCR-based diagnostic tests for the detection of HAdV in clinical specimens. However, qualitative detection of HAdV DNA in immunocompromised patients has a low predictive value because HAdV DNA can be detected occasionally in the blood of healthy, persistently infected individuals. Quantitation of HAdV DNA in blood samples may provide more informative data for predicting the risk of disseminated infection in immunocompromised patients. Heim *et al.* (2003) have recently described a 5′ nuclease LC assay for the sensitive and specific detection and quantitation of all 51 HAdV types in blood samples from immunocompromised patients. The assay targets a conserved region of the hexon gene and has a sensitivity of approximately 15 copies per reaction with a linear range of quantitation of 1.5×10^1 to 1.5×10^8 per reaction. Application of the assay to 234 clinical samples including blood, serum, eye swabs, and faeces demonstrated the LC assay to be more sensitive than a previously developed conventional in-house PCR assay. The HAdV VL were greater in the paediatric patients when compared to the adult patients and the controls with a small number of immunocompromised children having very high VL (up to 1.1×10^{10} copies/ml) which were associated with symptoms of disseminated disease. HAdV viraemia was detected in 4 of 27 paediatric and 8 of 93 adult stem cell transplant recipients but only in 5 of 306 healthy controls. The QPCR assay is a useful tool in the management of immunocompromised patients at risk of HAdV infection and may provide more simplified and rapid diagnosis of disseminated disease. Although there is no specific antiviral therapy for disseminated HAdV infection, cidofovir and

ribavirin have been used to treat severe infections in patients but no controlled clinical trial has been carried out. This QPCR assay may prove useful for monitoring HAdV VL in patients during clinical trials of antiviral agents for the treatment of disseminated infection.

Real-time PCR assays for the detection of antibiotic resistance, antiviral susceptibility and toxin genes

In addition to pathogen detection assays, real-time PCR methods have been applied to the detection of antibiotic resistance determinants and toxin genes in microbial pathogens. Some examples of real-time PCR assays for the detection of these gene targets are presented in Table 13.4. Assays for the detection of ciprofloxacin resistance have been described for *Campylobacter jejuni* (Wilson *et al.*, 2000), *Salmonella enterica* serotype *Typhimurium* DT104 (MR DT104) (Walker *et al.*, 2001), and *Yersinia pestis* (Lindler *et al.*, 2001). Resistance to quinolones such as ciprofloxacin is chromosomally mediated with mutations within the quinolone resistance-determining region (QRDR) of the DNA gyrase gene (*gyrA*) playing a major role in Gram-negative bacteria. A LC assay with three probes was developed for the detection and differentiation of the Asp-87-to-Asn, the Asp-87-to-Gly, and the Ser-83-to-Phe mutations within MR DT104. Strains homologous to the probes could be distinguished from strains that had different mutations by their probe–target melting temperatures. Application of the assay to 92 isolates of MR DT104 demonstrated that 86 of the isolates possessed one of the three mutations described above. Although the method could not determine the *gyrA* mutation for six strains it did provide much more rapid results for the other 86 isolates avoiding the need to sequence the *gyrA* to determine their mutation types.

A 5′ nuclease assay was described for the discrimination of ciprofloxacin-sensitive and ciprofloxacin-resistant strains of *C. jejuni* (Wilson *et al.*, 2000). The assay targeted the C to T transition in codon 86 in the QRDR of *gyrA* and was able to rapidly and reliably differentiate between of ciprofloxacin-sensitive and ciprofloxacin-resistant strains of *C. jejuni*. In the study of Lindler *et al.* (2001) ciprofloxacin

Table 13.4 Real-time PCR assays for the detection of antibiotic susceptibility, antibiotic resistance, antiviral susceptibility and toxin genes in clinical microbiology

Organism	Function of assay	Gene(s) targeted	Reference
Bordetella pertussis	Pertussis toxin gene variants	ptxS1A, ptxS1B, ptxS1D, ptxS1E	Makinen et al. (2002)
Campylobacter jejuni	Ciprofloxacin resistance	*gyrA*	Wilson et al. (2000)
Corynebacterium diptheriae	Toxin gene detection	A and B subunit toxin genes	Mothershed et al. (2002)
Coxiella burnetti	Antibiotic susceptibility testing	com1 gene	Brennan and Samuel (2003)
Enterococcus faecalis, Enterococcus faecium	Oxazolidinone resistance	G2576U rRNA mutation	Woodford et al. (2002)
Helicobacter pylori	Clarithromycin resistance	23S rRNA	Oleastro et al. (2003)
H. pylori	Clarithromycin resistance	23S rRNA	Matsumura et al. (2001)
H. pylori	Clarithromycin resistance	23S rRNA	Gibson et al. (1999)
H. pylori	Tetracycline resistance	16S rDNA	Lawson et al. (2005)
Mycobacterium tuberculosis	Rifampicin and isoniazid resistance	*rpoB* and *katG*	Garcia de Viedma et al. (2002)
M. tuberculosis	Rifampicin and isoniazid resistance	*rpoB* and *inhA*	Torres et al. (2000)
M. tuberculosis	Isoniazid resistance	*katG*	van Doorn et al. (2003)
M. tuberculosis	Rifampicin resistance	*rpoB*	Edwards et al. (2001b)
M. tuberculosis	Rifampicin and isoniazid resistance	*rpoB, katG* and *inhA*	Ruiz et al. (2004)
M. tuberculosis	Rifampicin, isoniazid and ethambutol resistance	*rpoB, katG* and *embB*	Wada et al. (2004)
M. tuberculosis	Rifampicin and isoniazid resistance	*rpoB* and *katG*	Marin et al. (2004)
M. tuberculosis	Rifampicin resistance	*rpoB*	Kocagoz et al. (2005)
Salmonella enterica serotype Typhimurium DT104	Ciprofloxacin resistance	*gyrA*	Walker et al. (2001)
Streptococcus pneumoniae	Penicillin susceptibility	ptp2b gene	Kearns et al. (2002a)
Staphylococcus species	Methicillin resistance	*mecA*	Killgore et al. (2000)
Staphylococcus aureus	Fluoroquinolone resistance	*grlA*	Lapierre et al. (2003)
S. aureus	Panton–Valentine leukocidin gene	*mecA, nuc, PVL*	McDonald et al. (2005)
S. aureus	Panton–Valentine leukocidin gene	*PVL*	Deurenberg et al. (2004)
Vancomycin-resistant enterococcus (VRE)	Vancomycin resistance	*vanA* and *vanB*	Palladino et al. (2003a)
VRE	Vancomycin resistance	*vanA* and *vanB*	Palladino et al. (2003b)
Yersinia pestis	Ciprofloxacin resistance	*gyrA*	Lindler et al. (2001)

resistance in *Y. pestis* was demonstrated to be due to a single nucleotide mutation in the QRDR of *gyrA*, with substitutions belonging to one of four groups. A LC assay was designed to differentiate between all four point mutations within the *gyrA* gene that resulted in ciprofloxacin resistance. The assay successfully discriminated between sensitive and resistant strains of *Y. pestis* on the basis of probe melting curve analysis following PCR amplification.

Clarithromycin resistance in *Helicobacter pylori* is associated with a single base mutation within the peptidyltransferase-encoding region of the 23S rRNA gene. The three mutations lead to the adenine residues at positions 2143 and 2144 being replaced with guanine (A2143G and A2144G) or cytosine (A2143C). Gibson *et al.* (1999) described a LC assay which could discriminate between clarithromycin sensitive and clarithromycin resistant strains using melting curve analysis of a probe complementary to the clarithromycin sensitive 23S rRNA sequence. In a similar study a hybridization probe LC assay was described for the detection of mutations within the 23S rRNA gene conferring resistance to clarithromycin (Matsumura *et al.*, 2001). The authors used three probes and melting curve analysis to successfully identify clarithromycin resistant strains directly in gastric biopsy samples. Similarly, in the study by Oleastro *et al.* (2003) a hybridization probe LC assay was designed to target the three point mutations (A2142C, A2142G, and A2143G) to determine clarithromycin susceptibility directly in gastric biopsy sample from patients with *H. pylori* infection. The assay differentiated between strains with the transition A to G in position 2142 or 2143, the transversion A to C in position 2142, and the wild-type sensitive genotype based on melting curve analysis. The assay was also shown to be able to detect resistance strains in mixed populations of *H. pylori* among sensitive wild-type strains. Rapid detection of resistance profiles of *H. pylori* directly in gastric biopsy specimens may prove useful in determining appropriate treatment regimens for the successful eradication of *H. pylori* from the patient.

The recommended treatment for *M. tuberculosis* is a combination of four drugs; rifampicin, isoniazid, pyrazinamide, and ethambutol with or without streptomycin. Multidrug-resistant (MDR) strains have arisen, most of which are resistant to rifampicin and at least one other drug. Detection of drug resistance in *M. tuberculosis* is usually performed by phenotypic testing; however, it requires considerable time to produce a result. Early detection of drug resistance in strains is important to ensure the timely initiation of effective treatment regimens and to help reduce the risk of transmission of MDR strains. The molecular basis of drug resistance is becoming better understood with more than 95% of rifampicin resistant strains being associated with mutations within an 81-bp region of the *rpoB* gene. Resistance to isoniazid has also been reported with between 60 and 70% of isoniazid resistant strains having mutations within the *katG* gene, with a specific mutation in codon 315 being responsible for many cases of resistance. To facilitate the rapid detection of rifampicin and isoniazid resistance in *M. tuberculosis*, a number of real-time assays have been developed. Edwards *et al.* (2001b) reported a bi-probe LC assay which detected all of the mutations present in the *rpoB* gene of 46 rifampicin resistant isolates of *M. tuberculosis*. A single-tube multiplex real-time LC assay for the detection of multiple rifampicin resistance mutations and high-level isoniazid resistance mutations in *M. tuberculosis* was reported by Garcia de Viedma *et al.* (2002). The assay was demonstrated to be able to detect 12 different mutations in eight codons within the whole *rpoB* core region, plus the most frequently occurring isoniazid resistance mutations.

Several other real-time PCR assays have been described for the detection of antibiotic resistance genes within a number of microbial pathogens. These include assays for the detection of the methicillin resistance gene, *mecA* the most frequent cause of methicillin resistance in staphylococci (Killgore *et al.*, 2000). The *vanA* and *vanB* genes of vancomycin-resistant enterococci were targeted in the assays of Palladino *et al.* (2003a,b). A real-time assay for the detection of fluoroquinolone resistance-associated mutations in the *grlA* gene of *Staphylococcus aureus* was described by Lapierre *et al.* (2003). The assay demonstrated good correlation (98.8%)

with the MICs of ciprofloxacin, levofloxacin, and gatifloxacin when applied to 85 S. aureus isolates with varying levels of fluoroquinolone resistance.

Real-time PCR assays have also been developed for the rapid detection of bacterial toxin genes. Corynebacterium diphtheriae is the aetiological agent of diphtheria. Laboratory confirmation of diphtheria relies on isolation of the organism from the patient and subsequent determination of the toxigenic potential of the isolate using the Elek test. Viable organisms that are present are not always detectable. This led to the development of PCR-based detection tests, which are often the only laboratory test available for the diagnosis of diphtheria in the absence of culturable organisms. A 5′ nuclease real-time PCR assay for the detection and differentiation of the A and B subunits of the toxin gene of C. diphtheriae has been described (Mothershed et al., 2002). The assays were demonstrated to be 750 times more sensitive than the conventional standard PCR assay currently considered the gold standard for molecular diagnosis of diphtheriae (LOD of 2 CFU compared to 1500 CFU/reaction). The assay had a 100% sensitivity and specificity when applied to 23 toxigenic strains of C. diphtheriae, nine non-toxigenic strains of C. diphtheriae, and 44 other strains representing a diverse range of other respiratory flora and respiratory pathogens. The assay was also applied to the direct detection of toxigenic C. diphtheriae in 36 clinical specimens from patients with confirmed clinical diphtheria (35 throat swabs and one throat pseudomembrane). The real-time assay detected one or both of the subunits of the toxin gene in 34 of the 36 specimens compared to only nine specimens being found to be positive in the conventional PCR. This assay provides a more sensitive and rapid alternative to conventional PCR for the detection of toxigenic strains of this important pathogen.

Future developments

This chapter has illustrated some of the areas in clinical microbiology in which the introduction of real-time methods have made a dramatic impact. The application of real-time methods to aid the diagnosis and management of infectious disease is one of the most rapidly growing fields

in PCR diagnostics with an increasing number of assays being reported in the literature every month. Many of these are 'in-house' assays making standardization of testing methods difficult and most have no internal controls for PCR inhibition which should be a standard requirement for any assay used as a diagnostic test. Incorporation of internal controls is fundamental to good laboratory practice and is the only way to ensure a negative result in an assay is correct and due to the sample not containing the target organism rather than through inhibition of the assay. For many of the pathogens described in the literature, any one of a number of different genes in the organism have been utilized as the target sequence for the assay; however, there have been very few reports of comparisons between the different gene targets to determine the most appropriate target in terms of sensitivity and specificity of detection. The development of commercially produced real-time PCR reagents and kits for diagnostics in clinical microbiology may reduce some of the problems associated with standardization. However, even as more of these kits and reagents become available they may not be used by every laboratory, as many laboratories may not be able to afford the additional costs associated with making the transition to using them in comparison to the relatively cheap alternative of 'in-house' assays. A number of companies have begun to produce kits for the real-time detection of a number of pathogens including; EBV, VZV, HSV, WNV, dengue, B. anthracis, E. histolytica, and Plasmodium (Qiagen, Germany); Group B Streptococcus (IDI-Strep B™ Assay) and L. pneumophila and M. pneumoniae (Minerva biolabs, Germany). In the future kits for other pathogens will be developed; however, most of these will target either the most frequently occurring pathogens found in the clinical laboratory or those which are most difficult to test for currently, ensuring the commercial success of such assays. This will mean that there will always be a need for in-house assays targeting more rarely occurring pathogens or for individual laboratory-specific interests or needs.

Prior to detection by real-time assays the nucleic acid in the clinical sample must be extracted in order for it to be detectable in the assay. For some specimens such as CSF this may be as simple as boiling the specimen; however, for more complex sample matrices such as pulmonary secretions more complex extraction procedures are necessary. Many types of clinical samples such as faeces and whole blood contain substances inhibitory to the PCR reaction and therefore the extraction method must also remove these to prevent inhibition of the PCR. Originally manual extraction methods were the only methods available to clinical laboratories; however, recently a number of automated or semi-automated platforms have become available for nucleic acid extraction from samples. These platforms and associated regents will help to standardize sample extraction methods; however, they will require validation studies to ensure they are suitable for the wide range of sample types found in clinical microbiology such as swabs, fluids, pulmonary secretions, tissues, CSF, urine, faeces, blood, and pus. The optimum type of sample for a particular assay and a particular pathogen also needs to be established and standardized. Clinical samples may contain very low numbers of pathogenic organisms against a background flora of many other microbes. Generic nucleic acid extraction methods produce nucleic acid extracts which contain a small amount of nucleic acid from the target pathogen against a background of nucleic acid from the rest of the mixed flora in the sample. A small volume of this sample is then tested in the PCR assay (usually between 1 and 10 µl) therefore the LOD is always compromised when there are only small numbers of target organism in the original sample. Pathogen-specific sample preparation methods which concentrate the numbers of target organisms before the nucleic acid extraction or specifically concentrate the target nucleic acid during extraction will help to increase the sensitivity of detection of assays. This will become increasingly important if the volume of the PCR reaction is further reduced through the development of miniature real-time PCR systems based on microfluidics and microchips (Liao et al. 2005).

Quality control will have an increasingly important role in the implementation of molecular diagnostic testing for the diagnosis of infectious disease. Quality control encompasses measures such as the inclusion of appropriate positive, negative, and inhibition controls in assay runs. The results of positive controls should be monitored over time to ensure the assay is performing consistently and that inter-assay reproducibility remains high. External quality control schemes will play a very crucial role to ensure high standards in molecular diagnostic in the future. The first external quality control scheme to be developed was the European Union Quality Control Concerted Action for Nucleic Acid Amplification in Diagnostic Virology. This temporary entity has been superseded by Quality Control for Molecular Diagnostics, a non-profit organization for the standardization and quality control of molecular diagnostics and genomic technologies (www.qcmd.org). This organization sends out proficiency panels of simulated clinical samples containing a wide range of viral and bacterial pathogens for molecular diagnostic assays. Over 100 laboratories from more than 60 countries regularly participate in the program which is endorsed by the European Society for Clinical Virology and the European Society for Microbiology and Infectious Disease. Laboratories providing molecular diagnostic testing should participate in this scheme to ensure quality of testing.

The introduction of real-time PCR methods in clinical microbiology has improved the detection of infectious disease agents and led to improvements in patient management and care. In the future new developments in real-time molecular diagnostics will lead to further benefits to the patient consolidating the role of real-time PCR as an essential tool in the clinical microbiology laboratory.

References

Aarts, H.J., Joosten, R.G., Henkens, M.H., Stegeman, H. and van Hoek, A.H. 2001. Rapid duplex PCR assay for the detection of pathogenic *Yersinia enterocolitica* strains. J. Microbiol. Methods 47, 209–17.

Abe, A., Inoue, K., Tanaka, T., Kato, J., Kajiyama, N., Kawaguchi, R., Tanaka, S., Yoshiba, M. and Kohara, M. 1999. Quantitation of Hepatitis B virus genomic DNA by real-time detection PCR. J. Clin. Microbiol. 37, 2899–2903.

Apfalter, P., Barousch, W., Nehr, M., Makristathis, A., Willinger, B., Rotter, M. and Hirschl, A.M. 2003. Comparison of a new quantitative ompA-based real-time PCR TaqMan assay for detection of *Chlamydia pneumoniae* DNA in respiratory specimens with four conventional PCR assays. J. Clin. Microbiol. 41, 592–600.

Aryee, E.A., Bailey, R.L., Natividad-Sancho, A., Kaye, S. and Holland, M.J. 2005. Detection, quantification and genotyping of Herpes Simplex Virus in cervicovaginal secretions by real-time PCR: a cross sectional survey. Virol J. 2, 61.

Ballard, A.L., Fry, N.K., Chan, L., Surman, S.B., Lee, J.V. Harrison T.G. and Towner K.J. 2000. Detection of *Legionella pneumophila* using a real-time PCR hybridisation assay. J. Clin. Microbiol. 38, 4215–4218.

Bélanger, S.D., Boissinot, M., Ménard, C., Picard, F.J. and Bergeron, M.G. 2002. Rapid detection of shiga toxin-producing bacteria in faeces by multiplex PCR with molecular beacons on the SmartCycler. J. Clin. Microbiol. 40, 1436–40.

Bélanger, S.D., Boissinot, M., Clairoux, N., Picard, F.J. and Bergeron, M.G. 2003. Rapid detection of *Clostridium difficile* in faeces by real-time PCR. J. Clin. Microbiol. 41, 730–4.

Bellin, T., Pulz, M., Matussek, A., Hempen, H.G. and Gunzer, F. 2001. Rapid detection of enterohaemorrhagic *Escherichia coli* by real-time PCR with fluorescent hybridisation probes. J. Clin. Microbiol. 39, 370–374.

Berger, A. and Preiser, W. 2002. Viral genome quantification as a tool for improving patient management: the example of HIV, HBV, HCV and CMV. J. Antimicrob. Chemother. 49, 713–21.

Boel, C.H., van Herk, C.M., Berretty, P.J., Onland, G.H. and van den Brule, A.J. 2005. Evaluation of conventional and real-time PCR assays using two targets for confirmation of results of the COBAS AMPLICOR *Chlamydia trachomatis/Neisseria gonorrhoeae* test for detection of *Neisseria gonorrhoeae* in clinical samples. J. Clin. Microbiol. 43, 2231–5.

Boivin, G., Cote, S., Dery, P., De Serres, G. and Bergeron, M.G. 2004. Multiplex real-time PCR assay for detection of influenza and human respiratory syncytial viruses. J. Clin. Microbiol. 42, 45–51.

Blessmann, J., Buss, H., Nu, P.A., Dinh, B.T., Ngo, Q.T., Van, A.L., Alla, M.D., Jackson, T.F., Ravdin, J.I. and Tannich, E. 2002. Real-time PCR for detection and differentiation of *Entamoeba histolytica* and *Entamoeba dispar* in faecal samples. J. Clin. Microbiol. 40, 4413–4417.

Brennan, R.E. and Samuel, J.E. 2003. Evaluation of *Coxiella burnetii* antibiotic susceptibilities by real-time PCR assay. J. Clin. Microbiol. 41, 1869–1874.

Bretagne, S., Durand, R., Olivi, M., Garin, J.F., Sulahian, A., Rivollet, D., Vidaud, M. and Deniau, M. 2001. Real-time PCR as a new tool for quantifying *Leishmania infantum* in liver in infected mice. Clin. Diagn. Lab. Immunol. 8, 828–31.

Broccolo, F., Locatelli, G., Sarmati, L., Piergiovanni, S., Veglia, F., Andreoni, M., Buttò, S., Ensoli, B., Lusso, P. and Malnati, M.S. 2002. Calibrated real-time PCR assay for quantitation of human Herpesvirus 8 DNA in biological fluids. J. Clin. Microbiol. 40, 4652–4658.

Burggraf, S., Reischl, U., Malik, N., Bollwein, M., Naumann, L. and Olgemöller, B. 2005a. Comparison of an internally controlled, large-volume LightCycler assay for detection of *Mycobacterium tuberculosis* in clinical samples with the COBAS AMPLICOR Assay. J. Clin. Microbiol. 43, 1564–1569.

Burggraf, S. and Olgemoller, B. 2005b. Straightforward procedure for internal control of real-time reverse transcription amplification assays. Clin. Chem. 51, 1508–10.

Callahan, J.D., Wu, S.J., Dion-Schultz, A., Mangold, B.E., Peruski, L.F., Watts, D.M., Porter, K.R., Murphy, G.R., Suharyono, W., King, C.C., Hayes, C.G. and Temenak, J.J. 2001. Development and evaluation of serotype- and group-specific fluorogenic reverse transcriptase PCR (TaqMan) assays for dengue virus. J. Clin. Microbiol. 39, 4119–24.

Cloud, J.L., Hymas, W.C., Turlak, A., Croft, A., Reischl, U., Daly, J.A. and Carroll, K.C. 2003. Description of a multiplex *Bordetella pertussis* and *Bordetella parapertussis* LightCycler PCR assay with inhibition control. Diagn. Microbiol. Infect. Dis. 46, 189–95.

Collot, S., Petit, B., Bordessoule, D., Alain, S., Touati, M., Denis, F. and Ranger-Rogez, S. 2002. Real-Time PCR for quantification of human Herpesvirus 6 DNA from lymph nodes and saliva. J. Clin. Microbiol. 40, 2445–2451.

Corless, C.E., Guiver, M., Borrow, R., Edwards-Jones, V., Fox, A.J. and Kaczmarski, E.B. 2001. Simultaneous detection of *Neisseria meningitidis*, *Haemophilus influenzae*, and *Streptococcus pneumoniae* in suspected cases of meningitis and septicaemia using real-time PCR. J. Clin. Microbiol. 39, 1553–8.

Costa, J.M., Pautas, C., Ernault, P., Foulet, F., Cordonnier, C. and Bretagne, S. 2000. Real-time PCR for diagnosis and follow-up of *Toxoplasma* reactivation after allogeneic stem cell transplantation using fluorescence resonance energy transfer hybridisation probes. J. Clin. Microbiol. 38, 2929–32.

Cote, S., Abed, Y. and Boivin, G. 2003. Comparative evaluation of real-time PCR assays for detection of the human Metapneumovirus. J. Clin. Microbiol. 41, 3631–5.

Cummings, K.L. and Tarleton, R.L. 2003. Rapid quantitation of *Trypanosoma cruzi* in host tissue by real-time PCR. Mol. Biochem. Parasitol. 129, 53–9.

Deguchi, T., Yoshida, T., Yokoi, S., Ito, M., Tamaki, M., Ishiko, H. and Maeda, S. 2002. Longitudinal quantitative detection by real-time PCR of *Mycoplasma genitalium* in first-pass urine of men with recurrent non-gonococcal urethritis. J. Clin. Microbiol. 40, 3854–6.

Desjardin, L.E., Chen, Y., Perkins, M.D., Teixeira, L., Cave, M.D. and Eisenach, K.D. 1998. Comparison of the ABI 7700 system (TaqMan) and competitive PCR for quantification of IS6110 DNA in sputum during treatment of tuberculosis. J. Clin. Microbiol. 36, 1964–8.

Deurenberg, R.H., Vink, C., Driessen, C., Bes, M., London, N., Etienne, J. and Stobberingh, E.E. 2004. Rapid detection of Panton–Valentine leukocidin from clinical isolates of *Staphylococcus aureus* strains by real-time PCR. FEMS. Microbiol. Lett. 240, 225–8.

Drosten, C., Göttig, S., Schilling, S., Asper, M., Panning, M., Schmitz, H. and Günther, S. 2002. Rapid detection and quantification of RNA of Ebola and Marburg viruses, Lassa virus, Crimean-Congo Haemorrhagic fever virus, Rift Valley fever virus, Dengue virus, and

Yellow fever virus by real-time reverse transcription-PCR. J. Clin. Microbiol. *40*, 2323–30.

Edwards, K.J., Kaufmann, M.E. and Saunders, N. A. 2001a. Rapid and accurate identification of coagulase-negative Staphylococci by real-time PCR. J. Clin. Microbiol. *39*, 3047–51.

Edwards, K.J., Metherel, L.A., Yates, M. and Saunders, N.A. 2001b. Detection of *rpoB* mutations in *Mycobacterium tuberculosis* by biprobe analysis. J. Clin. Microbiol. *39*, 3350–2.

Espy, M.J., Teo, R., Ross, T.K. Svien, K.A., Wold, A.D., Uhl, J.R. and Smith, T.F. 2000a. Diagnosis of Varicella-Zoster virus infections in the clinical laboratory by LightCycler PCR. J. Clin. Microbiol. *38*, 3187–9.

Espy, M.J. Uhl, J.R., Mitchell, P.S., Thorvilson, J.N., Svien, K.A., Wold, A.D. and Smith, T.F. 2000b. Diagnosis of Herpes Simplex virus infections in the clinical laboratory by LightCycler PCR. J. Clin. Microbiol. *38*, 795–9.

Espy, M.J., Cockerill, F.R., Meyer, R.F., Bowen, M.D., Poland, G.A., Hadfield, T.L. and Smith, T.F. 2002. Detection of Smallpox virus DNA by LightCycler PCR. J. Clin. Microbiol. *40*, 1985–1988.

Espy, M. J., Uhl, J. R., Sloan, L. M., Buckwalter, S. P., Jones, M. F., Vetter, E. A., Yao, J. D. C., Wengenack, N. L., Rosenblatt, J. E., Cockerill, III F. R. and Smith, T. F. 2006. Real-time PCR in clinical microbiology: Applications for routine laboratory testing. Clin. Microbiol. Rev. *19*, 165–256.

Fang, H. and Hedin, G. 2003. Rapid screening and identification of methicillin-resistant *Staphylococcus aureus* from clinical samples by selective-broth and real-time PCR assay. J. Clin. Microbiol. *41*, 2894–9.

Fenollar, F., Fournier, P.E., Raoult, D., Gerolami, R., Lepidi, H. and Poyart, C. 2002. Quantitative detection of *Tropheryma whipplei* DNA by real-time PCR. J. Clin. Microbiol. *40*, 1119–20.

Fernandez, C., Boutolleau, D., Manichanh, C., Mangeney, N., Agut, H. and Gautheret-Dejean, A. 2002. Quantitation of HHV-7 genome by real-time polymerase chain reaction assay using MGB probe technology. J. Virol. Methods *106*, 11–6.

Garcia de Viedma, D., del Sol Diaz Infantes, M., Lasala, F., Chaves, F., Alcala, L., Bouza, E. 2002. New real-time PCR able to detect in a single tube multiple rifampin resistance mutations and high-level isoniazid resistance mutations in *Mycobacterium tuberculosis*. J. Clin. Microbiol. *40*, 988–95.

Gault, E., Michel, Y., Dehée, A., Belabani, C., Nicolas, J.-C. and Garbarg-Chenon, A. 2001. Quantification of human Cytomegalovirus DNA by real-time PCR. J. Clin. Microbiol. *39*, 772–5.

Gautheret-Dejean, A., Manichanh, C., Thien-Ah-Koon, F., Fillet, A.M., Mangeney, N., Vidaud, M., Dhedin, N., Vernant, J.P. and Agut, H. 2002. Development of a real-time polymerase chain reaction assay for the diagnosis of human herpesvirus-6 infection and application to bone marrow transplant patients. J. Virol. Methods *100*, 27–35.

Geraats-Peters, C.W., Brouwers, M., Schneeberger, P.M., van der Zanden, A.G., Bruisten, S.M., Weers-Pothoff, G., Boel, C.H., van den Brule, A.J., Harmsen, H.G.

and Hermans, M.H. 2005. Specific and sensitive detection of *Neisseria gonorrhoeae* in clinical specimens by real-time PCR. J. Clin. Microbiol. *43*, 5653–9.

Geretti, A.M. and Brown, D.W. 2005. National survey of diagnostic services for genital herpes. Sex. Transm. Infect. *81*, 316–7.

Gibson, J.R., Saunders, N.A., Burke, B. and Owen, R.J. 1999. Novel method for rapid determination of clarithromycin sensitivity in *Helicobacter pylori*. J. Clin. Microbiol. *37*, 3746–8.

Greiner, O., Day, P.J., Bosshard, P.P., Imeri, F., Altwegg, M. and Nadal, D. 2001. Quantitative detection of *Streptococcus pneumoniae* in nasopharyngeal secretions by real-time PCR. J. Clin. Microbiol. *39*, 3129–34.

Greiner, O., Day, P.J., Altwegg, M. and Nadal, D. 2003. Quantitative detection of *Moraxella catarrhalis* in nasopharyngeal secretions by real-time PCR. J. Clin. Microbiol. *41*, 1386–90.

Griscelli, F., Barrois, M., Chauvin, S., Lastere, S., Bellet, D. and Bourhis, J.-H. 2001. Quantification of human Cytomegalovirus DNA in bone marrow transplant recipients by real-time PCR. J. Clin. Microbiol. *39*, 4362–9.

Grisold, A.J., Leitner, E., Mühlbauer, G., Marth, E. and Kessler, H.H. 2002. Detection of methicillin-resistant *Staphylococcus aureus* and simultaneous confirmation by automated nucleic acid extraction and Real-Time PCR. J. Clin. Microbiol. *40*, 2392–7.

Gueudin, M., Vabret, A., Petitjean, J., Gouarin, S., Brouard, J. and Freymuth, F. 2003. Quantitation of respiratory syncytial virus RNA in nasal aspirates of children by real-time RT-PCR assay. J. Virol. Methods *109*, 39–45.

Guiver, M., Borrow, R., Marsh, J., Gray, S.J., Kaczmarski, E.B., Howells, D., Boseley, P. and Fox A.J. 2000. Evaluation of the Applied Biosystems automated Taqman polymerase chain reaction system for the detection of meningococcal DNA. FEMS Immunol. Med. Microbiol. *28*, 173–179.

Guy, R.A., Xiao, C. and Horgen, P.A. 2004. Real-time PCR assay for detection and genotype differentiation of *Giardia lamblia* in stool specimens. J. Clin. Microbiol. *42*, 3317–20.

Hackett, S.J., Guiver M., Marsh, J., Sillis, J.A., Thomson, A.P., Kaczmarski, E.B. and Hart C.A. 2002. Meningococcal bacterial DNA load at presentation correlates with disease severity. Arch. Dis. Child. *86*, 44–46.

Halse, T.A., Musser, K.A. and Limberger, R.J. 2006. A multiplexed real-time PCR assay for rapid detection of *Chlamydia trachomatis* and identification of serovar L-2, the major cause of Lymphogranuloma venereum in New York. Mol. Cell. Probes. *20*, 290–7.

Hardegger, D., Nadal, D., Bossart, W., Altwegg, M. and Dutly, F. 2000. Rapid detection of *Mycoplasma pneumoniae* in clinical samples by real-time PCR. J. Microbiol. Methods *41*, 45–51.

Hardick, J., Yang, S., Lin, S., Duncan, D. and Gaydos, C. 2003. Use of the Roche LightCycler instrument in a real-time PCR for *Trichomonas vaginalis* in urine samples from females and males. J. Clin. Microbiol. *41*, 5619–22.

Hayden, R.T., Uhl, J.R., Qian, X., Hopkins, M.K., Aubry, M.C., Limper, A.H., Lloyd, R.V. and Cockerill, F.R. 2001. Direct detection of *Legionella* species from bronchoalveolar lavage and open lung biopsy specimens: comparison of LightCycler PCR, in situ hybridisation, direct fluorescence antigen detection and culture. J. Clin. Microbiol. 39, 2618–26.

He, Q., Wang, J.-P., Osato, M. and Lachman, L.B. 2002. Real-time quantitative PCR for detection of *Helicobacter pylori*. J. Clin. Microbiol. 40, 3720–28.

Heim, A., Ebnet, C., Harste, G. and Pring-Akerblom, P. 2003. Rapid and quantitative detection of human adenovirus DNA by real-time PCR. J. Med. Virol. 70, 228–39.

Hermsen, C.C., Telgt, D.S., Linders, E.H., van de Locht, L.A., Eling, W.M., Mensink, E.J. and Sauerwein, R.W. 2001. Detection of *Plasmodium falciparum* malaria parasites in vivo by real-time quantitative PCR. Mol. Biochem. Parasitol. 118, 247–51.

Higgins, J.A., Fayer, R., Trout, J.M., Xiao, L., Lal, A.A., Kerby, S. and Jenkins, M.C. 2001. Real-time PCR for the detection of *Cryptosporidium parvum*. J. Microbiol. Methods. 47, 323–37.

Hu, A., Colella, M., Tam, J.S., Rappaport, R. and Cheng, S.M. 2003. Simultaneous detection, subgrouping and quantitation of Respiratory Syncytial Virus A and B by real-time PCR. J. Clin. Microbiol. 41, 149–54.

Hui, R.K., Zeng, F., Chan, C.M., Yuen, K.Y., Peiris, J.S. and Leung, F.C. 2004. Reverse transcriptase PCR diagnostic assay for the coronavirus associated with severe acute respiratory syndrome. J. Clin. Microbiol. 42, 1994–9.

Huletsky, A., Giroux, R., Rossbach, V., Gagnon, M., Vaillancourt, M., Bernier, M., Gagnon, F., Truchon, K., Bastien, M., Picard, F.J., van Belkum, A., Ouellette, M., Roy, P.H. and Bergeron, M.G. 2004. New real-time PCR assay for rapid detection of methicillin-resistant *Staphylococcus aureus* directly from specimens containing a mixture of staphylococci. J. Clin. Microbiol. 42, 1875–84.

Jalal, H., Stephen, H., Curran, M.D., Burton, J., Bradley, M. and Carne, C. 2006. Development and validation of a rotor-gene real-time PCR assay for detection, identification, and quantification of *Chlamydia trachomatis* in a single reaction. J. Clin. Microbiol. 44:206–13.

Jauregui, L.H., Higgins, J., Zarlenga, D., Dubey, J.P. and Lunney, J.K. 2001. Development of a real-time PCR assay for detection of *Toxoplasma gondii* in pig and mouse tissues. J. Clin. Microbiol. 39, 2065–71.

Kageyama, T., Kojima, S., Shinohara, M., Uchida, K., Fukushi, S., Hoshino, F.B., Takeda, N. and Katayama, K. 2003. Broadly reactive and highly sensitive assay for Norwalk-like viruses based on real-time quantitative reverse transcription-PCR. J. Clin. Microbiol. 41, 1548–57.

Kearns, A.M., Turner, A.J., Taylor, C.E., George, P.W., Freeman, R. and Gennery, A.R. 2001a. LightCycler-based quantitative PCR for rapid detection of human herpesvirus 6 DNA in clinical material. J. Clin. Microbiol. 39, 3020–1.

Kearns, A.M., Guiver, M., James, V. and King, J. 2001b. Development of a real-time quantitative PCR for the detection of human cytomegalovirus. J. Virol. Methods 95, 121–31.

Kearns, A.M., Draper, B., Wipat, W., Turner, A.J., Wheeler, J., Freeman, R., Harwood, J., Gould, F.K. and Dark, J.H. 2001c. LightCycler-based quantitative PCR for detection of cytomegalovirus in blood, urine, and respiratory samples. J. Clin. Microbiol. 39, 2364–5.

Kearns, A.M., Graham, C., Burdess, D., Heatherington, J. and Freeman, R. 2002a. Rapid real-time PCR for determination of penicillin susceptibility in pneumococcal meningitis, including culture-negative cases. J. Clin. Microbiol. 40, 682–4.

Kearns, A.M., Turner, A.J., Eltringham, G.J. and Freeman, R. 2002b. Rapid detection and quantification of CMV DNA in urine using LightCycler-based real-time PCR. J. Clin. Virol. 24, 131–4.

Kessler, H.H., Mühlbauer, G., Rinner, B., Stelzl, E., Berger, A., Dörr, H-W., Santner, B., Marth, E. and Rabenau, H. 2000. Detection of Herpes Simplex Virus DNA by real-time PCR. J. Clin. Microbiol. 38, 2638–42.

Killgore, G.E., Holloway, B. and Tenover, F.C. 2000. A 5′ nuclease PCR (TaqMan) high-throughput assay for detection of the *mecA* gene in staphylococci. J. Clin. Microbiol. 38, 2516–9.

Kimura, H., Morita, M., Yabuta, Y., Kuzushima, K., Kato, K., Kojima, S., Matsuyama, T. and Morishima, T. 1999. Quantitative analysis of Epstein–Barr virus load by using a real-time PCR assay. J. Clin. Microbiol. 37, 132–6.

Kleiber, J., Walter, T., Haberhausen, G., Tsang, S., Babiel, R. and Rosenstraus, M. 2000. Performance characteristics of a quantitative, homogeneous TaqMan RT-PCR test for HCV RNA. J. Mol. Diagn. 2, 158–66.

Kocagoz, T., Saribas, Z. and Alp, A. 2005. Rapid determination of rifampin resistance in clinical isolates of *Mycobacterium tuberculosis* by real-time PCR. J. Clin. Microbiol. 43, 6015–9.

Koenig, M.G., Kosha, S.L., Doty, B.L. and Heath, D.G. 2004. Direct comparison of the BD ProbeTec ET system with in-house LightCycler PCR assays for detection of *Chlamydia trachomatis* and *Neisseria gonorrhoeae* from clinical specimens. J. Clin. Microbiol. 42, 5751–6.

Kösters, K., Riffelmann, M., Wirsing von Konig, C.H. 2001. Evaluation of a real-time PCR assay for detection of *Bordetella pertussis* and *B. parapertussis* in clinical samples. J. Med. Microbiol. 50, 436–40.

Kösters, K., Reischl, U., Schmetz, J., Riffelmann, M. and Wirsing von König, C.H. 2002. Real-time LightCycler PCR for detection and discrimination of *Bordetella pertussis* and *Bordetella parapertussis*. J. Clin. Microbiol. 40, 1719–1722.

Kraus, G., Cleary, T., Miller, N., Seivright, R., Young, A.K., Spruill, G. and Hnatyszyn, H.J. 2001. Rapid and specific detection of the *Mycobacterium tuberculosis* complex using fluorogenic probes and real-time PCR. Mol. Cell. Probes. 15, 375–83.

Kuoppa, Y., Boman, J., Scott, L., Kumlin, U., Eriksson, I. and Allard, A. 2002. Quantitative detection of respiratory *Chlamydia pneumoniae* infection by real-time PCR. J. Clin. Microbiol. 40, 2273–4.

Kuypers, J., Wright, N. and Morrow, R. 2004. Evaluation of quantitative and type-specific real-time RT-PCR assays for detection of respiratory syncytial virus in respiratory specimens from children. J. Clin. Virol. *31*, 123–9.

Kuypers, J., Wright, N., Corey, L. and Morrow, R. J. 2005. Detection and quantification of human metapneumovirus in paediatric specimens by real-time RT-PCR. Clin. Virol. *33*, 299–305.

Lai, K.K., Cook, L., Wendt, S., Corey, L. and Jerome, K.R. 2003. Evaluation of real-time PCR versus PCR with liquid-phase hybridisation for detection of Enterovirus RNA in cerebrospinal fluid. J. Clin. Microbiol. *41*, 3133–41.

Lanciotti, R.S., Kerst, A.J., Nasci, R.S., Godsey, M.S., Mitchell, C.J., Savage, H.M., Komar, N., Panella, N.A., Allen, B.C., Volpe, K.E., Davis, B.S. and Roehrig, J.T. 2000. Rapid detection of West Nile virus from human clinical specimens, field-collected mosquitoes, and avian samples by a TaqMan reverse transcriptase-PCR assay. J. Clin. Microbiol. *38*, 4066–71.

Lapierre, P., Huletsky, A., Fortin, V., Picard, F.J., Roy, P.H., Ouellette, M. and Bergeron, M.J. 2003. Real-time PCR assay for detection of fluoroquinolone resistance associated with *grlA* mutations in *Staphylococcus aureus*. J. Clin. Microbiol. *41*, 3246–51.

Lawson, A.J., Elviss, N.C. and Owen, R.J. 2005. Real-time PCR detection and frequency of 16S rDNA mutations associated with resistance and reduced susceptibility to tetracycline in *Helicobacter pylori* from England and Wales. J. Antimicrob. Chemother. *56*, 282–6.

Lee, M.-A., Tan, C.-H., Aw, L.-T., Tang, C.-S., Singh, M., Lee, S.-H., Chia, H.-P. and Yap, E.P.H. 2002. Real-time fluorescence-based PCR for detection of Malaria parasites. J. Clin. Microbiol. *40*, 4343–5.

Leruez-Ville, M., Ouachée, M., Delarue, R., Sauget, A.-S., Blanche, S., Buzyn, A. and Rouzioux, C. 2003. Monitoring Cytomegalovirus infection in adult and paediatric bone marrow transplant recipients by a real-time PCR assay performed with blood plasma. J. Clin. Microbiol. *41*, 2040–6.

Li, H., Dummer, J.S., Estes, W.R., Meng, S., Wright, P.F. and Tang, Y.W. 2003. Measurement of human Cytomegalovirus loads by quantitative real-time PCR for monitoring clinical intervention in transplant recipients. J. Clin. Microbiol. *41*, 187–91.

Liao, C.S., Lee, G.B., Liu, H.S., Hsieh, T.M. and Luo, C.H. 2005. Miniature RT-PCR system for diagnosis of RNA-based viruses. Nucleic Acids Res. *33*, e156.

Lindler, L.E., Fan, W. and Jahan, N. 2001. Detection of ciprofloxacin-resistant *Yersinia pestis* by fluorogenic PCR using the LightCycler. J. Clin. Microbiol. *39*, 3649–55.

Limor, J.R., Lal, A.A, and Xiao, L. 2002. Detection and differentiation of Cryptosporidium parasites that are pathogenic for humans by real-time PCR. J. Clin. Microbiol. *40*, 2335–8.

Lin, M.H., Chen, T.C., Kuo, T.T., Tseng, C.C., Tseng, C.P. 2000. Real-time PCR for quantitative detection of *Toxoplasma gondii*. J. Clin. Microbiol. *38*, 4121–5.

Locatelli, G., Santoro, F., Veglia, F., Gobbi, A., Lusso, P. and Malnati, M.S. 2000. Real-time quantitative PCR for human herpesvirus 6 DNA. J. Clin. Microbiol. *38*, 4042–8.

Logan, J. M. J., Edwards, K. J., Saunders, N. A. and Stanley, J. 2001. Rapid identification of *Campylobacter* spp. by melting peak analysis of biprobes in real-time PCR. J. Clin. Microbiol. *39*, 2227–32.

Machida, U., Kami, M., Fukui, T., Kazuyama, Y., Kinoshita, M., Tanaka, Y., Kanda, Y., Ogawa, S., Honda, H., Chiba, S., Mitani, K., Muto, Y., Osumi, K., Kimura, S., Hirai, H. 2000. Real-Time automated PCR for early diagnosis and monitoring of Cytomegalovirus infection after bone marrow transplantation. J. Clin. Microbiol. *38*, 2536–2542.

Mackay, I.M., Arden, K.E., Nitsche, A. Real-time PCR in virology. 2002. Nucleic Acids Res. *30*, 1292–305.

Mackay, I.M., Jacob, K.C., Woolhouse, D., Waller, K., Syrmis, M.W., Whiley, D.M., Siebert, D.J., Nissen, and M., Sloots, T.P. 2003. Molecular assays for detection of human metapneumovirus. J. Clin. Microbiol. *41*, 100–5.

Maertzdorf, J., Wang, C.K., Brown, J.B., Quinto, J.D., Chu, M., de Graaf, M., van den Hoogen, B.G., Spaete, R., Osterhaus, A.D. and Fouchier, R.A. 2004. Real-time reverse transcriptase PCR assay for detection of human metapneumoviruses from all known genetic lineages. J. Clin. Microbiol. *42*, 981–6.

Mahony, J.B., Jang, D., Chong, S., Luinstra, K., Sellors, J., Tyndall, M. and Chernesky, M. 1997. Detection of *Chlamydia trachomatis*, *Neisseria gonorrhoeae*, *Ureaplasma urealyticum*, and *Mycoplasma genitalium* in first-void urine specimens by multiplex polymerase chain reaction. Mol. Diagn. *2*, 161–168.

Makinen, J., Mertsola, J., Viljanen, M.K., Arvilommi, H. and He, Q. 2002. Rapid typing of *Bordetella pertussis* pertussis toxin gene variants by LightCycler real-time PCR and fluorescence resonance energy transfer hybridisation probe melting curve analysis. J. Clin. Microbiol. *40*, 2213–6.

Mangold, K.A., Manson, R.U., Koay, E.S., Stephens, L., Regner, M., Thomson, R.B. Jr, Peterson, L.R. and Kaul, K.L. 2005. Real-time PCR for detection and identification of *Plasmodium spp*. J. Clin. Microbiol. *43*, 2435–40.

Marin, M., Garcia de Viedma, D., Ruiz-Serrano, M.J. and Bouza, E. 2004. Rapid direct detection of multiple rifampin and isoniazid resistance mutations in *Mycobacterium tuberculosis* in respiratory samples by real-time PCR. Antimicrob. Agents Chemother. *48*, 4293–300.

Martell, M., Gómez, J., Esteban, J.L., Sauleda, S., Quer, J., Cabot, B., Esteban, R. and Guardia, J. 1999. High-throughput real-time reverse transcription-PCR quantitation of hepatitis C virus RNA. J. Clin. Microbiol. *37*, 327–32.

Marty, A., Greiner, O., Day, P.J., Gunziger, S., Muhlemann, K. and Nadal, D. 2004. Detection of *Haemophilus influenzae* type b by real-time PCR. J. Clin. Microbiol. *42*, 3813–5.

Matsumura, M., Hikiba, Y., Ogura, K., Togo, G., Tsukuda, I., Ushikawa. K., Shiratori, Y. and Omata, M. 2001. Rapid detection of mutations in the 23S rRNA gene of *Helicobacter pylori* that confers resistance to clarithromycin treatment to the bacterium. J. Clin. Microbiol. *39*, 691–5.

McDonald, R.R., Antonishyn, N.A., Hansen, T., Snook, L.A., Nagle, E., Mulvey, M.R., Levett, P.N. and Horsman, G.B. 2005. Development of a triplex real-time PCR assay for detection of Panton–Valentine leukocidin toxin genes in clinical isolates of methicillin-resistant *Staphylococcus aureus*. J. Clin. Microbiol. *43*, 6147–9.

McAvin, J.C., Reilly, P.A., Roudabush, R.M., Barnes, W.J., Salmen, A., Jackson, G.W., Beninga, K.K., Astorga, A., McCleskey, F.K., Huff, W.B., Niemeyer, D. and Lohman, K.L. 2001. Sensitive and specific method for rapid identification of *Streptococcus pneumoniae* using real-time fluorescence PCR. J. Clin. Microbiol. *39*, 3446–51.

McNees, A.L., White, Z.S., Zanwar, P., Vilchez, R.A. and Butel, J.S. Specific and quantitative detection of human polyomaviruses BKV, JCV, and SV40 by real time PCR. 2005. J. Clin. Virol. *34*, 52–62.

Mengelle, C., Sandres-Saune, K., Pasquier, C., Rostaing, L., Mansuy, J.M., Marty, M., Da Silva, I., Attal, M., Massip, P. and Izopet, J. 2003. Automated extraction and quantification of human cytomegalovirus DNA in whole blood by real-time PCR assay. J. Clin. Microbiol. *41*, 3840–5.

Menotti, J., Cassinat, B., Sarfati, C., Liguory, O., Derouin, F. and Molina, J.M. 2003. Development of a real-time PCR assay for quantitative detection of *Encephalitozoon intestinalis* DNA. J. Clin. Microbiol. *41*, 1410–3.

Miller, N., Cleary, T., Kraus, G., Young, A.K., Spruill, G. and Hnatyszyn H.J. 2002. Rapid and specific detection of *Mycobacterium tuberculosis* from acid-fast Bacillus smear-positive respiratory specimens and BacT/ALERT MP culture bottles by using fluorogenic probes and real-time PCR. J. Clin. Microbiol. *40*, 4143–7.

Mohamed, N., Elfaitouri, A., Fohlman, J., Friman, G. and Blomberg, J. 2004. A sensitive and quantitative single-tube real-time reverse transcriptase-PCR for detection of enteroviral RNA. J. Clin. Virol. *30*, 150–6.

Molling, P., Jacobsson, S., Backman, A. and Olcen, P. 2002. Direct and rapid identification and genogrouping of meningococci and *porA* amplification by LightCycler PCR. J. Clin. Microbiol. *40*, 4531–5.

Mommert, S., Gutzmer, R., Kapp, A. and Werfel, T. 2001. Sensitive detection of *Borrelia burgdorferi* sensu lato DNA and differentiation of *Borrelia* species by LightCycler PCR. J. Clin. Microbiol. *39*, 2663–7.

Monpoeho, S., Coste-Burel, M., Costa-Mattioli, M., Besse, B., Chomel, J.J., Billaudel, S. and Ferre, V. 2002. Application of a real-time polymerase chain reaction with internal positive control for detection and quantification of enterovirus in cerebrospinal fluid. Eur. J. Clin. Microbiol. Infect. Dis. *21*, 532–6.

Morre, S.A., Van Valkengoed, I.G.M., Moes, R.M., Boeke, A.J., Meijer, C.J.L.M. and Van den Brule, A.J.C. 1999. Determination of *Chlamydia trachomatis* prevalence in an asymptomatic screening population: performances of the LCx and COBAS Amplicor tests with urine specimens. J. Clin. Microbiol. *37*, 3092–6.

Morrison, T.B., Ma, Y., Weis, J.H. and Weis, J.J. 1999. Rapid and sensitive quantification of *Borrelia burgdorferi*-infected mouse tissues by continuous fluorescent monitoring of PCR. J. Clin. Microbiol. *37*, 987–92.

Mothershed, E.A. Cassiday, P.K., Pierson, K., Mayer, L.W. and Popovic, T. 2002. Development of a real-time fluorescence PCR assay for rapid detection of the diphtheria toxin gene. J. Clin. Microbiol. *40*, 4713–9.

Namvar, L., Olofsson, S., Bergstrom, T., Lindh, M. 2005. Detection and typing of Herpes Simplex virus (HSV) in mucocutaneous samples by TaqMan PCR targeting a gB segment homologous for HSV types 1 and 2. J. Clin. Microbiol. *43*, 2058–64.

Ng, C.T., Gilchrist, C.A., Lane, A., Roy, S., Haque, R. and Houpt, E.R. 2005. Multiplex real-time PCR assay using scorpion probes and DNA capture for genotype-specific detection of *Giardia lamblia* on faecal samples. J. Clin. Microbiol. *43*, 1256–60.

Ng, E.K., Cheng, P.K., Ng, A.Y., Hoang, T.L. and Lim, W.W. 2005. Influenza A H5N1 detection. Emerg. Infect. Dis. *11*, 1303–5.

Nicolas, L., Prina, E., Lang, T. and Milon, G. 2002. Real-time PCR for detection and quantitation of leishmania in mouse tissues. J. Clin. Microbiol. *40*, 1666–9.

Nielsen, E.M. and Andersen, M.T. 2003. Detection and characterization of verocytotoxin-producing *Escherichia coli* by automated 5' Nuclease PCR assay. J. Clin. Microbiol. *41*, 2884–93.

Niesters, H.G., van Esser, J., Fries, E., Wolthers, K.C., Cornelissen, J. and Osterhaus, A.D. 2000. Development of a real-time quantitative assay for detection of Epstein–Barr Virus. J. Clin. Microbiol. *38*, 712–5.

Niesters, H.G. 2001. Quantitation of viral load using real-time amplification techniques. Methods *25*, 419–29.

Niesters, H.G. 2002. Clinical virology in real time. J. Clin. Virol. 25 Suppl, 3, S3–12.

Nijhuis, M., van Maarseveen, N., Schuurman, R., Verkuijlen, S., de Vos, M., Hendriksen, K. and van Loon, A.M. 2002. Rapid and sensitive routine detection of all members of the genus enterovirus in different clinical specimens by real-time PCR. J. Clin. Microbiol. *40*, 3666–70.

Nitsche, A., Steuer, N., Schmidt, C.A., Landt, O., Ellerbrok, H., Pauli, G. and Siegert, W. 2000. Detection of human cytomegalovirus DNA by real-time quantitative PCR. J. Clin. Microbiol. *38*, 2734–7.

Nogva, H.K., Bergh, A., Holck, A. and Rudi, K. 2000. Application of the 5'-nuclease PCR assay in evaluation and development of methods for quantitative detection of *Campylobacter jejuni*. Appl. Environ. Microbiol. 66, 4029–36.

Oleastro, M., Ménard, A., Santos, A., Lamouliatte, H., Monteiro, L., Barthélémy, P. and Mégraud, F. 2003. Real-time PCR assay for rapid and accurate detection of point mutations conferring resistance to clarithromycin in *Helicobacter pylori*. J. Clin. Microbiol. *41*, 397–402.

Pahl, A., Kühlbrandt, U., Brune, K., Röllinghoff, M. and Gessner, A. 1999. Quantitative detection of *Borrelia burgdorferi* by real-time PCR. J. Clin. Microbiol. *37*, 1958–63.

Palladino, S., Kay, I.D., Costa, A.M., Lambert, E.J. and Flexman, J.P. 2003a. Real-time PCR for the rapid detection of *vanA* and *vanB* genes. Diagn. Microbiol. Infect. Dis. 45, 81–4.

Palladino, S., Kay, I.D., Flexman, J.P., Boehm, I., Costa, A.M.G., Lambert, E.J. and Christiansen, K.J. 2003b. Rapid detection of *vanA* and *vanB* genes directly from clinical specimens and enrichment broths by real-time multiplex PCR assay. J. Clin. Microbiol. *41*, 2483–6.

Palomares, C., Torres, M.J., Torres, A., Aznar, J. and Palomares, J.C. 2003. Rapid detection and identification of *Staphylococcus aureus* from blood culture specimens using real-time fluorescence PCR. Diagn. Microbiol. Infect. Dis. 45, 183–9.

Pang, X., Lee, B., Chui, L., Preiksaitis, J.K. and Monroe, S.S. 2004. Evaluation and validation of real-time reverse transcription-PCR assay using the LightCycler system for detection and quantitation of norovirus. J. Clin. Microbiol. *42*, 4679–85.

Pas, S.D., Fries, E., De Man, R.A., Osterhaus, A.D. and Niesters, H.G. 2000. Development of a quantitative real-time detection assay for Hepatitis B Virus DNA and comparison with two commercial assays. J. Clin. Microbiol. *38*, 2897–2901.

Patel, S., Zuckerman, M. and Smith, M. 2003. Real-time quantitative PCR of Epstein–Barr virus BZLF1 DNA using the LightCycler. J. Virol. Methods *109*, 227–33.

Payungporn, S., Chutinimitkul, S., Chaisingh, A., Damrongwantanapokin, S., Buranathai, C., Amonsin, A., Theamboonlers, A. and Poovorawan, Y. 2006. Single step multiplex real-time RT-PCR for H5N1 influenza A virus detection. J. Virol. Methods *131*, 143–7.

Pickett, M.A., Everson, J.S., Pead, P.J. and Clarke, I.N. 2005. The plasmids of *Chlamydia trachomatis* and *Chlamydophila pneumoniae* (N16): accurate determination of copy number and the paradoxical effect of plasmid-curing agents. Microbiology *151*, 893–903.

Piesman, J., Schneider, B.S. and Zeidner, N.S. 2001. Use of quantitative PCR to measure density of *Borrelia burgdorferi* in the midgut and salivary glands of feeding tick vectors. J. Clin. Microbiol. *39*, 4145–8.

Pietila, J., He, Q., Oksi, J. and Viljanen, M.K. 2000. Rapid differentiation of *Borrelia garinii* from *Borrelia afzelii* and *Borrelia burgdorferi* sensu stricto by LightCycler fluorescence melting curve analysis of a PCR product of the *recA* gene. J. Clin. Microbiol. *38*, 2756–9.

Rabenau, H.F., Clarici, A.M., Muhlbauer, G., Berger, A., Vince, A., Muller, S., Daghofer, E., Santner, B.I., Marth, E. and Kessler, H.H. 2002. Rapid detection of enterovirus infection by automated RNA extraction and real-time fluorescence PCR. J. Clin. Virol. 25, 155–64.

Ramaswamy, M., McDonald, C., Smith, M., Thomas, D., Maxwell, S., Tenant-Flowers, M. and Geretti, A.M. 2004. Diagnosis of genital herpes by real time PCR in routine clinical practice. Sex. Transm. Infect. *80*, 406–10.

Rantakokko-Jalava, K. and Jalava, J. 2001. Development of conventional and real-time PCR assays for detection of *Legionella* DNA in respiratory specimens. J. Clin. Microbiol. *39*, 2904–10.

Rauter, C., Oehme, R., Diterich, I., Engele, M. and Hartung, T. 2002. Distribution of clinically relevant *Borrelia* genospecies in ticks assessed by a novel, single-run, real-time PCR. J. Clin. Microbiol. *40*, 36–43.

Reischl, U., Linde, H-J., Metz, M., Leppmeier, B. and Lehn N. 2000. Rapid identification of methicillin-resistant *Staphylococcus aureus* and simultaneous species confirmation using real-time fluorescence PCR. J. Clin. Microbiol. *38*, 2429–33.

Reischl, U., Lehn, N., Sanden, G.N. and Loeffelholz, M.J. 2001. Real-time PCR assay targeting IS481 of *Bordetella pertussis* and molecular basis for detecting *Bordetella holmesii*. J. Clin. Microbiol. *39*, 1963–6.

Reischl, U., Youssef, M.T., Kilwinski, J., Lehn, N., Zhang, W.L., Karch, H. and Strockbine, N.A. 2002a. Real-time fluorescence PCR assays for detection and characterization of shiga toxin, intimin, and enterohemolysin genes from shiga toxin-producing *Escherichia coli*. J. Clin. Microbiol. *40*, 2555–65.

Reischl, U., Linde, H.J., Lehn, N., Landt, O., Barratt, K. and Wellinghausen, N. 2002b. Direct detection and differentiation of *Legionella* spp. and *Legionella pneumophila* in clinical specimens by dual-colour real-time PCR and melting curve analysis. J. Clin. Microbiol. *40*, 3814–7.

Reischl, U., Bretagne, S., Kruger, D., Ernault, P. and Costa, J.M. 2003a. Comparison of two DNA targets for the diagnosis of Toxoplasmosis by real-time PCR using fluorescence resonance energy transfer hybridisation probes. BMC Infect Dis. *3*, 7.

Reischl, U., Lehn, N., Simnacher, U., Marre, R. and Essig, A. 2003b. Rapid and standardised detection of *Chlamydia pneumoniae* using LightCycler real-time fluorescence PCR. Eur. J. Clin. Microbiol. Infect, Dis. *22*, 54–7.

Ruiz, M., Torres, M.J., Llanos, A.C., Arroyo, A., Palomares, J.C. and Aznar, J. 2004. Direct detection of rifampin and isoniazid-Resistant *Mycobacterium tuberculosis* in auramine-rhodamine-positive sputum specimens by real-time PCR. J. Clin. Microbiol. *42*, 1585–9.

Ryncarz, A.J., Goddard, J., Wald, A., Huang, M.L., Roizman, B. and Corey, L. 1999. Development of a high-throughput quantitative assay for detecting Herpes Simplex Virus DNA in clinical samples. J. Clin. Microbiol. *37*, 1941–7.

Safronetz, D., Humar, A. and Tipples, G.A. 2003. Differentiation and quantitation of human herpesviruses 6A, 6B and 7 by real-time PCR. J. Virol. Methods *112*, 99–105.

Sails, A.D., Fox, A.J., Bolton, F. J., Wareing, D.R.A. and Greenway, D.L.A. 2003. A real-time PCR assay for the detection of *Campylobacter jejuni* in foods after enrichment culture. Appl. Environ. Microbiol. *69*, 1383–90.

Sanchez, J.L. and Storch, G.A. 2002. Multiplex, quantitative, real-time PCR assay for cytomegalovirus and human DNA. J. Clin. Microbiol. *40*, 2381–6.

Scheltinga, S.A., Templeton, K.E., Beersma, M.F. and Claas, E.C. 2005. Diagnosis of human metapneumovirus and rhinovirus in patients with respiratory tract infections by an internally controlled multiplex real-time RNA PCR. J. Clin. Virol. *33*, 306–11.

Schulz, A., Mellenthin, K., Schönian, G., Fleischer, B. and Drosten, C. 2003. Detection, differentiation and quantitation of pathogenic *Leishmania* organisms by a fluorescence resonance energy transfer-based real-time PCR assay. J. Clin. Microbiol. *41*, 1529–35.

Scoular, A., Gillespie, G. and Carman, W.F. 2002. Polymerase chain reaction for diagnosis of genital herpes in a genitourinary medicine clinic. Sex. Transm. Infect. *78*:21–5.

Sharma, V.K., Dean-Nystrom, E.A. and Casey, T.A. 1999. Semi-automated fluorogenic PCR assays (TaqMan) for rapid detection of *Escherichia coli* O157, H7 and other shiga toxigenic *E. coli*. Mol. Cell. Probes. *13*, 291–302.

Sharma, V.K. and Dean-Nystrom, E.A. 2003. Detection of enterohaemorrhagic *Escherichia coli* O157, H7 by using a multiplex real-time PCR assay for genes encoding intimin and Shiga toxins. Vet Microbiol. *93*, 247–60.

Shrestha, N.K., Tuohy, M.J., Hall, G.S., Reischl, U., Gordon, S.M. and Procop, G.W. 2003. Detection and differentiation of *Mycobacterium tuberculosis* and non-tuberculous mycobacterial isolates by real-time PCR. J. Clin. Microbiol. *41*, 5121–6.

Shu, P.Y., Chang, S.F., Kuo, Y.C., Yueh, Y.Y., Chien, L.J., Sue, C.L., Lin, T.H. and Huang, J.H. 2003. Development of group- and serotype-specific one-step SYBR Green I based real-time reverse transcription-PCR assay for Dengue virus. J. Clin. Microbiol. *41*, 2408–16.

Si-Mohamed, A., Goff, J.L., Desire, N., Maylin, S., Glotz, D. and Belec, L. 2006. Detection and quantitation of BK virus DNA by real-time polymerase chain reaction in the LT-ag gene in adult renal transplant recipients. J. Virol. Methods. *131*, 21–7.

Sloan, L.M., Hopkins, M.K., Mitchell, P.S., Vetter, E.A. Rosenblatt, J.E., Harmsen, W.S., Cockerill, F.R. and Patel, R. 2002. Multiplex LightCycler PCR assay for detection and differentiation of *Bordetella pertussis* and *Bordetella parapertussis* in nasopharyngeal specimens. J. Clin. Microbiol. *40*, 96–100.

Smith, A.B., Mock, V., Melear, R., Colarusso, P. and Willis, D.E. 2003. Rapid detection of influenza A and B viruses in clinical specimens by LightCycler real time RT-PCR. J. Clin. Virol. *28*, 51–8.

Sofi Ibrahim, M., Kulesh, D.A., Saleh, S.S., Damon, I.K., Esposito, J.J., Schmaljohn, A.L. and Jahrling, P.B. 2003. Real-time PCR assay to detect smallpox virus. J. Clin. Microbiol. *41*, 3835–9.

Stamey, F.R., Patel, M.M., Holloway, B.P. and Pellett, P.E. 2001. Quantitative fluorogenic probe PCR assay for detection of human herpesvirus 8 DNA in clinical specimens. J. Clin. Microbiol. *39*, 3537–40.

Stark, D., Beebe, N., Marriott, D., Ellis, J. and Harkness, J. 2006. Evaluation of three diagnostic methods, including real-time PCR, for detection of *Dientamoeba fragilis* in stool specimens. J. Clin. Microbiol. *44*, 232–5.

Stevens, S.J., Verkuijlen, S.A., Brule, A.J., Middeldorp, J.M. 2002. Comparison of quantitative competitive PCR with LightCycler based PCR for measuring Epstein–Barr Virus DNA load in clinical specimens. J. Clin. Microbiol. *40*, 3986–92.

Stocher, M. and Berg, J. 2002. Normalized quantification of human cytomegalovirus DNA by competitive real-time PCR on the LightCycler instrument. J. Clin. Micro. *40*, 4547–53.

Stone, B., Burrows, J., Schepetiuk, S., Higgins, G., Hampson, A., Shaw, R. and Kok, T. 2004. Rapid detection and simultaneous subtype differentiation of influenza A viruses by real time PCR. J. Virol. Methods *117*, 103–12.

Sumino, K.C., Agapov, E., Pierce, R.A., Trulock, E.P., Pfeifer, J.D., Ritter, J.H., Gaudreault-Keener, M., Storch, G.A. and Holtzman, M.J. 2005. Detection of severe human metapneumovirus infection by real-time polymerase chain reaction and histopathological assessment. J. Infect. Dis. *192*, 1052–60.

Tan, T.Y., Corden, S., Barnes, R. and Cookson, B. 2001. Rapid Identification of methicillin-resistant *Staphylococcus aureus* from positive blood cultures by real-time fluorescence PCR. J. Clin. Microbiol. *39*, 4529–31.

Tanriverdi, S., Tanyeli, A., Baslamisli, F., Köksal, F., Kilinç, Y., Feng, X. Batzer, G., Tzipori, S. and Widmer, G. 2002. Detection and genotyping of oocysts of *Cryptosporidium parvum* by real-time PCR and melting curve analysis. J. Clin. Microbiol. *40*, 3237–3244.

Templeton, K.E., Scheltinga, S.A., van der Zee, A., Diederen, B.M., van Kruijssen, A.M., Goossens, H., Kuijper, E. and Claas, E.C. 2003. Evaluation of real-time PCR for detection of and discrimination between *Bordetella pertussis*, *Bordetella parapertussis*, and *Bordetella holmesii* for clinical diagnosis. J. Clin. Microbiol. 41:4121–6.

Tondella, M.L.C., Talkington, D.F., Holloway, B.P., Dowell, S.F., Cowley, K., Soriano-Gabarro, M., Elkind, M.S. and Fields, B.S. 2002. Development and evaluation of real-time PCR-based fluorescence assays for detection of *Chlamydia pneumoniae*. J. Clin. Microbiol. *40*, 575–83.

Torres, M.J., Criado, A., Palomares, J.C. and Aznar, J. 2000. Use of real-time PCR and fluorimetry for rapid detection of rifampin and isoniazid resistance-associated mutations in *Mycobacterium tuberculosis*. J. Clin. Microbiol. *38*, 3194–9.

Uhl, J.R., Adamson, S.C., Vetter, E.A., Schleck, C.D., Harmsen, W.S., Iverson, L.K., Santrach, P.J., Henry, N.K. and Cockerill, F.R. 2003. Comparison of LightCycler PCR, rapid antigen immunoassay, and culture for detection of group A streptococci from throat swabs. J. Clin. Microbiol. *41*, 242–9.

Ursi, D., Dirven, K., Loens, K., Leven, M. and Goossens, H. 2003. Detection of *Mycoplasma pneumoniae* in respiratory samples by real-time PCR using an inhibition control. J. Microbiol. Methods. 55, 149–53.

van den Hoogen, B.G., Herfst, S., Sprong, L., Cane, P.A., Forleo-Neto, E., de Swart, R.L., Osterhaus, A.D. and Fouchier, R.A. 2004. Antigenic and genetic variability of human metapneumoviruses. Emerg. Infect. Dis. 10:658–66.

van Doorn, H.R., Claas, E.C., Templeton, K.E., van der Zanden, A.G., te Koppele Vije, A., de Jong, M.D., Dankert, J. and Kuijper, E.J. 2003. Detection of a point mutation associated with high-level isoniazid resistance in *Mycobacterium tuberculosis* by using real-time PCR technology with 3′-minor groove binder-DNA probes. J. Clin. Microbiol. 41, 4630–5.

van Doornum, G.J.J., Schouls, L.M., Pijl, A., Cairo, I., Buimer, M. and Bruisten, S.. 2001. Comparison between the LCx probe system and the COBAS AMPLICOR system for detection of *Chlamydia trachomatis* and *Neisseria gonorrhoeae* infections in patients attending a clinic for treatment of sexually transmitted diseases in Amsterdam, The Netherlands. J. Clin. Microbiol. 39, 829–835.

van Doornum, G.J., Guldemeester, J., Osterhaus, A.D. and Niesters, H.G. 2003. Diagnosing herpesvirus infections by real-time amplification and rapid culture. J. Clin. Microbiol. 41, 576–80.

van Elden, L.J.R., Nijhuis, M., Schipper, P., Schuurman, R. and van Loon, A.M. 2001. Simultaneous detection of Influenza Viruses A and B using real-time quantitative PCR. J. Clin. Microbiol. 39, 196–200.

Varma, M., Hester, J.D., Schaefer 3rd, F.W., Ware, M.W. and Lindquist, H.D. 2003. Detection of *Cyclospora cayetanensis* using a quantitative real-time PCR assay. J. Microbiol. Methods 53, 27–36.

Verstrepen, W. A., Kuhn, S., Kockx, M.M., Van De Vyvere, M.E. and Mertens, A.H. 2001. Rapid detection of Enterovirus RNA in cerebrospinal fluid specimens with a novel single-tube real-time reverse transcription-PCR assay. J. Clin. Microbiol. 39, 4093–6.

Verweij, J.J., Schinkel, J., Laeijendecker, D., van Rooyen, M.A., van Lieshout, L. and Polderman, A.M. 2003. Real-time PCR for the detection of *Giardia lamblia*. Mol. Cell. Probes 17, 223–5.

Verweij, J.J., Blange, R.A., Templeton, K., Schinkel, J., Brienen, E.A., van Rooyen, M.A., van Lieshout, L. and Polderman, A.M. 2004. Simultaneous detection of *Entamoeba histolytica*, *Giardia lamblia*, and *Cryptosporidium parvum* in faecal samples by using multiplex real-time PCR. J. Clin. Microbiol. 42, 1220–3.

Wada, T., Maeda, S., Tamaru, A., Imai, S., Hase, A. and Kobayashi, K. 2004. Dual-probe assay for rapid detection of drug-resistant *Mycobacterium tuberculosis* by real-time PCR. J. Clin. Microbiol. 42, 5277–85.

Walker, R.A., Saunders, N., Lawson, A.J., Lindsay, E.A., Dassama, M., Ward, L.R., Woodward, M.J., Davies, R.H., Liebana, E. and Threlfall, E.J. 2001. Use of a LightCycler *gyrA* mutation assay for rapid identification of mutations conferring decreased susceptibility to ciprofloxacin in multiresistant *Salmonella enterica* serotype *Typhimurium* DT104 isolates. J. Clin. Microbiol. 39, 1443–8.

Weidmann, M., Meyer-König, U. and Hufert, F.T. 2003. Rapid detection of Herpes Simplex Virus and Varicella-Zoster Virus infections by real-time PCR. J. Clin. Microbiol. 41, 1565–8.

Wellinghausen, N., Frost, C. and Marre, R. 2001. Detection of *Legionellae* in hospital water samples by quantitative real-time LightCycler PCR. Appl. Environ. Microbiol. 67, 3985–93.

Welti, M., Jaton, K., Altwegg, M., Sahli, R., Wenger, A. and Bille, J. 2003. Development of a multiplex real-time quantitative PCR assay to detect *Chlamydia pneumoniae*, *Legionella pneumophila* and *Mycoplasma pneumoniae* in respiratory tract secretions. Diagn. Microbiol. Infect. Dis. 45, 85–95.

Whiley, D.M., LeCornec, G.M., Mackay, I.M., Siebert, D.J. and Sloots, T.P. A real-time PCR assay for the detection of *Neisseria gonorrhoeae* by LightCycler. 2002. Diagn. Microbiol. Infect. Dis. 42, 85–9.

Whiley, D.M., Buda, P.J., Bayliss, J., Cover, L., Bates, J. and Sloots, T.P. 2004. A new confirmatory *Neisseria gonorrhoeae* real-time PCR assay targeting the *porA* pseudogene. Eur. J. Clin. Microbiol. Infect. Dis. 23, 705–10.

Whiley, D.M., Buda, P.P., Freeman, K., Pattle, N.I., Bates, J. and Sloots, T.P. 2005. A real-time PCR assay for the detection of *Neisseria gonorrhoeae* in genital and extragenital specimens. Diagn. Microbiol. Infect. Dis. 52, 1–5.

Witney, A.A., Doolan, D.L., Anthony, R.M., Weiss, W.R., Hoffman, S.L. and Carucci, D.J. 2001. Determining liver stage parasite burden by real time quantitative PCR as a method for evaluating pre-erythrocytic malaria vaccine efficacy. Mol. Biochem. Parasitol. 118, 233–45.

Wilson, D.L., Abner, S.R., Newman, T.C., Mansfield, L.S. and Linz, J.E. 2000. Identification of ciprofloxacin-resistant *Campylobacter jejuni* by use of a fluorogenic PCR assay. J. Clin. Microbiol. 38, 3971–8.

Wilson, D.A., Yen-Lieberman, B., Reischl, U., Gordon, S.M. and Procop, G.W. 2003. Detection of *Legionella pneumophila* by real-time PCR for the *mip* Gene. J. Clin. Microbiol. 41, 3327–30.

Woodford, N., Tysall, L., Auckland, C., Stockdale, M.W., Lawson, A.J., Walker, R.A. and Livermore, D.M. 2002. Detection of oxazolidinone-resistant *Enterococcus faecalis* and *Enterococcus faecium* strains by real-time PCR and PCR-restriction fragment length polymorphism analysis. J. Clin. Microbiol. 40, 4298–300.

Wolk, D.M., Schneider, S.K., Wengenack, N.L., Sloan, L.M., Rosenblatt, J.E. 2002. Real-time PCR method for detection of *Encephalitozoon intestinalis* from stool specimens. J. Clin. Microbiol. 40, 3922–8.

Yang, J.H., Lai, J.P., Douglas, S.D., Metzger, D., Zhu, X.H. and Ho, W.Z. 2002. Real-time RT-PCR for quantitation of hepatitis C virus RNA. J. Virol. Methods 102, 119–28.

Yang, S., Lin, S., Khalil, A., Gaydos, C., Nuemberger, E., Juan, G., Hardick, J., Bartlett, J.G., Auwaerter, P.G.

and Rothman, R.E. 2005. Quantitative PCR assay using sputum samples for rapid diagnosis of pneumococcal pneumonia in adult emergency department patients. J. Clin. Microbiol. *43*, 3221–6.

Yoshida, T., Deguchi, T., Ito, M., Maeda, S., Tamaki, M. and Ishiko, H. 2002. Quantitative detection of *Mycoplasma genitalium* from first-pass urine of men with urethritis and asymptomatic men by real-time PCR. J. Clin. Microbiol. *40*, 1451–5.

Zeaiter, Z., Fournier, P.E., Greub, G., Raoult, D. 2003. Diagnosis of *Bartonella* endocarditis by a real-time nested PCR assay using serum. J. Clin. Microbiol. *41*, 919–25.

Zeidner, N.S., Schneider, B.S., Dolan, M.C. and Piesman, J. 2001. An analysis of spirochete load, strain, and pathology in a model of tick-transmitted Lyme borreliosis. Vector Borne Zoonotic Dis. 1, 35–44.

Diagnosis of Invasive Fungal Infections

David S. Perlin

14

Abstract

The mounting prevalence of invasive fungal disease in immunocompromised patients is exacerbated by inadequate methods for pathogen detection. PCR-based amplification approaches have been developed to address this problem because conventional methods for pathogen identification lack sensitivity, specificity and speed, and some infectious organisms are difficult to culture. PCR amplification of ribosomal genes and their internal transcribed spacer regions coupled with sequence-specific detection probes are the most reliable approaches for fungal identification. Real-time self-reporting probes capable of single nucleotide allelic discrimination have expanded PCR applications to target mechanisms of drug resistance. Clinical applications of PCR are expanding for diagnosing invasive fungal diseases in blood and respiratory specimens at an early stage to improve treatment outcomes for high-risk patients.

Introduction

Opportunistic fungal infections are widespread in immunosuppressed individuals such as AIDS and cancer patients, and represent a significant cause of morbidity and mortality for such severely ill patients (Almirante et al., 2005; Hajjeh et al., 2004; Idemyor, 2003; Marr, 2004; Richards et al., 2000). The growing incidence of life-threatening invasive fungal infections is often a consequence of advanced medical intervention for various cancers, acute leukaemia, burns, gastrointestinal disease, and premature birth.

Fungal infections are the fourth leading cause of blood stream infections in U.S. hospitals (Pfaller et al., 1998a,; Richards et al., 1999). Numerous yeasts and moulds contribute to clinical disease with Candida spp. and Aspergillus spp. representing the most common life-threatening infections. C. albicans accounts for ~50% of all fungal blood stream infections, although infections due to other Candida spp. (e.g. C. glabrata, C. tropacalis, C. krusei and C. parapsilosis) are significant and in certain clinical settings can be more important (Colombo et al., 2003; Pfaller et al., 1999; Safdar et al., 2001). Invasive fungal infections are difficult to prevent, diagnose and treat among patients with prolonged neutropenia and severe graft-versus-host disease resulting in high mortality (Fukuda et al., 2003). Invasive infections due to moulds are particularly difficult to manage with invasive aspergillosis (IA) being the most common filamentous fungal infection observed in immunocompromised patients, and a leading cause of fungal mortality (Hope et al., 2004; Marr et al., 2002). A. fumigatus is the most dominant species causing invasive mould disease, although A. flavus is also important as are other Aspergillus spp. (e.g. A. niger, A. terreus). Recent in vitro and in vivo studies indicate that A. terreus is distinguished from other Aspergillus spp. as being more resistant to certain antifungal drugs (amphotericin B) (Warn et al., 2003), which necessitates the use of alternative antifungal agents along with reconstituting host immune factors (Steinbach and Stevens, 2003).

The diagnostic quandary: conventional approaches are inadequate

Early diagnosis of fungaemia remains a problem for the management of immunocompromised patients, since signs and symptoms of diseases are non-specific, blood cultures are commonly negative, and patients are often unable to undergo invasive diagnostic procedures (Stevens, 2002). Positive confirmation of early infection is difficult to obtain by conventional procedures like blood culture or those that depend upon a functioning immune system, and too often, deep-seated fungal infections are diagnosed only at autopsy (Bodey *et al.*, 1992). Although progress in detecting fungaemia has recently been made, conventional fungal blood cultures still require 7–10 days of incubation for candidaemias and are mostly negative for aspergillosis. An evaluation of morphological features, reproductive structures, and biochemical properties useful for identifying fungi may take days to weeks to develop in culture, and evaluation of these characteristics requires expertise in mycology. Once colonies are confirmed, typical drug susceptibility testing to determine *in vitro* MIC values, as defined by the Clinical and Laboratory Standards Institute (CLSI) require 48 h of growth or longer (Standards, 1997; Standards, 1998). The clinical diagnosis of IA is most often made based on clinical and radiographic findings, which are non-specific. Therapies are initiated empirically, often without ever establishing the diagnosis (Wingard and Leather, 2001). Rarely is the susceptibility of the infecting organism determined. Predictably, the case definition for IA is difficult. IA is either considered proven, probable or possible, as defined according to the joint guidelines of the European Organization for Research and Treatment of Cancer (EORTC) and the National Institutes of Health (NIH) Mycoses Study Group (MSG) (Denning *et al.*, 1998). For critically ill patients, such time delays mandate that empirical therapy in the absence of a confirmed organism will be initiated. With both yeasts and moulds, drug susceptibility determinations are valuable to understand how the infecting organism may be adapting. In many cases, susceptibility testing is not performed. Early detection of fungi in blood can improve the survival of patients with invasive disease by accelerating the initiation of appropriate antifungal treatment while the fungal loads are still low.

A new generation of diagnostics: antigen-based assays

A strong recognition by clinical mycologists that fungal diagnostics have lagged behind detection of other pathogens and diseases has led to the strong embrace of molecular testing for antigens as surrogate markers of active infection and PCR-based nucleic acid testing for rapid identification of an infecting pathogen (Verweij, 2005). Fungal cell wall components such as glucans and galactomannans, which are actively shed during growth and development, are the basis of antigen assays for rapid diagnostic testing (Andriole, 1993; De Repentigny *et al.*, 1994; Fujita and Hashimoto, 1992; Latge, 1995, 1999; Patterson *et al.*, 1995; Paugam *et al.*, 1998; Pinel *et al.*, 2003; Sarfati *et al.*, 1995; Sendid *et al.*, 1999; Siemann and Koch-Dorfler, 2001; Wheat, 2003). These assays are under wide-scale clinical evaluation and include commercial products such as the Glucatell (1→3)-beta-D-glucan (BG) detection assay (Associates of Cape Cod) (Odabasi *et al.*, 2004) or the Platelia *Aspergillus* galactomannan antigenaemia assay (Alexander, 2002; Mennink-Kersten *et al.*, 2005; Pinel *et al.*, 2003). These new antigen-based detection assays assist physicians in making decisions with other clinical and diagnostic factors. Yet, their value is limited by the potential for false-positive and false-negative results due to an assortment of factors (Alexander, 2002; Maertens *et al.*, 2004; Marr *et al.*, 2004; Mennink-Kersten *et al.*, 2005; Pinel *et al.*, 2003; Scotter and Chambers, 2005; Sulahian *et al.*, 2003; Viscoli *et al.*, 2004), which include interference due to contaminating antigen resulting from treatment with piperacillin-tazobactam (Adam *et al.*, 2004; Sulahian *et al.*, 2003; Viscoli *et al.*, 2004). An unanswered question for antigen-based assays is whether knowledge of results can alter the use of empirical therapy to treat febrile patients at high risk of invasive fungal disease.

Nucleic acid-based assays

Perhaps the area of most impact for fungal diagnostics has been the development of PCR-based assays for ultra-sensitive detection of fungal pathogens in blood and respiratory specimens (Elie *et al.*, 1998; Hopfer *et al.*, 1993; Loffler *et al.*, 1998; Makimura *et al.*, 1994; Morace *et al.*, 1997; Muncan and Wise, 1996; Reiss and Morrison, 1993; Reiss *et al.*, 1998; Sandhu *et al.*, 1995; Shin *et al.*, 1999; Talluri *et al.*, 1998; Wahyuningsih *et al.*, 2000). Genomic differences between fungi offer an alternative to culture for detection and identification of fungi by provid-ing pan-fungal, genera- and species-specific DNA targets that can be accurately resolved by molecular methods (Elie *et al.*, 1998; Perlin and Park, 2001; Reiss *et al.*, 1998; Turenne *et al.*, 1999). Nucleic acid-based amplification assays for the detection of fungal nucleic acids may be the optimal diagnostic approach because they are more rapid and sensitive than current culture-based and biochemical methods, and encompass a full range of fungal genera and specimen types (Ellepola *et al.*, 2003; Hinrikson *et al.*, 2005a; Hui *et al.*, 2000; Perlin and Park, 2001; Reiss *et al.*, 2000; Skladny *et al.*, 1999; Verweij, 2005). Such assays can identify *Aspergillus* in blood when conventional culture-based approaches are negative (Hinrikson *et al.*, 2005a; Loeffler *et al.*, 2002a; Millon *et al.*, 2005). PCR primers designed to conserved regions of fungal rRNA genes are used to amplify sequence-variable frag-ments of genes or intervening non-coding regions such as ITS1 and ITS2 (Chen *et al.*, 2000; Elie *et al.*, 1998; Lott *et al.*, 1998; Park *et al.*, 2000b; Reiss *et al.*, 1998; Turenne *et al.*, 1999) or 18s (Einsele *et al.*, 1997; Lu *et al.*, 1995; Makimura *et al.*, 1994; Reiss *et al.*, 1998; Skladny *et al.*, 1999; Vanittanakom *et al.*, 2002). These targets have been validated on more than two dozen different fungi and can be developed more generically for pan-fungal detection (Hendolin *et al.*, 2000).

PCR detection formats
Standard PCR-based amplification is often insufficient for identification because of rela-tively high levels of false-positive results, since in the various target regions some species vary by as little a single nucleotide (Hinrikson *et al.*, 2005b). Techniques were developed to delineate minor differences between fungal species and/ or genera such as single-strand conformational polymorphism (SSCP) (Walsh *et al.*, 1995),

random amplified polymorphic DNA (RAPD) PCR (RAPD-PCR) (Brandt *et al.*, 1998; Perlin and Park, 2001), reverse-hybridization line probe assay (LiPA) (Martin *et al.*, 2000) and PCR-restriction enzyme analysis (Morace *et al.*, 1999). Improved sensitivity and specificity was achieved following target amplification with species-specific DNA probes in enzyme immunoassay (PCR-EIA) or enzyme-linked immunosorbent assay PCR-ELISA format to distinguish among *Candida* and *Aspergillus* spp., as well as *Fusarium* and *Penicillium* spp., and other fungal genera (Einsele *et al.*, 1997; Elie *et al.*, 1998; Ellepola *et al.*, 2003; Loeffler *et al.*, 1998, 2002a; Scotter *et al.*, 2005; Scotter and Chambers, 2005; Shin *et al.*, 1997). Probe–target hybridization is highly temperature dependent, and depending on the nucleotide composition of the probe, random annealing can be a problem. PCR amplification used in concert with high-fidelity hybridization probes has helped to improve accuracy. The introduction of self-reporting fluorescent probes in real-time PCR assays (RT-PCR) have suc-cessfully overcome this problem by a mechanistic requirement for high-fidelity binding to target sequences for detection. Such high-fidelity probes have the additional advantage that both PCR amplification and detection can be performed in a sealed tube, which reduces the possibility of contamination. This is important for clinical microbiology laboratories because PCR amplifi-cation has the potential to amplify small amounts of target DNA from contaminating organisms or even human DNA (Loeffler *et al.*, 1999; Vanee-choutte and Van Eldere, 1997). RT-PCR does not require post-PCR sample handling, which also prevents contamination due to product carry-over, resulting in higher throughput assays. Furthermore, RT-PCR using high-fidelity self-reporting probes are quantitative and have a large dynamic range. LightCycler^{TM} FRET probes (Kim, 2001), TaqMan^{TM} (Heid *et al.*, 1996), and Molecular Beacons (Tyagi and Kramer, 1996) are high-fidelity hybridization probe detection systems that have been used to rapidly identify fungal pathogens in a real-time assays (Brandt *et al.*, 1998; Buchheidt and Hummel, 2005; Buchheidt *et al.*, 2004; Kim, 2001; Loeffler *et al.*, 2000b; Park *et al.*, 2000a,b; White *et al.*, 2003) in either single target or multiplex format

(Balashov *et al.*, 2005; Hendolin *et al.*, 2000; Luo and Mitchell, 2002).

Sensitivity

Sensitivity is an important issue for fungal identification because early recognition of disease requires detection when fungal burdens are low. To achieve lower limit detection (0.1 to 1 pg DNA or 1–10 cfu per ml blood or gram of tissue), two-step nested PCR approaches have been used (Buchheidt *et al.*, 2002; Chryssanthou *et al.*, 1994; Halliday *et al.*, 2005; Kawamura *et al.*, 1999; Mathis *et al.*, 1997; Rabodonirina *et al.*, 1997; Skladny *et al.*, 1999; Yamakami *et al.*, 1998). PCR-based diagnostic assays utilizing real-time self reporting probes are capable of detecting 1–5 organisms per ml of blood in tracheal aspirates, bronchiolar lavage or tissue specimens (Costa *et al.*, 2002; Loeffler *et al.*, 2002a; Morace *et al.*, 1999; Perlin and Park, 2001; Spiess *et al.*, 2003). These approaches provide an opportunity to diagnose fungal infections at an early stage and in a timeframe that can influence patient management. Yet, molecular methods must take into account prior colonization with commensal organisms such as *C. albicans* or persistent respiratory challenge with environmental spores of *Aspergillus*. Such baseline factors require reliable quantitative data and clear threshold levels to be established for this technology. Ultimately, validation must be correlated with clinical outcome.

Multiplex PCR assays

Ideally, a single reaction assay capable of identifying multiple pathogens is a more efficient and cost-effective approach for a clinical microbiology laboratory. Multiplex assays require probes representing different targets that can be reliably resolved in the same reaction tube or well. When dealing with fluorophores, the spectral properties of the probe–target hybrid must be significantly different from the unbound probes to permit unambiguous probe identification. FRET probes, TaqMan[TM], and Molecular Beacon probes are all suitable for multiplex assays, and numerous fluorophore options are available for multiplex applications including 6-carboxy-fluorescein (FAM), tetrachloro-6-carboxy-fluorescein (TET), or hexachloro-6-carboxy-fluorescein (HEX) (Marras *et al.*, 2006). However, universal

donor FRET probes are limited by the receptor molecule, which must have significant spectral overlap with the donor light emission. Universal acceptor probe systems, such as molecular beacons may be more suited for multiplex applications because their quenching largely occurs through a mechanism involving a direct transfer of energy from fluorophore to quencher (Bonnet *et al.*, 1999; Marras *et al.*, 1999; Vet *et al.*, 1999). This property enables a common quencher molecule to be used with beacons, which vastly increases the number of possible fluorophores that can be used as reporters. Limiting the number of amplification primers during PCR aids in multiplex applications. The internal transcribed spacer region of the ribosomal RNA (rRNA) gene is ideal for multiplexing because it allows a wide range of fungal genera and species to be identified from a common pair of primers (Elie *et al.*, 1998; Lott *et al.*, 1993; Reiss *et al.*, 1998) (Fig. 14.1). Probes to the ITS regions have been used to detect three different *Candida* spp. in a single TaqMan assay (Shin *et al.*, 1999), while molecular beacons directed at the same region have been used to detect four separate *Candida* and *Aspergillus* spp. in single assays (Park *et al.*, 2000a). A tandem multiplex assay that relied upon amplicon size was used to distinguish between *A. fumigatus*, *C. albicans*, *C. glabrata*, *C. parapsilosis*, *C. tropicalis*, and *C. neoformans* (Luo and Mitchell, 2002). Finally, a procedure based on panfungal PCR and multiplex liquid hybridization was developed for the detection of fungi in tissue specimens. After capture with specific probes, eight common fungal pathogens (*A. flavus*, *A. fumigatus*, *C. albicans*, *C. krusei*, *C. glabrata*, *C. parapsilosis*, *C. tropicalis*, and *C. neoformans*) were identified according to the size of the amplification product on an automated DNA sequencer (Hendolin *et al.*, 2000).

Rapid identification of drug resistance

The molecular mechanisms responsible for azole resistance in *C. albicans* are common to other *Candida* spp., as well as to less related fungi (Vanden Bossche *et al.*, 1998). These mechanisms include point mutations in the drug target gene, *ERG11*, which alter the drug-binding domain reducing affinity for drug. Reducing intracel-

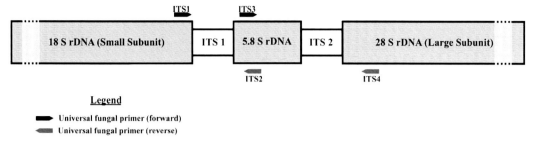

Figure 14.1 Schematic diagram of fungal ribosomal genes and internal transcribed spacer (ITS) regions. Location of amplification primers are indicated (Perlin and Park, 2001).

lular levels of drug by over-expressing membrane bound drug efflux pumps is another prominent mechanism. The two families of efflux pumps include ATP binding cassette transporters (ABC) genes *CDR1* and *CDR2* and multifacilitator genes *MDR1* or *BEN^r* and more recently FLU1 (Sanglard *et al.*, 1995; Sanglard and Odds, 2002). Azole resistance is mostly correlated with the up-regulation of genes of a single family, although there are examples of genes of both families being co-regulated in the same azole-resistant isolate (White *et al.*, 2002). Resistance has also been linked to overexpression of the target, often linked to a chromosomal anomaly. The evolution of drug resistance in a susceptible strain may involve a single mechanism or may reflect a step-wise accumulation of mutations resulting in target site affinity changes and/or induction of various types of drug efflux transporters (Lopez-Ribot *et al.*, 1999; White *et al.*, 2002)

As molecular technology for the detection of fungal pathogens matures, there is an opportunity to develop in parallel molecular probing technology that can be used in conjunction with these systems to evaluate drug resistance. The key to this approach are platforms that reliably distinguish single nucleotide changes (single nucleotide allele-discrimination), are quantitative, and are dynamic enough to be used in a multiplexed assay to reduce costs. Ideally, such an approach, when used in conjunction with genus and/or species-specific probes, should identify a specific fungal pathogen and determine whether it is drug resistant. Rapid identification of fungal pathogens that show primary resistance or reduced susceptibility to antifungal agents (e.g. *A. terreus* with amphotericin b or *C. glabrata*

with fluconazole) is straight-forward. However, a majority of the common yeasts and moulds develop secondary resistance mechanisms resulting from mutations that alter the drug target site, such as mutations in *Erg11* (*Cyp51A* in *Aspergillus*), which confer azole resistance (Diaz-Guerra *et al.*, 2003; Edlind *et al.*, 2001; Loffler *et al.*, 1997; Nascimento *et al.*, 2003; Sanglard *et al.*, 2003; Sanglard and Odds, 2002; White *et al.*, 2002) or *FKS1*, which confer resistance to echinocandin drugs (Park *et al.*, 2005) and/or to over-expression of multi-drug resistance (mdr) ABC- and MFS-type efflux pumps in yeasts (Sanglard, 2002; Sanglard and Odds, 2002; White *et al.*, 2002) and moulds (Nascimento *et al.*, 2003; Slaven *et al.*, 2002). Traditional Northern hybridization methods have given way to more precise real-time PCR methods for profiling mdr pump over-expression linked to azole resistance. Such profiling has identified over-expression of *CDR1*, *CDR2* and *MDR1* as being closely linked to fluconazole resistance in *C. albicans* (Frade *et al.*, 2004; Park and Perlin, 2005) and itraconazole resistance in *A. fumigatus* (Nascimento *et al.*, 2003). A wide array of mutations in *Erg11* have been linked to fluconazole resistance in the *Candida* spp. (Marichal *et al.*, 1999). Although subsets of mutations closely linked to resistance phenotypes are emerging as useful for molecular diagnostics (Park and Perlin, 2005), which can be profiled by real-time PCR (Loeffler *et al.*, 2000a). In *A. fumigatus*, resistance to the azole itraconazole involves limited target site mutations at selected loci, Gly54 and Met220, within Cyp51A (Diaz-Guerra *et al.*, 2003; Mellado *et al.*, 2004; Nascimento *et al.*, 2003). The narrow range of resistance-associated mutations allowed the use of real-time PCR with a limited primer amplification set and either allele-specific molecular beacon probes (Balashov et al., 2005) or pyrosequencing (Trama et al., 2005) for mutation profiling. The use of allele-specific molecular beacons enabled eight separate alleles to be profiled in a single multiplexed PCR reaction (Balashov et al., 2005). The reaction is ideal for high-throughput applications and was designed either to identify a specific allele by assigning separate fluorophores to each mutation or to identify overall resistance by assigning a single fluorophore to all resistance alleles and a separate fluorophore to the wild type susceptible allele (Fig. 14.2). In this format, high-throughput evaluation of resistance could be made as either susceptible or resistant.

False-positives due to contamination

Early PCR was hampered by its tendency to create positive results due to contamination. This problem was solved by the introduction of the uracil-N-glycosylase (UNG) anti-contamination enzyme. PCR also amplifies DNA from dead organisms rendering a result diagnostically correct as positive, but clinically as false-positive if the infection was historic (Burkardt, 2000). The frequency and risk of contamination in fungal PCR were analysed in collaboration among five European centres involving 2800 samples. It was found that 3.3% of the negative controls

were contaminated during the DNA extraction, and 4.7% of the PCR mixtures were contaminated during the amplification process. Common contaminants included DNA arising from *A. fumigatus, Saccharomyces cerevisiae*, and *Acremonium* spp. Commercial products used in nucleic acid isolation such as zymolyase powder or PCR buffers may contain such fungal DNA. However, it was concluded that the risk of contamination is not higher in fungal PCR assays than in other diagnostic PCR-based assays (Loeffler et al., 1999).

Nucleic acid extraction

The success of any PCR-based diagnostic assay relies upon efficient isolation of fungal nucleic acid from a specimen. Isolation of nucleic acid can vary depending on the type of organism (yeasts or moulds), its physiological state (vegetative vs. spore) and specimen type (blood, respiratory or tissue). To assess differences in common nucleic acid extraction procedures, six commercial methods were compared and found to produce markedly differing yields of fungal DNA depending on the organism and its physiological state; no single extraction method was optimal for all organisms (Fredricks et al., 2005). For these reasons, most groups tend to rely upon an empirical laboratory-specific standard approach for isolating nucleic acids from fungi. The incorporation of known quantities of carrier DNA or non-target fungal cells provides an

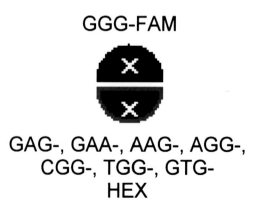

Figure 14.2 Single-tube multiplex assay format for itraconazole resistance. Graphic output for the single tube assay is shown by the split circle. The top semicircle is highlighted when the FAM signal is observed reporting the presence of the wild type drug susceptible *cyp51A* allele. The bottom semicircle is highlighted when the HEX signal is observed reporting the presence of any of seven mutant *cyp51A* alleles conferring resistance to itraconazole (Balashov *et al.* 2005).

imperfect but important measure of nucleic acid extraction efficiency. Automated processing of samples is important and such an assay has been described for the extraction of DNA from clinically important fungi with a MagNA Pure LC instrument (Loeffler et al., 2002b). The efficiency of the test was evaluated by isolating DNA from 23 species of yeast and filamentous fungi and demonstrating amplification by real-time PCR. Sixty-three of sixty seven clinical samples showed identical results by automated and manual methods (Loeffler et al., 2002b).

PCR and invasive fungal disease

Molecular-based diagnostics have been most intensely focused on invasive fungal disease due to Aspergillus spp. Invasive aspergillosis (IA) remains a devastating disease in neutropenic patients with haematological malignancies largely because it is difficult to identify infecting organisms early in the diseases course in high risk patients. The promise of PCR diagnostic approaches was that improved sensitivity and specificity could overcome a lag in early detection and improve clinical outcome (Verweij, 2005; Verweij et al., 2000). Animal models of invasive disease utilizing immunosuppressed rats and mice have been used to compare A. fumigatus-specific PCR with antigen detection. In a rat model, nested PCR was superior to those of galactomannan detection assay and (1→3)-beta-D-glucan detection assay (Scotter and Chambers, 2005). PCR-ELISA and quantitative LightCycler PCR were compared to culture in an assay of blood and organ specimens in experimentally infected mice and rabbits. PCR was superior to culture. Importantly, 74% of infected lung tissue was found to be PCR positive, while only 62% of these tissues were culture positive. A total of 64 (of 145) culture-negative lungs were positive by PCR (Loeffler et al., 2002a). Similarly, Aspergillus DNA was detected in culture-negative blood (Loeffler et al., 2002a). Early PCR applications suggested that it could aid in detection of invasive fungal disease (Chryssanthou et al., 1994; Hebart et al., 2000a; Hebart et al., 2000b; Morace et al., 1999; Spreadbury et al., 1993; Tang et al., 1993). In a study involving 93 patients at risk for invasive fungal disease, there was a 100% correlation obtained between PCR and positive

histology, culture, or high-resolution computed tomography results with a test specificity of 89% (Skladny et al., 1999). Numerous studies with patients have now demonstrated the usefulness of nested, PCR-ELISA or real-time PCR in detecting Aspergillus in blood relative to antigen detection (Kawamura, 1999; Yamakami, 1998; Musher, 2004; Buchheidt, 2004; Millon, 2005; Sanguinetti, 2003; Lass-Florl, 2001; Rimek, 2002). A large prospective study involving 263 serial blood samples from 25 haematological patients at risk of IA compared PCR-ELISA and galactomannan antigen assays. Optimal results for PCR detection and GM were: 100% and 60% sensitivity with 85% and 95% specificity (Scotter et al., 2005). PCR was determined to be very sensitive for the diagnosis of IA, but it was associated with a moderate rate of false positives (Scotter et al., 2005). In another prospective study, nested PCR was used to evaluate a large haematological population involving 197 bronchoalveolar lavage samples from 176 subjects, including 141 neutropenic, febrile patients with lung infiltrates. The sensitivity and specificity values of PCR were 93.9% and 94.4% with a positive predictive value of 83.8%; a negative predictive value of 98.1% was also obtained (Buchheidt et al., 2002). PCR is an important tool for evaluating invasive fungal disease from blood and respiratory samples; although it's high sensitivity does result in false positives. For this reason, PCR is best used clinically in conjunction with other markers of disease. Its use in following disease states following the initiation of therapy are more varied (Lass-Florl et al., 2001; Lass-Florl et al., 2005), and is evolving.

Conclusion

Target development and the application of highly sensitive and specific probing technology have enabled PCR to have a profound impact on the identification and ultra-sensitive detection of fungi. The value of PCR in clinical diagnosis of invasive fungal disease is gaining wider acceptance. Currently, it is being used as an adjunct for diagnosis along with clinical symptoms and other diagnostic markers such as antigen testing and radiographic findings. The new challenge for PCR-based diagnostics is whether it can be used reliably to assess drug resistance and to follow clinical response to therapy.

References

Adam, O., Auperin, A., Wilquin, F., Bourhis, J.H., Gachot, B. and Chachaty, E. 2004. Treatment with piperacillin-tazobactam and false-positive *Aspergillus* galactomannan antigen test results for patients with hematological malignancies. Clin. Infect. Dis. 38, 917–920.

Alexander, B.D. 2002. Diagnosis of fungal infection: new technologies for the mycology laboratory. Transpl. Infect. Dis.4 Suppl. 3, 32–37.

Almirante, B., Rodriguez, D., Park, B. J., Cuenca-Estrella, M., Planes, A. M., Almela, M., Mensa, J., Sanchez, F., Ayats, J., Gimenez, M., *et al.* 2005. Epidemiology and predictors of mortality in cases of Candida bloodstream infection: results from population-based surveillance, Barcelona, Spain, from 2002. to 2003. J. Clin. Microbiol. 43, 1829–1835.

Andriole, V.T. (1993). Infections with *Aspergillus* species. Clin. Infect. Dis. *17* Suppl. 2, S481–486.

Balashov, S.V., Gardiner, R., Park, S. and Perlin, D.S. 2005. Rapid, high-throughput, multiplex, real-time PCR for identification of mutations in the cyp51A gene of *Aspergillus fumigatus* that confer resistance to itraconazole. J. Clin. Microbiol. 43, 214–222.

Bodey, G., Bueltmann, B., Duguid, W., Gibbs, D., Hanak, H., Hotchi, M., Mall, G., Martino, P., Meunier, F., Milliken, S. *et al.* (1992). Fungal infections in cancer patients: an international autopsy survey. Eur J. Clin. Microbiol. Infect. Dis. *11*, 99–109.

Bonnet, G., Tyagi, S., Libchaber, A. and Kramer, F.R. (1999). Thermodynamic basis of the enhanced specificity of structured DNA probes. Proc. Natl. Acad. Sci. USA 96, 6171–6176.

Brandt, M.E., Padhye, A.A., Mayer, L.W. and Holloway, B.P. (1998). Utility of random amplified polymorphic DNA PCR and TaqMan automated detection in molecular identification of *Aspergillus fumigatus*. J. Clin. Microbiol. 36, 2057–2062.

Buchheidt, D., Baust, C., Skladny, H., Baldus, M., Brauninger, S. and Hehlmann, R. 2002. Clinical evaluation of a polymerase chain reaction assay to detect *Aspergillus* species in bronchoalveolar lavage samples of neutropenic patients. Br. J. Haematol. *116*, 803–811.

Buchheidt, D. and Hummel, M. 2005. *Aspergillus* polymerase chain reaction (PCR) diagnosis. Med. Mycol.43 Suppl 1, S139–145.

Buchheidt, D., Hummel, M., Schleiermacher, D., Spiess, B., Schwerdtfeger, R., Cornely, O.A., Wilhelm, S., Reuter, S., Kern, W., Sudhoff, T., *et al.* 2004. Prospective clinical evaluation of a LightCycler-mediated polymerase chain reaction assay, a nested-PCR assay and a galactomannan enzyme-linked immunosorbent assay for detection of invasive aspergillosis in neutropenic cancer patients and haematological stem cell transplant recipients. Br. J. Haematol. *125*, 196–202.

Burkardt, H.J. 2000. Standardization and quality control of PCR analyses. Clin. Chem. Lab. Med. 38, 87–91.

Chen, Y.C., Eisner, J.D., Kattar, M.M., Rassoulian-Barrett, S.L., LaFe, K., Yarfitz, S.L., Limaye, A.P. and Cookson, B. T. 2000. Identification of medically

important yeasts using PCR-based detection of DNA sequence polymorphisms in the internal transcribed spacer 2 region of the rRNA genes. J. Clin. Microbiol. 38, 2302–2310.

Chryssanthou, E., Andersson, B., Petrini, B., Lofdahl, S. and Tollemar, J. 1994. Detection of *Candida albicans* DNA in serum by polymerase chain reaction. Scand. J. Infect. Dis. 26, 479–485.

Colombo, A.L., Perfect, J., DiNubile, M., Bartizal, K., Motyl, M., Hicks, P., Lupinacci, R., Sable, C. and Kartsonis, N. 2003. Global distribution and outcomes for Candida species causing invasive candidiasis: results from an international randomized double-blind study of caspofungin versus amphotericin B for the treatment of invasive candidiasis. Eur J. Clin. Microbiol. Infect. Dis. 22, 470–474.

Costa, C., Costa, J.M., Desterke, C., Botterel, F., Cordonnier, C. and Bretagne, S. 2002. Real-time PCR coupled with automated DNA extraction and detection of galactomannan antigen in serum by enzyme-linked immunosorbent assay for diagnosis of invasive aspergillosis. J. Clin. Microbiol. 40, 2224–2227.

De Repentigny, L., Kaufman, L., Cole, G.T., Kruse, D., Latge, J.P. and Matthews, R.C. (1994). Immunodiagnosis of invasive fungal infections. J. Med. Vet. Mycol. 32, 239–252.

Denning, D.W., Marinus, A., Cohen, J., Spence, D., Herbrecht, R., Pagano, L., Kibbler, C., Kcrmery, V., Offner, F., Cordonnier, C., *et al.* (1998). An EORTC multicentre prospective survey of invasive aspergillosis in haematological patients: diagnosis and therapeutic outcome. EORTC Invasive Fungal Infections Cooperative Group. J. Infect. 37, 173–180.

Diaz-Guerra, T. M., Mellado, E., Cuenca-Estrella, M. and Rodriguez-Tudela, J.L. 2003. A point mutation in the 14alpha-sterol demethylase gene cyp51A contributes to itraconazole resistance in *Aspergillus fumigatus*. Antimicrob. Agents Chemother. 47, 1120–1124.

Edlind, T.D., Henry, K. W., Metera, K.A. and Katiyar, S.K. 2001. *Aspergillus fumigatus* CYP51 sequence: potential basis for fluconazole resistance. Med. Mycol.39, 299–302.

Einsele, H., Hebart, H., Roller, G., Loffler, J., Rothenhofer, I., Muller, C.A., Bowden, R.A., van Burik, J., Engelhard, D., Kanz, L. and Schumacher, U. (1997). Detection and identification of fungal pathogens in blood by using molecular probes. J. Clin. Microbiol. 35, 1353–1360.

Elie, C.M., Lott, T.J., Reiss, E. and Morrison, C.J. (1998). Rapid identification of *Candida* species with species-specific DNA probes. J. Clin. Microbiol. 36, 3260–3265.

Ellepola, A.N., Hurst, S. F., Elie, C.M. and Morrison, C.J. 2003. Rapid and unequivocal differentiation of *Candida dubliniensis* from other *Candida* species using species-specific DNA probes: comparison with phenotypic identification methods. Oral Microbiol. Immunol. 18, 379–388.

Frade, J.P., Warnock, D.W. and Arthington-Skaggs, B.A. 2004. Rapid quantification of drug resistance gene expression in *Candida albicans* by reverse transcriptase

LightCycler PCR and fluorescent probe hybridization. J. Clin. Microbiol. *42*, 2085–2093.

Fredricks, D.N., Smith, C. and Meier, A. 2005. Comparison of six DNA extraction methods for recovery of fungal DNA as assessed by quantitative PCR. J. Clin. Microbiol. *43*, 5122–5128.

Fujita, S. and Hashimoto, T. (1992). Detection of serum *Candida* antigens by enzyme-linked immunosorbent assay and a latex agglutination test with anti-*Candida albicans* and anti-*Candida krusei* antibodies. J. Clin. Microbiol. *30*, 3132–3137.

Fukuda, T., Boeckh, M., Carter, R.A., Sandmaier, B.M., Maris, M.B., Maloney, D.G., Martin, P.J., Storb, R.F. and Marr, K.A. 2003. Risks and outcomes of invasive fungal infections in recipients of allogeneic hematopoietic stem cell transplants after nonmyeloablative conditioning. Blood *102*, 827–833.

Hajjeh, R.A., Sofair, A.N., Harrison, L.H., Lyon, G.M., Arthington-Skaggs, B.A., Mirza, S.A., Phelan, M., Morgan, J., Lee-Yang, W., Ciblak, M. A., et al. 2004. Incidence of bloodstream infections due to Candida species and in vitro susceptibilities of isolates collected from 1998 to 2000 in a population-based active surveillance program. J. Clin. Microbiol. *42*, 1519–1527.

Halliday, C., Wu, Q.X., James, G. and Sorrell, T. 2005. Development of a nested qualitative real-time PCR assay to detect *Aspergillus* species DNA in clinical specimens. J. Clin. Microbiol. *43*, 5366–5368.

Hebart, H., Loffler, J., Meisner, C., Serey, F., Schmidt, D., Bohme, A., Martin, H., Engel, A., Bunje, D., Kern, W.V., et al. 2000a). Early detection of aspergillus infection after allogeneic stem cell transplantation by polymerase chain reaction screening. J. Infect. Dis. *181*, 1713–1719.

Hebart, H., Loffler, J., Reitze, H., Engel, A., Schumacher, U., Klingebiel, T., Bader, P., Bohme, A., Martin, H., Bunjes, D., et al. 2000b. Prospective screening by a panfungal polymerase chain reaction assay in patients at risk for fungal infections: implications for the management of febrile neutropenia. Br. J. Haematol. *111*, 635–640.

Heid, C.A., Stevens, J., Livak, K.J. and Williams, P.M. (1996). Real time quantitative PCR. Genome Res. 6, 986–994.

Hendolin, P.H., Paulin, L., Koukila-Kahkola, P., Anttila, V.J., Malmberg, H., Richardson, M. and Ylikoski, J. 2000. Panfungal PCR and multiplex liquid hybridization for detection of fungi in tissue specimens. J. Clin. Microbiol. *38*, 4186–4192.

Hinrikson, H.P., Hurst, S.F., De Aguirre, L. and Morrison, C.J. 2005a. Molecular methods for the identification of *Aspergillus* species. Med. Mycol.*43* Suppl *1*, S129–137.

Hinrikson, H.P., Hurst, S.F., Lott, T.J., Warnock, D.W. and Morrison, C. J. 2005b. Assessment of ribosomal large-subunit D1-D2, internal transcribed spacer 1, and internal transcribed spacer 2 regions as targets for molecular identification of medically important *Aspergillus* species. J. Clin. Microbiol. *43*, 2092–2103.

Hope, W.W., Padwell, A., Guiver, M. and Denning, D.W. 2004. Invasive pulmonary aspergillosis with sponta-

neous resolution and the diagnostic utility of PCR from tissue specimens. J. Infect. *49*, 136–140.

Hopfer, R.L., Walden, P., Setterquist, S. and Highsmith, W.E. (1993). Detection and differentiation of fungi in clinical specimens using polymerase chain reaction (PCR) amplification and restriction enzyme analysis. J. Med. Vet. Mycol. *31*, 65–75.

Hui, M., Ip, M., Chan, P.K., Chin, M.L. and Cheng, A.F. 2000. Rapid identification of medically important candida to species level by polymerase chain reaction and single-strand conformational polymorphism [In Process Citation]. Diagn. Microbiol. Infect. Dis. *38*, 95–99.

Idemyor, V. 2003. Emerging opportunistic fungal infections: where are we heading? J. Natl. Med. Assoc. *95*, 1211–1215.

Kawamura, S., Maesaki, S., Noda, T., Hirakata, Y., Tomono, K., Tashiro, T. and Kohno, S. (1999). Comparison between PCR and detection of antigen in sera for diagnosis of pulmonary aspergillosis. J. Clin. Microbiol. *37*, 218–220.

Kim, D. W. 2001. Real time quantitative PCR. Exp. Mol. Med. *33*, 101–109.

Lass-Florl, C., Aigner, J., Gunsilius, E., Petzer, A., Nachbaur, D., Gastl, G., Einsele, H., Loffler, J., Dierich, M. P. and Wurzner, R. 2001. Screening for *Aspergillus* spp. using polymerase chain reaction of whole blood samples from patients with haematological malignancies. Br. J. Haematol. *113*, 180–184.

Lass-Florl, C., Gunsilius, E., Gastl, G., Freund, M., Dierich, M. P. and Petzer, A. 2005. Clinical evaluation of Aspergillus-PCR for detection of invasive aspergillosis in immunosuppressed patients. Mycoses *48* Suppl. *1*, 12–17.

Latge, J.P. 1995. Tools and trends in the detection of *Aspergillus fumigatus*. Curr. Top. Med. Mycol. 6, 245–281.

Latge, J.P. 1999. *Aspergillus fumigatus* and aspergillosis. Clin. Microbiol. Rev.*12*, 310–350.

Loeffler, J., Hagmeyer, L., Hebart, H., Henke, N., Schumacher, U. and Einsele, H. 2000a. Rapid detection of point mutations by fluorescence resonance energy transfer and probe melting curves in *Candida* species. Clin. Chem. *46*, 631–635.

Loeffler, J., Hebart, H., Bialek, R., Hagmeyer, L., Schmidt, D., Serey, F. P., Hartmann, M., Eucker, J. and Einsele, H. 1999. Contaminations occurring in fungal PCR assays. J. Clin. Microbiol. *37*, 1200–1202.

Loeffler, J., Henke, N., Hebart, H., Schmidt, D., Hagmeyer, L., Schumacher, U. and Einsele, H. 2000b. Quantification of fungal DNA by using fluorescence resonance energy transfer and the LightCycler system. J. Clin. Microbiol. *38*, 586–590.

Loeffler, J., Kloepfer, K., Hebart, H., Najvar, L., Graybill, J.R., Kirkpatrick, W. R., Patterson, T.F., Dietz, K., Bialek, R. and Einsele, H. 2002a. Polymerase chain reaction detection of aspergillus DNA in experimental models of invasive aspergillosis. J. Infect. Dis *185*, 1203–1206.

Loeffler, J., Schmidt, K., Hebart, H., Schumacher, U. and Einsele, H. 2002b. Automated extraction of genomic DNA from medically important yeast species and filamentous fungi by using the MagNA Pure LC system. J. Clin. Microbiol. 40, 2240–2243.

Loffler, J., Hebart, H., Sepe, S., Schumcher, U., Klingebiel, T. and Einsele, H. 1998. Detection of PCR-amplified fungal DNA by using a PCR-ELISA system. Med. Mycol. 36, 275–279.

Loffler, J., Kelly, S. L., Hebart, H., Schumacher, U., Lass-Florl, C. and Einsele, H. 1997. Molecular analysis of cyp51 from fluconazole-resistant Candida albicans strains. FEMS Microbiol. Lett. 151, 263–268.

Lopez-Ribot, J.L., McAtee, R.K., Perea, S., Kirkpatrick, W.R., Rinaldi, M.G. and Patterson, T.F. 1999. Multiple resistant phenotypes of Candida albicans co-exist during episodes of oropharyngeal candidiasis in human immunodeficiency virus- infected patients [In Process Citation]. Antimicrob. Agents Chemother. 43, 1621–1630.

Lott, T.J., Burns, B.M., Zancope-Oliveira, R., Elie, C.M. and Reiss, E. 1998. Sequence analysis of the internal transcribed spacer 2 (ITS2) from yeast species within the genus Candida. Curr. Microbiol. 36, 63–69.

Lott, T.J., Kuykendall, R.J. and Reiss, E. 1993. Nucleotide sequence analysis of the 5.8S rDNA and adjacent ITS2 region of Candida albicans and related species. Yeast 9, 1199–1206.

Lu, J.J., Chen, C.H., Bartlett, M.S., Smith, J. W. and Lee, C.H. (1995). Comparison of six different PCR methods for detection of Pneumocystis carinii. J. Clin. Microbiol. 33, 2785–2788.

Luo, G. and Mitchell, T.G. 2002. Rapid identification of pathogenic fungi directly from cultures by using multiplex PCR. J. Clin. Microbiol. 40, 2860–2865.

Maertens, J., Theunissen, K., Verhoef, G. and Van Eldere, J. 2004. False-positive Aspergillus galactomannan antigen test results. Clin. Infect. Dis. 39, 289–290.

Makimura, K., Murayama, S.Y. and Yamaguchi, H. (1994). Detection of a wide range of medically important fungi by the polymerase chain reaction. J. Med. Microbiol. 40, 358–364.

Marichal, P., Koymans, L., Willemsens, S., Bellens, D., Verhasselt, P., Luyten, W., Borgers, M., Ramaekers, F. C., Odds, F. C. and Bossche, H. V. (1999). Contribution of mutations in the cytochrome P450 14alpha-demethylase (Erg11p, Cyp51p) to azole resistance in Candida albicans. Microbiology 145 (Pt 10), 2701–2713.

Marr, K. A. 2004. Invasive Candida infections: the changing epidemiology. Oncology (Huntingt) 18, 9–14.

Marr, K. A., Balajee, S. A., McLaughlin, L., Tabouret, M., Bentsen, C. and Walsh, T. J. 2004. Detection of galactomannan antigenemia by enzyme immunoassay for the diagnosis of invasive aspergillosis: variables that affect performance. J. Infect. Dis 190, 641–649.

Marr, K.A., Patterson, T. and Denning, D. 2002. Aspergillosis. Pathogenesis, clinical manifestations, and therapy. Infect. Dis. Clin. North Am. 16, 875–894, vi.

Marras, S.A., Kramer, F.R. and Tyagi, S. (1999). Multiplex detection of single-nucleotide variations using molecular beacons. Genet. Anal. 14, 151–156.

Marras, S.A., Tyagi, S. and Kramer, F.R. 2006. Real-time assays with molecular beacons and other fluorescent

nucleic acid hybridization probes. Clin. Chim. Acta 363, 48–60.

Martin, C., Roberts, D., van Der Weide, M., Rossau, R., Jannes, G., Smith, T. and Maher, M. 2000. Development of a PCR-based line probe assay for identification of fungal pathogens. J. Clin. Microbiol. 38, 3735–3742.

Mathis, A., Weber, R., Kuster, H. and Speich, R. (1997). Simplified sample processing combined with a sensitive one-tube nested PCR assay for detection of Pneumocystis carinii in respiratory specimens. J. Clin. Microbiol. 35, 1691–1695.

Mellado, E., Garcia-Effron, G., Alcazar-Fuoli, L., Cuenca-Estrella, M. and Rodriguez-Tudela, J.L. 2004. Substitutions at methionine 220 in the 14alpha-sterol demethylase (Cyp51A) of Aspergillus fumigatus are responsible for resistance in vitro to azole antifungal drugs. Antimicrob. Agents Chemother. 48, 2747–2750.

Mennink-Kersten, M.A., Ruegebrink, D., Klont, R.R., Warris, A., Gavini, F., Op den Camp, H.J. and Verweij, P. E. 2005. Bifidobacterial lipoglycan as a new cause for false-positive platelia Aspergillus enzyme-linked immunosorbent assay reactivity. J. Clin. Microbiol. 43, 3925–3931.

Millon, L., Piarroux, R., Deconinck, E., Bulabois, C.E., Grenouillet, F., Rohrlich, P., Costa, J.M. and Bretagne, S. 2005. Use of real-time PCR to process the first galactomannan-positive serum sample in diagnosing invasive aspergillosis. J. Clin. Microbiol. 43, 5097–5101.

Morace, G., Pagano, L., Sanguinetti, M., Posteraro, B., Mele, L., Equitani, F., D'Amore, G., Leone, G. and Fadda, G. (1999). PCR-restriction enzyme analysis for detection of Candida DNA in blood from febrile patients with hematological malignancies. J. Clin. Microbiol. 37, 1871–1875.

Morace, G., Sanguinetti, M., Posteraro, B., Lo Cascio, G. and Fadda, G. (1997). Identification of various medically important Candida species in clinical specimens by PCR-restriction enzyme analysis. J. Clin. Microbiol. 35, 667–672.

Muncan, P. and Wise, G.J. 1996. Early identification of candiduria by polymerase chain reaction in high risk patients. J. Urol. 156, 154–156.

Nascimento, A.M., Goldman, G.H., Park, S., Marras, S.A., Delmas, G., Oza, U., Lolans, K., Dudley, M. N., Mann, P. A. and Perlin, D.S. 2003. Multiple resistance mechanisms among Aspergillus fumigatus mutants with high-level resistance to itraconazole. Antimicrob. Agents Chemother. 47, 1719–1726.

Odabasi, Z., Mattiuzzi, G., Estey, E., Kantarjian, H., Saeki, F., Ridge, R. J., Ketchum, P.A., Finkelman, M.A., Rex, J.H. and Ostrosky-Zeichner, L. 2004. Beta-D-glucan as a diagnostic adjunct for invasive fungal infections: validation, cut-off development, and performance in patients with acute myelogenous leukemia and myelodysplastic syndrome. Clin. Infect. Dis. 39, 199–205.

Park, S. and Perlin, D.S. 2005. Establishing surrogate markers for fluconazole resistance in Candida albicans. Microb. Drug Resist. 11, 232–238.

Park, S., Marras, S.A.E., Kiehn, T. E., Chaturvedi, V., Tyagi, S. and Perlin, D.S. 2000a. Rapid Detection of *Candida* and *Aspergillus* spp. Using Molecular Beacons. Paper presented at: In Program and abstracts of the 40th Interscience Conference on Antimicrobial Agents and Chemotherapy. American Society for Microbiology (Washington, DC).

Park, S., Wong, M., Marras, S.A., Cross, E.W., Kiehn, T.E., Chaturvedi, V., Tyagi, S. and Perlin, D.S. 2000b. Rapid identification of *Candida dubliniensis* using a species-specific molecular beacon. J. Clin. Microbiol. 38, 2829–2836.

Patterson, T.F., Miniter, P., Patterson, J.E., Rappeport, J.M. and Andriole, V.T. 1995. Aspergillus antigen detection in the diagnosis of invasive aspergillosis. J. Infect. Dis 171, 1553–1558.

Park, S., Kelly, R., Kahn, J. N., Robles, J., Hsu, M. J., Register, E., Li, W., Vyas, V., Fan, H., Abruzzo, G., *et al.* 2005. Specific substitutions in the echinocandin target Fks1p account for reduced susceptibility of rare laboratory and clinical *Candida* sp. isolates. Antimicrob. Agents Chemother. 49, 3264–3273.

Paugam, A., Sarfati, J., Romieu, R., Viguier, M., Dupouy-Camet, J. and Latge, J.P. 1998. Detection of *Aspergillus* galactomannan: comparison of an enzyme-linked immunoassay and a europium-linked time-resolved fluoroimmunoassay. J. Clin. Microbiol. 36, 3079–3080.

Perlin, D.S. and Park, S. 2001. Rapid identification of fungal pathogens: molecular approaches for a new millennium. Rev Med Microbiol 12 (Suppl.), S13–20.

Pfaller, M.A., Jones, R.N., Messer, S.A., Edmond, M.B. and Wenzel, R.P. 1998a. National surveillance of nosocomial blood stream infection due to *Candida albicans*: frequency of occurrence and antifungal susceptibility in the SCOPE Program. Diagn. Microbiol. Infect. Dis. 31, 327–332.

Pfaller, M.A., Jones, R.N., Messer, S.A., Edmond, M.B. and Wenzel, R.P. 1998b. National surveillance of nosocomial blood stream infection due to species of Candida other than *Candida albicans*: frequency of occurrence and antifungal susceptibility in the SCOPE Program. SCOPE Participant Group. Surveillance and Control of Pathogens of Epidemiologic. Diagn. Microbiol. Infect. Dis. 30, 121–129.

Pfaller, M.A., Messer, S.A., Hollis, R.J., Jones, R.N., Doern, G.V., Brandt, M.E. and Hajjeh, R.A. 1999. Trends in species distribution and susceptibility to fluconazole among blood stream isolates of *Candida* species in the United States. Diagn. Microbiol. Infect. Dis. 33, 217–222.

Pinel, C., Fricker-Hidalgo, H., Lebeau, B., Garban, F., Hamidfar, R., Ambroise-Thomas, P. and Grillot, R. 2003. Detection of circulating *Aspergillus fumigatus* galactomannan: value and limits of the Platelia test for diagnosing invasive aspergillosis. J. Clin. Microbiol. 41, 2184–2186.

Rabodonirina, M., Raffenot, D., Cotte, L., Boibieux, A., Mayencon, M., Bayle, G., Persat, F., Rabatel, F., Trepo, C., Peyramond, D. and Piens, M.A. 1997. Rapid detection of *Pneumocystis carinii* in bronchoalveolar lavage specimens from human immunodeficiency virus-infected patients: use of a simple DNA extraction procedure and nested PCR. J. Clin. Microbiol. 35, 2748–2751.

Reiss, E. and Morrison, C. J. 1993. Nonculture methods for diagnosis of disseminated candidiasis. Clin. Microbiol. Rev.6, 311–323.

Reiss, E., Obayashi, T., Orle, K., Yoshida, M. and Zancope-Oliveira, R.M. 2000. Non-culture based diagnostic tests for mycotic infections. Med. Mycol. 38 Suppl. 1, 147–159.

Reiss, E., Tanaka, K., Bruker, G., Chazalet, V., Coleman, D., Debeaupuis, J.P., Hanazawa, R., Latge, J.P., Lortholary, J., Makimura, K., *et al.* (1998). Molecular diagnosis and epidemiology of fungal infections. Med. Mycol.36, 249–257.

Richards, M.J., Edwards, J. R., Culver, D.H. and Gaynes, R.P. 1999. Nosocomial infections in medical intensive care units in the United States. National Nosocomial Infections Surveillance System. Crit. Care Med. 27, 887–892.

Richards, M.J., Edwards, J. R., Culver, D.H. and Gaynes, R.P. 2000. Nosocomial infections in combined medical-surgical intensive care units in the United States. Infect. Control Hosp. Epidemiol. 21, 510–515.

Rimek, D. and Kappe, R. 2002. Invasive aspergillosis: results of an 8-year study. Mycoses 45 Suppl., 18–21.

Safdar, A., Chaturvedi, V., Cross, E.W., Park, S., Bernard, E.M., Armstrong, D. and Perlin, D.S. 2001. Prospective study of Candida species in patients at a comprehensive cancer center. Antimicrob. Agents Chemother. 45, 2129–2133.

Sandhu, G.S., Kline, B.C., Stockman, L. and Roberts, G.D. 1995. Molecular probes for diagnosis of fungal infections. J. Clin. Microbiol. 33, 2913–2919.

Sanglard, D. 2002. Resistance of human fungal pathogens to antifungal drugs. Curr. Opin. Microbiol. 5, 379–385.

Sanglard, D., Ischer, F., Parkinson, T., Falconer, D. and Bille, J. 2003. *Candida albicans* mutations in the ergosterol biosynthetic pathway and resistance to several antifungal agents. Antimicrob. Agents Chemother. 47, 2404–2412.

Sanglard, D., Kuchler, K., Ischer, F., Pagani, J.L., Monod, M. and Bille, J. (1995). Mechanisms of resistance to azole antifungal agents in *Candida albicans* isolates from AIDS patients involve specific multidrug transporters. Antimicrob. Agents Chemother. 39, 2378–2386.

Sanglard, D. and Odds, F. C. 2002. Resistance of Candida species to antifungal agents: molecular mechanisms and clinical consequences. Lancet Infect Dis. 2, 73–85.

Sanguinetti, M., Posteraro, B., Pagano, L., Pagliari, G., Fianchi, L., Mele, L., La Sorda, M., Franco, A. and Fadda, G. 2003. Comparison of real-time PCR, conventional PCR, and galactomannan antigen detection by enzyme-linked immunosorbent assay using bronchoalveolar lavage fluid samples from hematology patients for diagnosis of invasive pulmonary aspergillosis. J. Clin. Microbiol. 41, 3922–5.

Sarfati, J., Boucias, D.G. and Latge, J.P. 1995. Antigens of *Aspergillus fumigatus* produced in vivo. J. Med. Vet. Mycol. 33, 9–14.

Scotter, J.M., Campbell, P., Anderson, T.P., Murdoch, D.R., Chambers, S.T. and Patton, W.N. 2005. Comparison of PCR-ELISA and galactomannan detection for the diagnosis of invasive aspergillosis. Pathology 37, 246–253.

Scotter, J.M. and Chambers, S.T. 2005. Comparison of galactomannan detection, PCR-enzyme-linked immunosorbent assay, and real-time PCR for diagnosis of invasive aspergillosis in a neutropenic rat model and effect of caspofungin acetate. Clin. Diagn. Lab. Immunol. 12, 1322–1327.

Sendid, B., Tabouret, M., Poirot, J.L., Mathieu, D., Fruit, J. and Poulain, D. 1999. New enzyme immunoassays for sensitive detection of circulating Candida albicans mannan and antimannan antibodies: useful combined test for diagnosis of systemic candidiasis. J. Clin. Microbiol. 37, 1510–1517.

Shin, J.H., Nolte, F.S., Holloway, B.P. and Morrison, C.J. 1999. Rapid identification of up to three Candida species in a single reaction tube by a 5′ exonuclease assay using fluorescent DNA probes. J. Clin. Microbiol. 37, 165–170.

Shin, J.H., Nolte, F.S. and Morrison, C.J. 1997. Rapid identification of Candida species in blood cultures by a clinically useful PCR method. J. Clin. Microbiol. 35, 1454–1459.

Siemann, M. and Koch-Dorfler, M. 2001. The Platelia Aspergillus ELISA in diagnosis of invasive pulmonary aspergilosis (IPA). Mycoses 44, 266–272.

Skladny, H., Buchheidt, D., Baust, C., Krieg-Schneider, F., Seifarth, W., Leib-Mosch, C. and Hehlmann, R. 1999. Specific detection of Aspergillus species in blood and bronchoalveolar lavage samples of immunocompromised patients by two-step PCR. J. Clin. Microbiol. 37, 3865–3871.

Slaven, J.W., Anderson, M. J., Sanglard, D., Dixon, G.K., Bille, J., Roberts, I.S. and Denning, D.W. 2002. Increased expression of a novel Aspergillus fumigatus ABC transporter gene, atrF, in the presence of itraconazole in an itraconazole resistant clinical isolate. Fungal Genet. Biol. 36, 199–206.

Spiess, B., Buchheidt, D., Baust, C., Skladny, H., Seifarth, W., Zeilfelder, U., Leib-Mosch, C., Morz, H. and Hehlmann, R. 2003. Development of a LightCycler PCR assay for detection and quantification of Aspergillus fumigatus DNA in clinical samples from neutropenic patients. J. Clin. Microbiol. 41, 1811–1818.

Spreadbury, C., Holden, D., Aufauvre-Brown, A., Bainbridge, B. and Cohen, J. 1993. Detection of Aspergillus fumigatus by polymerase chain reaction. J. Clin. Microbiol. 31, 615–621.

Standards, N.C.f.C L. 1997. Reference method for broth dilution antifungal susceptibility testing of yeasts. Approved standard M27-A. In National Committee for Clinical Laboratory Standards. (Wayne, PA.).

Standards, N.C.f.C.L. 1998. Reference method for broth dilution antifungal susceptibility testing of conidium-forming filamentous fungi; proposed standard M38-P. In National Committee for Clinical Laboratory Standards. (Wayne, PA.).

Steinbach, W.J. and Stevens, D.A. 2003. Review of newer antifungal and immunomodulatory strategies for invasive aspergillosis. Clin. Infect. Dis. 37 Suppl. 3, S157–187.

Stevens, D.A. 2002. Diagnosis of fungal infections: current status. J. Antimicrob. Chemother. 49 Suppl. 1, 11–19.

Sulahian, A., Touratier, S. and Ribaud, P. 2003. False positive test for aspergillus antigenemia related to concomitant administration of piperacillin and tazobactam. N. Engl. J. Med. 349, 2366–2367.

Talluri, G., Mangone, C., Freyle, J., Shirazian, D., Lehman, H. and Wise, G.J. (1998). Polymerase chain reaction used to detect candidemia in patients with candiduria. Urology 51, 501–505.

Tang, C.M., Holden, D.W., Aufauvre-Brown, A. and Cohen, J. 1993. The detection of Aspergillus spp. by the polymerase chain reaction and its evaluation in bronchoalveolar lavage fluid. Am. Rev. Respir. Dis 148, 1313–1317.

Trama, J.P., Mordechai, E. and Adelson, M.E. 2005. Detection of Aspergillus fumigatus and a mutation that confers reduced susceptibility to itraconazole and posaconazole by real-time PCR and pyrosequencing. J. Clin. Microbiol. 43, 906–908.

Turenne, C.Y., Sanche, S.E., Hoban, D.J., Karlowsky, J.A. and Kabani, A.M. 1999. Rapid identification of fungi by using the ITS2 genetic region and an automated fluorescent capillary electrophoresis system. J. Clin. Microbiol. 37, 1846–1851.

Tyagi, S. and Kramer, F.R. 1996. Molecular beacons: probes that fluoresce upon hybridization. Nat. Biotechnol. 14, 303–308.

Vanden Bossche, H., Dromer, F., Improvisi, I., Lozano-Chiu, M., Rex, J.H. and Sanglard, D. 1998. Antifungal drug resistance in pathogenic fungi. Med. Mycol. 36, 119–128.

Vaneechoutte, M. and Van Eldere, J. 1997. The possibilities and limitations of nucleic acid amplification technology in diagnostic microbiology. J. Med. Microbiol. 46, 188–194.

Vanittanakom, N., Vanittanakom, P. and Hay, R.J. 2002. Rapid identification of Penicillium marneffei by PCR-based detection of specific sequences on the rRNA gene. J. Clin. Microbiol. 40, 1739–1742.

Verweij, P.E. 2005. Advances in diagnostic testing. Med. Mycol. 43 Suppl. 1, S121–124.

Verweij, P.E., Figueroa, J., Van Burik, J., Holdom, M.D., Dei-Cas, E., Gomez, B.L. and Mendes-Giannini, M.J. 2000. Clinical applications of non-culture based methods for the diagnosis and management of opportunistic and endemic mycoses. Med. Mycol. 38 Suppl. 1, 161–171.

Vet, J.A., Majithia, A.R., Marras, S.A., Tyagi, S., Dube, S., Poiesz, B.J. and Kramer, F.R. 1999. Multiplex detection of four pathogenic retroviruses using molecular beacons. Proc. Natl. Acad. Sci. USA 96, 6394–6399.

Viscoli, C., Machetti, M., Cappellano, P., Bucci, B., Bruzzi, P., Van Lint, M.T. and Bacigalupo, A. 2004. False-positive galactomannan platelia Aspergillus test results for patients receiving piperacillin-tazobactam. Clin. Infect. Dis. 38, 913–916.

Wahyuningsih, R., Freisleben, H. J., Sonntag, H.G. and Schnitzler, P. 2000. Simple and rapid detection of

Candida albicans DNA in serum by PCR for diagnosis of invasive candidiasis. J. Clin. Microbiol. *38*, 3016–3021.

Walsh, T.J., Francesconi, A., Kasai, M. and Chanock, S.J. (1995). PCR and single-strand conformational polymorphism for recognition of medically important opportunistic fungi. J. Clin. Microbiol. *33*, 3216–3220.

Warn, P.A., Morrissey, G., Morrissey, J. and Denning, D.W. 2003. Activity of micafungin (FK463) against an itraconazole-resistant strain of *Aspergillus fumigatus* and a strain of *Aspergillus terreus* demonstrating in vivo resistance to amphotericin B. J Antimicrob Chemother *51*, 913–919.

Wheat, L.J. 2003. Rapid diagnosis of invasive aspergillosis by antigen detection. Transpl. Infect. Dis.5, 158–166.

White, P.L., Shetty, A. and Barnes, R.A. 2003. Detection of seven Candida species using the Light-Cycler system. J. Med. Microbiol. *52*, 229–238.

White, T.C., Holleman, S., Dy, F., Mirels, L.F. and Stevens, D.A. 2002. Resistance mechanisms in clinical isolates of *Candida albicans*. Antimicrob. Agents Chemother. *46*, 1704–1713.

Wingard, J.R. and Leather, H. L. 2001. Empiric antifungal therapy for the neutropenic patient. Oncology (Williston Park) *15*, 351–363; discussion 363–354, 367–359.

Yamakami, Y., Hashimoto, A., Yamagata, E., Kamberi, P., Karashima, R., Nagai, H. and Nasu, M. 1998. Evaluation of PCR for detection of DNA specific for *Aspergillus* species in sera of patients with various forms of pulmonary aspergillosis. J. Clin. Microbiol. *36*, 3619–3623.

Biodefence

Christina Egan, Nick M. Cirino and Kimberlee A. Musser

15

The world is a dangerous place to live, not just because of the people who are evil, but because of the people who don't do anything about it.

Albert Einstein

Abstract

With the public's reawakened concern regarding use of biological agents as weapons, the rapid detection, discrimination, and identification of pathogenic organisms and toxins has become a priority for state and federal government agencies. High-confidence, cost-effective, and near real-time diagnostic methods are essential to protecting national health security whether the target is public health, agriculture, commodities, or water supply infrastructures. While culture-based methods have been, and will likely remain, the gold standard for microbiological diagnostics, PCR-based tests offer significant advantages in sensitivity, specificity, speed and data richness that make them invaluable to diagnostic laboratories. In this chapter, we will describe the application of real-time PCR methods in biodefence. We will discuss the use of real-time PCR in biodefence in terms of general workflow and processing considerations, clinical diagnostic applications, environmental diagnostic applications, and multiplex screening. Real-time PCR assays can be either quantitative (qPCR) or qualitative, depending on whether a standard curve is included with the analytical run. Most diagnostic and biodefence applications utilize the qualitative nature of real-time PCR as a detection platform; this chapter will focus on the benefits of these types of assays. Finally, we will consider the future uses and anticipated advances in real-time PCR applications as related to biodefence.

Introduction

Biodefence is a field that has readily embraced the diagnostic capabilities of real-time PCR, given that as the ability to rapidly and specifically detect a pathogen involved in an event of bioterrorism (BT) is of utmost importance. Real-time PCR is the method of choice for initial detection because it is rapid, highly sensitive, specific, and the equipment can be field portable (Edwards et al., 2006). Real-time PCR is utilized in many aspects of biodefence response and preparedness, including testing in the field, surveillance of water or air, environmental sampling of powders and mail (Canter et al., 2005; Higgins et al., 2003), and testing of clinical specimens. This chapter will describe the application of real-time PCR and highlight the benefits of utilizing the technology in the many different areas of biodefence.

The use of real-time PCR is recognized by many military, federal, state, and local entities that are involved on one or many levels in the detection of biological threat agents. These biothreats are defined as microorganisms or toxins that can be used as biological weapons. Table 15.1 lists microorganisms that pose a genuine risk of use as a weapon against humans, animals, and plants. Similarly, Table 15.2 shows biological toxins which could be employed as bioweapons. Detection of the microorganisms and toxins is essential to support public health and national security. The US Laboratory Response Network (LRN) (Website: http://www.bt.cdc.gov/lrn/) was established by the Centers for Disease Control and Prevention (CDC) to provide

Table 15.1 List of select agents

(A) Health and human services select agents and toxins

Cercopithecine herpesvirus 1 (herpes B virus)

Coccidioides posadasii

Crimean-Congo haemorrhagic fever virus

Ebola viruses

Lassa fever virus

Marburg virus

Monkeypox virus

Rickettsia prowazekii

Rickettsia rickettsii

South American haemorrhagic fever viruses

Flexal

Guanarito

Junin

Machupo

Sabia

Tick-borne encephalitis complex viruses (flaviviruses)

Central European tick-borne encephalitis

Far Eastern tick-borne encephalitis

Kyasanur forest disease

Omsk haemorrhagic fever

Russian spring and summer encephalitis

Variola major virus (smallpox virus)

Variola minor virus (alastrim)

Yersinia pestis

Abrin (plant toxin)

Conotoxins (mycotoxin)

Diacetoxyscirpenol

Ricin

Saxitoxin

Shiga-like ribosome-inactivating proteins

Tetrodotoxin

(B) USDA non-overlap high consequence livestock pathogens and toxins

African horse sickness virus

African swine fever virus

Akabane virus

Avian influenza virus (highly pathogenic)

Blue tongue virus (exotic)

Bovine spongiform encephalopathy agent

Camel pox virus

Classical swine fever virus

Cowdria ruminantium (heartwater)

Foot and mouth disease virus

Goat pox virus

Table 15.1 *continued*

Lumpy skin disease virus

Japanese encephalitis virus

Malignant catarrhal fever virus (exotic)

Menangle virus

Mycoplasma capricolum/M.F38/*M. mycoides* var. *Capri*

Mycoplasma mycoides mycoides

Newcastle disease virus (velogenic)

Peste Des Petits Ruminants virus

Rinderpest virus

Sheep pox virus

Swine vesicular disease virus

Vesicular stomatitis virus (exotic)

(C) HHS/USDA overlap high-consequence livestock pathogens and toxins/select agents

Bacillus anthracis

Brucella abortus

Brucella melitensis

Brucella suis

Burkholderia mallei (formerly *Pseudomonas mallei*)

Burkholderia pseudomallei (formerly *Pseudomonas pseudomallei*)

Botulinum neurotoxin-producing species of *Clostridium*

Coccidioides immitis

Coxiella burnetii

Eastern equine encephalitis virus

Francisella tularensis

Hendra virus

Nipah virus

Rift Valley fever virus

Venezuelan equine encephalitis virus

Botulinum neurotoxin

Clostridium perfringens epsilon toxin

Shigatoxin

D-2

Staphylococcal enterotoxin

T-2 toxin

(D) Listed plant pathogens

Liberobacter africanus

Liberobacter asiaticus

Peronosclerospora philippinensis

Ralstonia solanacearum race 3, biovar 2

Schlerophthora rayssiae var. *zeae*

Synchytrium endobioticum

Xanthomonas oryzae

Xylella fastidiosa (citrus variegated chlorosis strain)

Table 15.2 Maximal toxin amounts permissible per principal investigator

Toxin	Amount (mg)
Abrin	100
Botulinum neurotoxin	0.5
Clostridium perfringens epsilon toxin	100
Conotoxin	100
Diacetoxyscirpenol (DAS)	1000
Ricin	100
Saxitoxin	100 mg
Shigatoxin	100
Shiga-like ribosome inactivating proteins	100
Staphylococcal enterotoxins	5
T-2 toxin	1000
Tetrodotoxin	100

validated real-time PCR protocols and reagents to public health laboratories for the detection of potential bioterrorist agents. The reagents used in these advanced diagnostic methods, including real-time PCR, are provided by the LRN and are restricted to laboratories that are reference laboratories, usually state and county public health laboratories. The rationale to restrict these tests is based on many factors, including safety of laboratory personnel, availability of expertise and dedicated personnel necessary to run and maintain facilities for these tests, access to high-quality reagents and maintenance of proficiency standards (Synder *et al.*, 2004).

Many of the agents considered the primary threats for a bioterrorism attack, such as *Francisella tularensis, Yersinia pestis,* and *Clostridium botulinum,* are slow-growing and fastidious, and may require a specialized growth medium or environment. Additionally, testing for any of these agents necessitates the availability of specialized facilities minimally biosafety level (BSL)-2, with many needing to be handled in the more limited BSL-3 and BSL-4 facilities, where gold-standard tests, including culture, must be carried out. Examples like *Bacillus anthracis,* which can be produced in highly aerosolizable forms that are especially conducive to laboratory-acquired infections, must be handled at BSL-3. Other agents such as variola virus (smallpox) and the haemorrhagic fever viruses such as Ebola and Marburg, are classified as BSL-4 agents and can-

not be handled in most state or local public health laboratories. Real-time PCR is an ideal method to screen for these agents without growing them in culture. Employing real-time PCR, the analyst first inactivates the agent(s) in a biothreat sample and then tests for multiple organisms or multiple targets. Thus, this methodology provides a measure of safety, because samples can be inactivated prior to analysis (Nitsche *et al.*, 2004).

Real-time PCR assays have been developed for some biothreat-classified agents such as B. *anthracis, Y. pestis, F. tularensis, C. botulinum,* orthopoxviruses, and viral haemorrhagic fever viruses. To date, the number of real-time PCR detection assays applied to the diagnosis of infectious diseases is relatively small in comparison to the published literature utilizing this technology. In the last few years, however, the number of real-time PCR assays for detection of biothreat agents has grown rapidly (Fig. 15.1A). As expected, there was an increase after 2001 in publications describing assays to detect B. *anthracis.* In the last 2–3 years, assays have been published that can identify other BT agents such as *Y. pestis* or orthopoxvirus (Fig. 15.1B). Most promising is the small but growing number of assays that can detect multiple agents. Since the threat of bioagents other than *B. anthracis* exists, assays that can detect multiple agents are critical to a rapid public health response, if potential exposures are to be minimized. Multiplex real-time PCR is a technology that can be utilized to

A
Real-time PCR publications

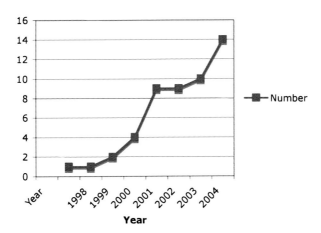

B
Real-time PCR Publications

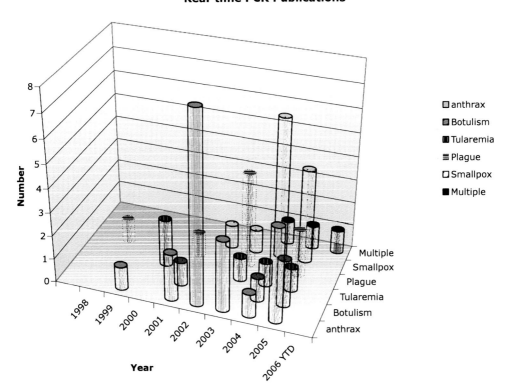

Figure 15.1 Recent literature published on biothreat agent detection by real-time PCR. (A) Publications on real-time PCR designed for the detection of select agents by year. (B) Publications on real-time PCR designed for the detection of select agents by year and organism. This analysis was performed by searching the http://www.ncbi.nlm.nih.gov web site using the Pubmed database. Queries were performed by combining the keywords: [real-time PCR OR qPCR OR 5′ nuclease OR lightcycler OR beacon OR scorpion OR taqman] and the bacterial genus and species and the disease caused by the agent [*Bacillus anthracis* OR anthrax; *Yersinia pestis* OR plague; *Francisella tularensis* OR tularaemia; *Clostridium botulinum* OR botulism; variola OR orthopox OR smallpox]. The information gathered by the authors and then evaluated by reading the abstract for each publication listed and either including or excluding the article from the data. Publications that described the detection of more than one select agent were included only in the multiple agent category.

detect more than one agent in a single tube. The use of multiplex real-time PCR in biodefence will be discussed in this chapter.

Assay validation

The validation of protocols is a critical area of method development for real-time PCR assays in the field of biodefence. Assay validation should take into account any applicable state and federal regulations as well as sample considerations such as the range of concentrations of bacteria or DNA expected in a particular clinical or environmental sample (Cirino *et al.*, 2006). The Select Agent Rule designed to regulate laboratories that work with agents with biothreat potential has imposed additional restrictions on laboratories that develop and/or use real-time PCR assays. Stringent record keeping for the use, storage, and destruction of culture stocks of agents must be maintained by the laboratory; thereby adding an additional layer of complexity when select agent assays are developed or used.

Because there are many real-time PCR chemistries, instruments and probe dyes available, users must take care to fully validate an existing protocol for their particular facility's equipment and the manner in which the assay is used such as sample types that may be tested. Validation studies must include sensitivity, specificity and reproducibility data for the assay using the specific matrix that will be tested. The validation process, while important for any real-time PCR assay, is crucial for BT diagnostics, because many decisions regarding vaccination, quarantine and treatment will potentially be made on the basis of this initial information, before full confirmatory work involving culture, biochemical characterization and antibiotic susceptibility profiles has been carried out and the results reported to the appropriate authorities.

Other elements important to the success of real-time PCR for biodefence applications are sampling and collection protocols appropriate for the event or agent of interest. Sample processing can be critical for effective diagnostics. This is especially true for any PCR-based method. DNA extraction is an important, initial step in the evaluation of the performance of real-time PCR on biothreat samples. Fortunately, many commercially available reaction components (e.g. mastermixes) for real-time PCR include additives that reduce the effects of potential PCR-specific inhibitors that are commonly found in environmental and clinical sample matrices (Halse *et al.*, 2006) typical of biothreat events.

As for any technique applied to the diagnosis of pathogens, the capability to operate using only limited sample volume and the ability to assess more than one pathogen or gene in that volume (i.e., multiplex capability) are essential. Real-time PCR requires minimal sample volume and can be used on sample types commonly analysed in laboratories for detection of potential biothreat agents (Ryu *et al.*, 2003; Cockerill, 2003). The clinical diagnostic value of traditional diagnostic methods such as microscopy and culture can be limited, compared to molecular methods, by a lack of sensitivity and specificity and by prior or ongoing antimicrobial treatment of the patient (for human specimens) (Mackay, 2004; Whelen and Persing, 1996). Real-time PCR compared to gold-standard methods such as culture, can achieve a high sensitivity, down to detection of a single microbial organism. Additionally, real-time PCR is effective for the detection of fastidious or non-cultivable organisms and for application to nonviable or mixed samples for which culture or other conventional methods are not possible. Melo and co-workers found multiplex PCR to be more sensitive than culture for the diagnosis of plague, for both older retrospective and more recent samples (Melo *et al.*, 2003).

Laboratories performing real-time PCR on potential agents of BT must also take into account many factors that may not be necessary for other applications of real-time PCR. These include chain-of-custody protocols and biosecurity measures such as secure facilities, surveillance cameras and locked incubators and freezers with minimal access, consistent with federal guidelines such as the Select Agent Rule. Additionally, laboratory workers must have knowledge of the appropriate protocols expected to be performed and the location where a specimen will need to be forwarded for final confirmatory testing if identified as a select agent. Finally, a laboratory must have in place appropriate protocols, equipment and reagents for testing specimens by real-time PCR in the facility in which the specimen is processed (i.e. BSL-3, BSL-4) or approved protocols for removing DNA extracts from this facility following sterility determinations. The latter option is of interest to many laboratories that have only limited number of real-time PCR instruments and need to maximize their diagnostic capabilities for other agents processed in BSL-2 laboratories to avoid having real-time PCR platforms, reagent clean facilities and additional personnel in high containment space.

As with any validated assay performed, a proficiency testing (PT) program must be established to maintain the competency of the laboratory staff. This can be carried out in several ways. With many of the BT assays that have been developed, there are no commercially available PT programs (Synder *et al.*, 2004). Each laboratory should establish a blinded PT program and each analyst should be tested on a routine basis.

PCR workflow

As for any area of microbiology, PCR laboratory design and workflow are important to successful testing. The use of proper PCR flow, as in standard PCR applications, is a necessity, as is the use of proper extraction and amplification controls (Mifflin, 2003). Although real-time PCR is a closed-system approach, PCR flow should still be carefully considered. It is necessary to maintain separate areas for reagents and mastermix considered 'PCR clean', as well as areas for specimen processing and extraction, amplification and analysis (Mitchell *et al.*, 2004; Persing, 1991). Laboratories should maintain this PCR flow as unidirectional, such that staff are not be allowed to enter a PCR Clean area once they have entered PCR amplification areas on a given day. Many large laboratories have dedicated personnel for each of the different areas of PCR analysis. Recent published literature contains laboratory flow schemes that are excellent references for those contemplating the addition of real-time PCR into their laboratory testing algorithms (Millar *et al.*, 2002; Mitchell *et al.*, 2004). Laboratories should consider, when possible, dedicated equipment such as pipettes, centrifuges, vortex mixers, refrigerators and freezers in each area as well as protocols for the use of each area. Protocols should include proper cleaning of the area and equipment with 10% bleach to reduce the potential for amplicon or DNA contamination.

Additionally, racks utilized in the real-time PCR process can be soaked in 10% bleach (Mitchell *et al.*, 2004). As a standard the wearing of gloves, shoe covers, lab coats and other personal protective equipment should also be established. At the very least, dedicated pipettes and adequate glove use should be maintained.

Routine environmental sampling of rooms and/or areas should also be considered. This testing process can involve the swabbing of areas such as doorknobs to rooms, refrigerators/freezers, centrifuges, racks and pipettes. A swab can be moistened in water or buffer and used to sample an area, after which it is returned to a microfuge tube containing the water or buffer and properly labelled. This tube can then undergo DNA extraction or direct testing in a PCR assay either for a specific target or a broad-range target like 16S rDNA that would detect all bacteria.

It is also critical in the diagnosis of biothreat agents that false positives and false negatives are minimized. To decrease the risk of false positives, nested primers should be avoided whenever possible, since these are a major contributor to contamination (Wannamaker *et al.*, 1989). Additionally, proper training of analysts is crucial to the success of this method. Real-time PCR instrumentation is extremely complex and requires routine maintenance for background determination, for calibration of dyes and for ensuring that heating blocks and elements are functioning properly. The occurrence of false negatives can be greatly decreased by timely attention to these details.

Clinical assays

Test results for specimens from the clinically ill must be obtained as rapidly as possible to provide the physician with beneficial information crucial for patient treatment. This is especially true in the field of infectious disease where treatment must be given as soon as possible. PCR fills an important gap in the testing for infectious agents because culture-based testing may take days to produce a result, whereas PCR requires a few hours. PCR assays have vastly improved clinical diagnostics of detection for biothreat agents and real-time PCR has added benefits over traditional PCR analysis. The time necessary to perform a real-time PCR assay is much less than that needed for conventional PCR and it may provide an extra layer of specificity by the inclusion of a specific oligonucleotide probe. Quick turn-around time for test results, while essential in a clinical setting becomes of utmost importance when a biothreat agent such as *B. anthracis* or Variola major is suspected. The initiation of an epidemiological or criminal investigation hinges on rapid identification of the agent.

Many naturally occurring infections have been identified that were caused by organisms that are classified as biothreat agents, such as *Brucella* spp. or *F. tularensis*. Through an epidemiological investigation it can be determined whether an infection was acquired in an area endemic for a particular pathogen, from a contaminated food source, from an accidental infection in a laboratory setting, or from an actual BT event. Real-time PCR has enabled rapid detection of agents that may be used in BT and it is this rapid determination that starts an epidemiologic investigation to either rule in or rule out bioterrorism causes. In a biothreat event, rapid determination is essential to begin both criminal and public health investigations to minimize the number of individuals exposed to the agents and to begin prophylaxis for those already exposed.

The power of both conventional and real-time PCR methods is the exquisite specificity that is obtained in each assay. This is a critical component in testing for biothreat agents. Many biothreat agents possess genetic similarity. For example, *B. anthracis* shows >99% genetic homology with *B. thuringiensis* and *B. cereus* within certain areas of the genome (Ash and Collins, 1992; Ash *et al.*, 1991). For a properly designed real-time PCR assay, the addition of an additional fluorescent oligonucleotide sequence referred to as a probe adds significant specificity to the assay. This type of PCR assay was utilized by the LRN in the investigation of the intentional releases of anthrax in the US postal system in 2001. The CDC had developed real-time PCR assays for *B. anthracis* and utilized these assays to analyse large numbers of both clinical and environmental samples (Hoffmaster *et al.*, 2002b). These real-time PCR assays were extremely useful when laboratories received hundreds of items a day and a mechanism was needed to quickly rule-out

B. anthracis in these samples. This allowed the laboratories to focus further additional molecular and culture testing on samples that produced positive results in the initial real-time PCR assay. Valuable information was quickly reported back to law enforcement personnel and epidemiologists to aid in their investigation. As reported by Hoffmaster and co-workers (2002b), every clinical specimen that tested positive at CDC in the real-time PCR assays subsequently was found to be positive by culture (Hoffmaster *et al.*, 2002a). Because of the highly sensitive and specific nature of the real-time PCR technique, it proved to be an invaluable tool in the 2001 anthrax investigations. Future bioterrorism-related investigations will also rely on it for its invaluable capacity as a screening tool.

Real-time PCR has also been combined with melting point analysis, in order to aid in the diagnosis of closely related organisms. Bystrom and co-workers described the use of real-time PCR to distinguish among closely related *F. tularensis* strains, in order to detect tularaemia from lymph node tissue (Bystrom *et al.*, 2005). Utilizing melting point analysis after real-time PCR, they obtained subspecies information that could be extremely useful in a bioterrorism event for determining whether an intentional release had occurred; further, it could provide epidemiological evidence to link a potential source to infected patients and identify origin.

Real-time PCR for biothreat agents other than *B. anthracis* have been developed and utilized in recent years (Christensen *et al.*, 2006; Debeaumont *et al.*, 2005; Jones *et al.*, 2005; Newby *et al.*, 2003; Tomaso *et al.*, 2003). While many of these assays were designed for diagnosis of agents such as *Brucella* spp., *F. tularensis*, and *Y. pestis* in a biothreat situation, they have been particularly useful for the diagnosis of naturally occurring infections of these agents in areas endemic for the bacteria. After a primary clinical diagnosis of infection with a biothreat agent, it may be difficult to determine whether the source of an infection is a true intentional bioterrorist event. The following case highlights the importance of a rapid diagnosis in order to confirm a clinical diagnosis and begin an epidemiological investigation. In 2003, in New York City, a male and female presented at an emergency room with symptoms consistent with bubonic plague (Anonymous, 2003). Specimens were taken, and *Y. pestis* was quickly identified by the LRN real-time PCR assay. Because plague is not endemic in New York State there was a suspicion of bioterrorism. An epidemiological investigation quickly determined that the infection was of natural origin. The couple had recently arrived in New York from New Mexico, where plague is endemic. *Y. pestis* had been isolated from fleas trapped in the couple's backyard in New Mexico and additional detailed genetic analyses were used to determine that the pathogen that caused the infection in the couple had originated in the flea population behind their home.

Several real-time PCR assays have been reported in the literature for use with clinical specimen types for biothreat detection (Table 15.3). One example of the necessity for rapid diagnosis is intoxication by *C. botulinum* toxin. Botulism causes a descending paralysis that can occur extremely quickly, especially in infants. Akbulut and colleagues recently reported the development of a real-time PCR assay for the detection *C. botulinum* toxins A, B, and E from serum or tissue samples or wound swabs (Akbulut *et al.*, 2004). The use of real-time PCR is essential when *C. botulinum* toxin is suspected because the gold-standard test, the mouse bioassay, can require several days to produce a result. Further, the laboratory may not be supplied with the appropriate amount of specimen particularly for infant cases. Real-time PCR can serve as a screen for presumptive toxin identification as well as a screen for associated environmental samples to determine the source of the infection. Our laboratory has determined that real-time PCR analysis for *C. botulinum* toxin is invaluable in this regard, given that many sample types, such as stool, food, water and soil samples, can contain competing organisms and other *Clostridium* species. Real-time PCR can be used as a screening tool and positive isolates can then be identified for testing in the mouse bioassay. This strategy significantly reduces the number of mice needed for the bioassay, as well as providing cost savings, since only a subset of the original bacterial colonies must be tested.

While there has been an increase in the number of real-time PCR assays for biothreats in last few years, overall in the past five years, there have been very few papers published on the use of these real-time PCR assays on different clinical specimen types. Often, significant research needs to be done in order to develop PCR protocols for extraction of DNA from various specimen types. The identification of the need to develop assays for real-time PCR within the last five years is a positive sign. However, there needs to be initiatives and published literature to provide data showing that these assays can be used for testing with different clinical specimens. To date, there is only one FDA approved real-time PCR assay for a biothreat agent (avian influenza virus) in clinical specimen types (Anonymous, 2006). This paucity of assays could have serious consequences if or when another biothreat attack occurs. Often, the development of such protocols and the validation of appropriate numbers of specimens requires significant allocation of resources that would be difficult to free up during a BT event when there will be large numbers of specimens that must be analysed. Commercial kits and reagents have become available for use in biothreat detection (Table 15.4). Many of these assays were developed for use with environmental samples and are not yet validated for clinical specimens, leaving a critical gap in a biothreat event.

For each of the pathogens thought to be possible BT agents, there is at least one published real-time PCR assay evaluated for at least one clinical specimen type. One example is the detection of *Y. pestis* in sputum (Loiez *et al.*, 2003). Sputum is known to contain substances that inhibit the PCR reaction, a fact which caused the authors to modify their extraction procedure. This and other published assays highlight the importance of optimization of real-time PCR assays as well as the inclusion of a step to monitor inhibition in each sample tested.

Commercial kits and reagents have become available for use in biothreat detection (Table 15.4). The majority of the available kits and assays were designed to be used with environmental samples and have not been validated on clinical specimens, leaving a critical gap in diagnostic availability of reagents in a biothreat event.

Table 15.3 Real-rime PCR assays published for clinical specimen types

Agent	Specimen	Target	Chemistry	Sensitivity	Instrument	Reference
Bacillus anthracis	Serum, Tissue, Pleural fluid, Sputum	pXO1, pXO2, Chromosome	TaqMan	167 cells (1 pg DNA)	Light Cycler, ABI 7700, Smart Cycler	Hoffmaster et al. (2002b)
Clostridium botulinum toxin types A,B,E	Serum, wound swabs aspirates, pus	Toxin gene	TaqMan	16–141 GE	Light Cycler, ABI 7700	Akbulut et al. (2004)
Brucella spp.	Serum	*bcsp31*	SybrGreen	1 GE	LightCycler	Queipo-Ortuno et al. (2005)
Yersinia pestis	Sputum	*pla* gene	MGB TaqMan	1.5 CFU/mLl	ABI7000	Loïez et al. (2003)
Francisella tularensis	Lymph node	rRNA, *lpnA*	SybrGreen	10–100CFU, 100–150 GE	iCycler	Byström et al. (2005)
Orthopoxvirus	Viral isolates	Haemagglutinin gene	TaqMan	25 gene copies	LightCycler, SmartCycler	Sofi Ibrahim et al. (2003)
Burkholderia pseudomallei	Blood	*orf2*	Taqman	8.4x103 CFU/ml	SmartCycler	Novak et al. (2006)
Coxiella burnetii	Serum	*sod* gene	SybrGreen	1 gene copy	LightCycler	Fournier et al. (2003)
Ebola/Marburg virus	Serum	L gene	SyberGreen	2.677 GE/ml plasma	LightCycler	Drosten et al. (2002)
Lassa virus	Serum	L gene	SyberGreen	2.445 GE/ml plasma	LightCycler	Drosten et al. (2002)

Abbreviations: GE, genome equivalents;CFU, colony-forming units.

Table 15.4 Available commercial real-time PCR kits for biothreat detection

Real-time PCR Kit	Manufacturer	Agent(s) detected	RT-PCR platform	Number of targets	Specimen type	Notes
BA 4-Plex Assay	Cepheid	B. anthracis	GeneXpert	2+2 controls	Environmental	
CycleavePCR Bacillus anthracis Detection Kit	TaKaRa	B. anthracis	Smart Cycler	2	Environmental, clinical	Uses cycling probe technology
LightCycler Bacillus anthracis Detection Kit	Roche	B. anthracis	LightCycler	2	Environmental, clinical	
PathAlert™ Detection Kit	Invitrogen	B. anthracis, Y. pestis, orthopoxvirus, F. tularensis	Agilent 2100 Bioanalyser, 5100 Automated Lab on a Chip Platform	Four multiplex kits	Environmental (air, food, water)	Combines multiplex real-time PCR with a microfluidic chip
Certified Lux ™Primer Sets for infectious agents	Invitrogen	B. anthracis, Y. pestis, Orthopoxvirus, F. tularensis, C. botulinum toxin Types B,E	Compatible with most real-time PCR instruments	Ba (3), Yp (3), Or (1), Ft (2), CbotB (1), CbotE (1)	Linear DNA	
artus™ B. anthracis PCR kit, artus™ Orthopox PCR kit	Qiagen	B. anthracis, Orthopoxvirus	LightCycler, ABI Prism 7000,7900HT, Rotor-gene	2+2 controls	Spores and vegetative cells from diverse sample types (tissues, blood, stool, soil)	
BioThreat Screening Kit	Idaho Technology	B. anthracis, F. tularensis, Y. pestis, Brucella spp.	LightCycler, Rapid Cycler, Razor, R.A.P.I.D.	1 target each	Environmental	Reagents are freeze-dried
Pathogen BioReagents	Idaho Technology	B. anthracis, F. tularensis, Y. pestis, Brucella spp., Variola, C. botulinum Type A	Razor, R.A.P.I.D.	Ba (3), Ft (2), Yp (2), Br (1) Va (1), Cbot (1)	Environmental	Reagents are freeze-dried
BioSeeq	Smiths Detection	B. anthracis. F. tularensis	BioSeeq	NA	Environmental (powders)	Hand-held unit
BioSeeq Mail Sentry	Smiths Detection	B. anthracis, F. tularensis,	Mail Sentry	NA	Environmental (mail)	On-site mail screening system

Environmental diagnostic applications

The processing and analysis of environmental samples are multifaceted tasks. While there exist a limited number of typical clinical matrices (e.g. blood, serum, urine, sputum, tissue), there is a myriad of highly diverse, complex matrix types in the environment. From the simplest matrix (potable water) to the most complex matrices (raw sewage, soil, or foods) the primary sample processing and extraction will be critical to the success of real-time PCR analyses. Each matrix type has particular characteristics that necessitate specific protocols to achieve efficient recovery of nucleic acids, while excluding confounding materials (Lewis *et al.*, 2000). For example, humic and fulvic acids commonly found in soil will hinder PCR amplification (Kreader, 1996; Miller, 2001) as will metal chelators and heavy metals in certain foods (Ijzerman *et al.*, 1997); (Kreader, 1996) and haem in blood (Akane *et al.*, 1994). Wilson provides an excellent review of the factors that are known to inhibit PCR amplification (Wilson, 1997). Unfortunately, it is impossible to validate extraction procedures on every possible environmental sample matrix, so most kits or methods are marketed for one or two general sample matrix types like soil, blood, or stool. Therefore, when new or atypical matrices are being processed it is essential to include positive controls in each extraction procedure.

Each class of these PCR-confounding materials (e.g. polymerase inhibitors, nucleases, extreme pH modifiers) requires implementation of defined steps to isolate the amplifiable nucleic acids from the inhibitory compounds (de Franchis *et al.*, 1988; Filion *et al.*, 2003; Kuske *et al.*, 1998; Sutlovic *et al.*, 2005; Tsai and Olson, 1992). Fortunately, the basic methods for nucleic acid extraction have been well characterized, and solid-phase extraction procedures are used routinely in most research and diagnostic laboratories (Yamamoto, 2002). The many extraction kits available for nucleic acid processing provide differing levels of PCR-inhibitor mitigation. In addition, certain pathogens, like spore-forming bacteria, require harsher or additional processing steps to release genetic material for PCR-based detection. For all of these reasons, the process-

ing and PCR analysis of environmental samples require significant preliminary testing and evaluation, in addition to inclusion of processing controls. To facilitate better understanding by the reader of the various applications of real-time PCR, this section is organized according to end-product applications: liquid testing, food testing, solids testing (e.g. powders, soils) and air testing. In addition, the areas of forensic PCR and automated PCR for environmental monitoring will be discussed.

Real-time PCR analysis of liquids and foods

Water quality testing is the most common environmental testing enterprise in developed countries and one of the highest priority public health concerns in underdeveloped countries (Pedley and Pond, 2003). Comprehensive assessment of water quality is complex, and many diagnostic methods are required to ensure a safe potable water supply. Because water quality testing is a monitoring/surveillance activity meant to ensure integrity of the product, it is not routinely performed in response to a defined incident. PCR-based analysis constitutes only a small proportion of the diagnostic tests used to assess water quality. Determinations of turbidity, pH, metal ion levels and general bioburden far outweigh analysis by PCR for specific pathogens in the water (Shepard *et al.*, 2006). Initial application of PCR to determine bioburden was found to be less cost-effective than the use of classical methods, so use of PCR for this assessment has not been extensive. Instead of generic bioburden determination, levels of certain indicator organisms have been shown to be predictive of water quality and PCR screens for these indicator organisms have been developed (Lebuhn *et al.*, 2004). Because of dilution factors in municipal water supplies and because of the fact that many of the select agents can be found naturally in the environment somewhere in the U.S., PCR-based analysis for many select agents (Table 15.1) has not been performed routinely, even though real-time PCR methods are available for these targets (Bode *et al.*, 2004; Christensen *et al.*, 2006; Ivnitski *et al.*, 2003; McAvin *et al.*, 2003; McAvin *et al.*, 2004 Chase *et al.*, 2005; Tomaso *et al.*, 2003.

The most common and widespread application of PCR to water testing is PCR monitoring for trace levels of enteropathogenic *Escherichia coli* O157:H7 either at the water source or after post-processing at waste treatment facilities (Campbell *et al.*, 2001; Ibekwe *et al.*, 2002; Ibekwe *et al.*, 2004). Additional applications to detect viruses (Carducci *et al.*, 2003) and parasites (Fontaine and Guillot, 2002) are becoming more common, as is screening of wastewater for livestock run-off for bacteria carrying antibiotic resistance genes (Volkmann *et al.*, 2004).

Solid (e.g. beef, chicken, fresh produce) and liquid (e.g. milk, carbonated beverages, baby formula) foods are also amenable to analysis by real-time PCR, although the method is not routinely employed for them because of cost considerations (Perelle *et al.*, 2004). Similar to water testing, the most common application of real-time PCR to food testing is the detection of enteropathogenic *E. coli* O157:H7 (Demarco and Lim, 2002; Nguyen *et al.*, 2004) typically in beef. Additional tests for *Listeria* (Nguyen *et al.*, 2004) and *Staphylococcus aureus* in beef (Alarcon *et al.*, 2006) and *Salmonella* in meat products (Perelle *et al.*, 2004), have been reported. These directed diagnostics were developed because the pathogens of interest are commonly associated with certain livestock or dairy products and are known to cause some level of morbidity and mortality in consumers. Clearly, the ability to screen food for pathogenic bacteria is a major focus of food biosecurity (McKillip and Drake, 2004) as is the ability to detect plant pathogens on crops (Schaad *et al.*, 2003).

Botulism food poisoning is caused by ingestion of a toxin secreted by the bacterium *C. botulinum* and this toxin is one of the most toxic substances known. Botulism is typically associated with home-canned foods, although the bacteria can occur naturally in raw honey and the soil. Botulinum toxin is classified as a select agent because of its extreme toxicity, ease of acquisition and production. Therefore, detection of the toxin or the bacterial DNA in solid (Yoon *et al.*, 2005) or liquid (Perdue, 2003) foods is considered a high priority for public health and national security. Several real-time PCR methods exist to detect these in food and environmental matrices (Szabo *et al.*, 1994).

Real-time PCR analysis of soils and powders

Whereas real-time PCR analysis of water and of liquid and solid foods is done for surveillance, analysis of soil or powders is typically done in response to a suspected or known event, be it a natural outbreak or intentional contamination. Hence, many of the real-time PCR assays that have been developed for detection of biothreat agents are employed for soil and/or powder matrices (Ivnitski *et al.*, 2003). Not surprisingly, identification of *B. anthracis* from powders associated with mail has become a routine procedure for many public health laboratories since 2001 (Teshale *et al.*, 2002). Real-time PCR assays for *B. anthracis* have also been validated for samples in soil matrices (Alam *et al.*, 2003); (Beyer *et al.*, 1995; Cheun *et al.*, 2003; Ryu *et al.*, 2003), since *B. anthracis* spores can survive in soils for decades. Following any future bioterrorism event, reclamation of buildings and the surrounding environment is likely to entail a process of repeated testing for remaining viable spores. Other assays for detection of biothreats in soils/powders include detection of *Burkholderia* spp. (Miller *et al.*, 2002; Salles *et al.*, 2002), *Y. pestis* (McAvin *et al.*, 2003), *F. tularensis* (Versage *et al.*, 2003), and *E. coli* O157:H7 (Campbell *et al.*, 2001; Ibekwe *et al.*, 2004).

Real-time PCR analysis of air samples

Before 2001, air monitoring was performed to detect chemical contaminants or pollution, but it rarely was employed to test for biological organisms. Since that time however, air monitoring for biothreats has become a national biodefence focus (Lim *et al.*, 2005). Now, operating throughout the U.S. Postal Service system is the Biohazard Detection System (BDS), which currently screens all mailed letters for *B. anthracis* and will soon be adding additional biothreat targets to its assays. The BDS system collects air from 'pinchpoints' in the mail processing stream and subjects the collected material to real-time PCR analysis. The system has been operating in postal facilities for over a year with exemplary results (Jaffer, 2004). Additionally, the BioWatch system is operating in many major metropolitan areas in the US. The BioWatch program collects air samples and analyses them for several biothreat agents. Both BioWatch and BDS operate in near-real time and have emergency response protocols incorporated in case there is a biothreat detection. Other automated air sampling systems are in development and are expected to reduce analysis time and operational costs, while expanding the number of pathogens detected. One such system, the autonomous pathogen detection system (APDS), has been deployed in select US cities (Hindson *et al.*, 2005; McBride *et al.*, 2003).

Multiplex real-time PCR in biodefence

The forgoing sections have highlighted the value of real-time PCR in situations when a biothreat agent is suspected. In certain cases, the causative agent may be unclear, or it may be one among a list of bacterial or viral agents. In the early stages of the anthrax events of 2001, anthrax could potentially be mis-diagnosed in patients as influenza due to similar symptomatology (Anonymous, 2001). Many biothreat agents initially present with symptoms similar to those of the flu. Given this lack of symptomatic specificity, it is likely that a particular specimen will require testing for a number of agents. Multiplex real-time PCR can provide the necessary tool in this situation. At present, multiplexed assays can be used to detect four or five target analytes in one tube. The latest generation instruments such as the InSyte from Biobank and the Rotorgene 6000 from Corbett research are reported to be able to detect at least six or seven different analytes by using multiple channels that can distinguish different wavelengths in a single PCR assay (Kubista *et al.*, 2006).

Multiplexed real-time PCR can be used for detection of multiple agents, to confirm pathogen isolates by using multiple targets, or to differentiate an organism from closely related organisms at the species level. Multiplex PCR allows for the simultaneous detection of multiple genes in one organism such as the two plasmids (pX01 and pX02) necessary for virulence in *B. anthracis* to distinguish it from its closely related near neighbours. Several real-time PCR kits that have been developed that are syndromic in design, (based on patient symptomatology); however, most of the commercially available kits for detection of biothreat agents have been designed and vali-

dated using environmental samples (Table 15.4). Multiplex PCR assays have been developed for select agents in clinical specimens such as *B. anthracis* (Bell *et al.*, 2002; Ramisse *et al.*, 1996; Ryu *et al.*, 2003), *Y. pestis* (Melo *et al.*, 2003; Tomaso *et al.*, 2003), *F. tularensis* (Grunow *et al.*, 2000; Sjostedt *et al.*, 1997; Versage *et al.*, 2003), *C. botulinum* toxins (Lindstrom *et al.*, 2001), *S. aureus* enterotoxin type B (SEB) (Letertre *et al.*, 2003), orthopoxviruses (Kulesh *et al.*, 2004; Panning *et al.*, 2004) and viral haemorrhagic fever viruses (Bronzoni *et al.*, 2004). For a review of multiplex assays for biothreat agents, see Cirino *et al.* (2004). The literature published on multiplex real-time PCR assays is small and few of these assays have been applied to the diagnosis of infectious diseases (Mackay, 2004), with the focus on areas that already test for multiple organisms such as respiratory pathogens (Templeton *et al.*, 2004) or food-borne bacteria (Fout *et al.*, 2003). In the last few years, however, more assays have been reported, most notably for biothreat agents.

Future prospects

A breakthrough technology is defined as one that rapidly displaces other technologies and renders them obsolete. PCR was one such breakthrough technology for genetic manipulation and nucleic acid detection. The last four years have seen an explosive growth in the number of new technologies being developed for diagnostics and the next five years should see some of these nascent technologies becoming widely employed for detection of biothreat agents.

Real-time PCR technology has existed for almost 10 years and is likely to be displaced within the next decade, although the ability to improve multiplexing capability on this elegant platform may keep it viable for a longer period. Availability of new dyes, better methods for optical discrimination and more powerful software will be critical to the longevity of real-time PCR. Microsphere arrays, microarrays and biosensors are still in their early evolutionary phases but are likely to become significant technologies within the next five years. The limitations of the technology, such as expense and time for development, will be decreased as these platforms are increasingly combined with other technologies such as magnetic separation, biofilms, microfluidics and fibreoptics. It is likely that each successful technology will develop and expand into a defined analytical niche.

Advances in reporter technologies will also play a significant role in determining what types of methodologies become widely employed. Novel fluorophores and fluorescent quenchers are being synthesized routinely and these will be critical in fluorescent or FRET-based diagnostic multiplexing. Similarly, up-converting phosphor reporters, which have already been applied to monitor changes in RNA expression levels (van De Rijke *et al.*, 2001), may prove useful in the detection of biothreat agents. More diverse reporter compounds will improve the sensitivity of various platforms and will offer significant advantages for multiplexing.

Another area in which breakthroughs are likely to occur in the next five years is sample processing. The various detection platforms require very diverse pre-analysis processing, a fact which makes a unified diagnostic algorithm difficult to develop. Better software, robotics, and understanding of the field of diagnostics on a global basis should facilitate continued improvements in sample processing methodologies. Current lab-on-a-chip technology should evolve into a broadly integrated component of diagnostics.

Although real-time PCR was developed over a decade ago, it has only recently evolved to become sufficiently mature and cost-effective for integration into the routine workflow of diagnostic and public health laboratories. Real-time PCR has been used for some time in the pharmaceutical industry for drug target quantitation, genetic screening and tracking of genetically modified organisms (GMOs). With recent advances in real-time PCR, the field of biodefence should acquire assays in coming years that will be able to determine numbers of spores or cells in a sample type the biothreat agent present and even determine the viability of the organisms.

The ongoing improvements in real-time PCR analytical systems and software, combined with reduced cost and the increasing ability to multiplex analytical assays, should ensure real-time PCR in diagnostic laboratories and in biodefence research and applications for another decade or more.

References

Anonymous. 2001. Considerations for distinguishing influenza-like illness from inhalational anthrax. MMWR Morb. Mortal Wkly Rep. *50*, 984–986.

Anonymous. 2003. Imported plague – New York City, 2002. MMWR Morb Mortal Wkly Rep. *52*, 725–728.

Anonymous. 2006. New laboratory assay for diagnostic testing of avian influenza A/H5 (Asian Lineage). MMWR Morb Mortal Wkly Rep. *55*, 127.

Akane, A., Matsubara, K., Nakamura, H., Takahashi, S. and Kimura, K. 1994. Identification of the haem compound copurified with deoxyribonucleic acid (DNA) from bloodstains, a major inhibitor of polymerase chain reaction (PCR) amplification. J. Forensic Sci. *39*, 362–372.

Akbulut, D., Grant, K. and McLauchlin, J. 2004. Development and application of Real-Time PCR assays to detect fragments of the *Clostridium botulinum* types A, B, and E neurotoxin genes for investigation of human foodborne and infant botulism. Foodborne Pathog. Dis. *1*, 247–257.

Alam, S.I., Agarwal, G., Kamboj, D., Rai, G. and Singh, L. 2003. Detection of spores of *Bacillus anthracis* from environment using polymerase chain reaction. Ind. J. Exp. Biol. *41*, 177–180.

Alarcon, B., Vicedo, B. and Aznar, R. 2006. PCR-based procedures for detection and quantification of *Staphylococcus aureus* and their application in food. J. Appl. Microbiol. *100*, 352–364.

Ash, C., Collins, M. (1992). Comparative analysis of 23S ribosomal RNA gene sequences of *Bacillus anthracis* and emetic *Bacillus cereus* determined by PCR-direct sequencing. FEMS Microbiol. Lett. *73*, 75–80.

Ash, C., Farrow, J., Dorsch, M., Stackebrandt, E. and Collins, M. (1991). Comparative analysis of *Bacillus anthracis*, *Bacillus cereus*, and related species on the basis of reverse transcriptase sequencing of 16S rRNA. Int. J. Syst. Bacteriol. *41*, 343–346.

Bell, C. A., Uhl, J. R., Hadfield, T., David, J., Meyer, R., Smith, T. and Cockerill, F., 3rd 2002. Detection of *Bacillus anthracis* DNA by LightCycler PCR. J. Clin. Microbiol. *40*, 2897–2902.

Beyer, W., Glockner, P., Otto, J. and Bohm, R. 1995. A nested PCR method for the detection of *Bacillus anthracis* in environmental samples collected from former tannery sites. Microbiol. Res. *150*, 179–186.

Bode, E., Hurtle, W. and Norwood, D. 2004. Real-time PCR assay for a unique chromosomal sequence of *Bacillus anthracis*. J. Clin. Microbiol. *42*, 5825–5831.

Bronzoni, R.V., Moreli, M.L., Cruz, A.C. and Figueiredo, L.T. 2004. Multiplex nested PCR for Brazilian Alphavirus diagnosis. Trans. R. Soc. Trop. Med. Hyg. *98*, 456–461.

Bystrom, M., Bocher, S., Magnusson, A., Prag, J. and Johansson, A. 2005. Tularemia in Denmark: identification of a *Francisella tularensis* subsp. *holarctica* strain by real-time PCR and high-resolution typing by multiple-locus variable-number tandem repeat analysis. J. Clin. Microbiol. *43*, 5355–5358.

Campbell, G.R., Prosser, J., Glover, A. and Killham, K. 2001. Detection of *Escherichia coli* O157,H7 in soil and water using multiplex PCR. J. Appl. Microbiol. *91*, 1004–1010.

Canter, D.A., Gunning, D., Rodgers, P., O'Connor, L., Traunero, C. and Kempter, C. 2005. Remediation of *Bacillus anthracis* contamination in the U.S. Department of Justice mail facility. Biosecur. Bioterror. *3*, 119–127.

Carducci, A., Casini, B., Bani, A., Rovini, E., Verani, M., Mazzoni, F. and Giuntini, A. 2003. Virological control of groundwater quality using biomolecular tests. Water Sci. Technol. *47*, 261–266.

Chase, C.J., Ulrich, M., Wasieloski, L., Jr., Kondig, J., Garrison, J., Lindler, L. and Kulesh, D. 2005. Real-Time PCR assays targeting a unique chromosomal sequence of *Yersinia pestis*. Clin. Chem. *51*, 1778–1785.

Cheun, H.I., Makino, S., Watarai, M., Erdenebaatar, J., Kawamoto, K. and Uchida, I. 2003. Rapid and effective detection of anthrax spores in soil by PCR. J. Appl. Microbiol. *95*, 728–733.

Christensen, D.R., Hartman, L., Loveless, B., Frye, M., Shipley, M., Bridge, D., Richards, M., Kaplan, R., Garrison, J., Baldwin, C., *et al.* 2006. Detection of biological threat agents by real-time PCR: comparison of assay performance on the R.A.P.I.D., the LightCycler, and the Smart Cycler platforms. Clin. Chem. *52*, 141–145.

Cirino, N. M., Musser, K. and Egan, C. 2004. Multiplex diagnostic platforms for detection of biothreat agents. Expert Rev. Mol. Diagn. *4*, 841–857.

Cirino, N.M., Shail, M. and Egan, C. 2006. Bioterrorism: Prevention, Preparedness, and Protection. J.V. Borrelli Editor. New York: Nova Science Publishing, Inc.).

Cockerill, F.R., 3rd 2003. Application of rapid-cycle real-time polymerase chain reaction for diagnostic testing in the clinical microbiology laboratory. Arch. Pathol. Lab. Med. *127*, 1112–1120.

de Franchis, R., Cross, N., Foulkes, N. and Cox, T. 1988. A potent inhibitor of Taq polymerase copurifies with human genomic DNA. Nucleic Acids Res. *16*, 10355.

Debeaumont, C., Falconnet, P. and Maurin, M. 2005. Real-time PCR for detection of *Brucella* spp. DNA in human serum samples. Eur. J. Clin. Microbiol. Infect. Dis. *24*, 842–845.

Demarco, D.R. and Lim, D. 2002. Detection of *Escherichia coli* O157,H7 in 10- and 25-gram ground beef samples with an evanescent-wave biosensor with silica and polystyrene waveguides. J. Food Prot. *65*, 596–602.

Drosten, C., Göttig, S., Schilling, S., Asper, M., Panning, M., Schmitz, H. and Günther, S. 2002. Rapid detection and quantification of RNA of Ebola and Marburg viruses, Lassa virus, Crimean-Congo Hemorrhagic Fever virus, Rift Valley Fever virus, Dengue virus, and Yellow Fever virus by real-time reverse transcription PCR. J. Clin. Microbiol. *40*, 2323–2330.

Edwards, K.A., Clancy, H. and Baeumner, A. 2006. *Bacillus anthracis*: toxicology, epidemiology and current rapid-detection methods. Anal. Bioanal. Chem. *384*, 73–84.

Filion, M., St-Arnaud, M. and Jabaji-Hare, S. 2003. Direct quantification of fungal DNA from soil substrate using real-time PCR. J. Microbiol. Methods *53*, 67–76.

Fontaine, M. and Guillot, E. 2002. Development of a TaqMan quantitative PCR assay specific for *Cryptosporidium parvum*. FEMS Microbiol. Lett. *214*, 13–17.

Fournier, P. and Raoult, D. 2003. Comparison of PCR and serology for the diagnosis of acute Q fever. J. Clin. Microbiol. *41*, 5094–5098.

Fout, G.S., Martinson, B., Moyer, M. and Dahling, D. 2003. A multiplex reverse transcription-PCR method for detection of human enteric viruses in groundwater. Appl. Environ. Microbiol. *69*, 3158–3164.

Grunow, R., Splettstoesser, W., McDonald, S., Otterbein, C., O'Brien, T., Morgan, C., Aldrich, J., Hofer, E., Finke, E. and Meyer, H. 2000. Detection of *Francisella tularensis* in biological specimens using a capture enzyme-linked immunosorbent assay, an immunochromatographic handheld assay, and a PCR. Clin. Diagn. Lab. Immunol. *7*, 86–90.

Halse, T.A., Musser, K. and Limberger, R. 2006. A multiplexed real-time PCR assay for rapid detection of *Chlamydia trachomatis* and identification of serovar L-2, the major cause of lymphogranuloma venereum in New York. Mol Cell. Probes *20*, 290–297.

Higgins, J.A., Cooper, M., Schroeder-Tucker, L., Black, S., Miller, D., Karns, J., Manthey, E., Breeze, R. and Perdue, M. 2003. A field investigation of *Bacillus anthracis* contamination of U.S. Department of Agriculture and other Washington, D.C., buildings during the anthrax attack of October 2001. Appl. Environ. Microbiol. *69*, 593–599.

Hindson, B.J., Makarewicz, A., Setlur, U., Henderer, B., McBride, M. and Dzenitis, J. 2005. APDS: the autonomous pathogen detection system. Biosens Bioelectron. *20*, 1925–1931.

Hoffmaster, A.R., Fitzgerald, C., Ribot, E., Mayer, L. and Popovic, T. 2002a. Molecular subtyping of *Bacillus anthracis* and the 2001. bioterrorism-associated anthrax outbreak, United States. Emerg. Infect. Dis. *8*, 1111–1116.

Hoffmaster, A.R., Meyer, R., Bowen, M., Marston, C., Weyant, R., Thurman, Messenger, S., Minor, E., Winchell, J., Rassmussen, M., *et al.* 2002b. Evaluation and validation of a real-time polymerase chain reaction assay for rapid identification of *Bacillus anthracis*. Emerg. Infect. Dis. *8*, 1178–1182.

Ibekwe, A.M., Watt, P.M., Grieve, C.M., Sharma, V K. and Lyons, S.R. 2002. Multiplex fluorogenic real-time PCR for detection and quantification of *Escherichia coli* O157,H7 in dairy wastewater wetlands. Appl. Environ. Microbiol. *68*, 4853–4862.

Ibekwe, A..M., Watt, P., Shouse, P. and Grieve, C. 2004. Fate of *Escherichia coli* O157,H7 in irrigation water on soils and plants as validated by culture method and real-time PCR. Can. J. Microbiol. *50*, 1007–1014.

Ijzerman, M.M., Dahling, D.R. and Fout, G.S. (1997). A method to remove environmental inhibitors prior to the detection of waterborne enteric viruses by reverse transcription-polymerase chain reaction. J. Virol. Methods *63*, 145–153.

Ivnitski, D., O'Neil, D., Gattuso, A., Schlicht, R., Calidonna, M. and Fisher, R. 2003. Nucleic acid approaches for detection and identification of biological warfare and infectious disease agents. Biotechniques 35, 862–869.

Jaffer, A.S. 2004. Biohazard detection system deployment resumes. US Postal Service Postal News 38.

Jones, S.W., Dobson, M., Francesconi, S., Schoske, R. and Crawford, R. 2005. DNA assays for detection, identification, and individualisation of select agent microorganisms. Croat. Med. J. 46, 522–529.

Kreader, C.A. 1996. Relief of amplification inhibition in PCR with bovine serum albumin or T4 gene 32 protein. App. Environ. Microbiol. 62, 1102–1106.

Kubista, M., Andrade, J.M., Bengtsson, M., Forootan, A., Jonak, J., Lind, K., Sindelka, R., Sjoback, R., Sjogreen, B., Strombom, L., et al. 2006. The real-time polymerase chain reaction. Mol. Aspects Med. 27, 95–125.

Kulesh, D.A., Baker, R.O., Loveless, B.M., Norwood, D., Zwiers, S.H., Mucker, E., Hartmann, C., Herrera, R., Miller, D., Christensen, D., et al. 2004. Smallpox and pan-orthopox virus detection by real-time 3'-minor groove binder TaqMan assays on the roche LightCycler and the Cepheid SmartCycler platforms. J. Clin. Microbiol. 42, 601–609.

Kuske, C.R., Banton, K.L., Adorada, D.L., Stark, P.C., Hill, K.K. and Jackson, P.J. (1998). Small-scale DNA sample preparation method for field PCR detection of microbial cells and spores in soil. Appl. Environ. Microbiol. 64, 2463–2472.

Lebuhn, M., Effenberger, M., Garces, G., Gronauer, A. and Wilderer, P.A. 2004. Evaluating real-time PCR for the quantification of distinct pathogens and indicator organisms in environmental samples. Water Sci. Technol. 50, 263–270.

Letertre, C., Perelle, S., Dilasser, F. and Fach, P. 2003. A strategy based on 5' nuclease multiplex PCR to detect enterotoxin genes sea to sej of Staphylococcus aureus. Mol. Cell Probes 17, 227–235.

Lewis, G.D., Molloy, S.L., Greening, G.E. and Dawson, J. 2000. Influence of environmental factors on virus detection by RT-PCR and cell culture. J. Appl. Microbiol. 88, 633–640.

Lim, D.V., Simpson, J.M., Kearns, E.A. and Kramer, M.F. 2005. Current and developing technologies for monitoring agents of bioterrorism and biowarfare. Clin. Microbiol. Rev. 18, 583–607.

Lindstrom, M., Keto, R., Markkula, A., Nevas, M., Hielm, S. and Korkeala, H. 2001. Multiplex PCR assay for detection and identification of Clostridium botulinum types A, B, E, and F in food and fecal material. Appl. Environ. Microbiol. 67, 5694–5699.

Loïez, C., Herwegh, S., Wallet, F., Armand, S., Guinet, F. and Courcol, R J. 2003. Detection of Yersinia pestis in sputum by real-time PCR. J. Clin. Microbiol. 41, 4873–4875.

Mackay, I.M. 2004. Real-time PCR in the microbiology laboratory. Clin. Microbiol. Infect. 10, 190–212.

McAvin, J.C., McConathy, M.A., Rohrer, A.J., Huff, W.B., Barnes, W.J. and Lohman, K.L. 2003. A real-time fluorescence polymerase chain reaction assay for the identification of Yersinia pestis using a field-deployable thermocycler. Mil. Med. 168, 852–855.

McAvin, J.C., Morton, M.M., Roudabush, R.M., Atchley, D.H. and Hickman, J.R. 2004. Identification of Francisella tularensis using real-time fluorescence polymerase chain reaction. Mil. Med. 169, 330–333.

McBride, M.T., Masquelier, D., Hindson, B.J., Makarewicz, A. J., Brown, S., Burris, K., Metz, T., Langlois, R.G., Tsang, K.W., Bryan, R., et al. 2003. Autonomous detection of aerosolized Bacillus anthracis and Yersinia pestis. Anal. Chem. 75, 5293–5299.

McKillip, J.L. and Drake, M. 2004. Real-time nucleic acid-based detection methods for pathogenic bacteria in food. J. Food Prot. 67, 823–832.

Melo, A.C., Almeida, A.M. and Leal, N.C. 2003. Retrospective study of a plague outbreak by multiplex-PCR. Lett. Appl. Microbiol. 37, 361–364.

Mifflin, T.E. 2003. Setting up a PCR Laboratory. PCR Primer: A Laboratory Manual. New York: Cold Spring Harbor Laboratory Press.

Millar, B.C., Xu, J. and Moore, J.E. 2002. Risk assessment models and contamination management: implications for broad-range ribosomal DNA PCR as a diagnostic tool in medical bacteriology. J. Clin. Microbiol. 40, 1575–1580.

Miller, D.N. 2001. Evaluation of gel filtration resins for the removal of PCR-inhibitory substances from soils and sediments. J. Microbiol. Methods 44, 49–58.

Miller, S.C., LiPuma, J.J. and Parke, J.L. 2002. Culture-based and non-growth-dependent detection of the Burkholderia cepacia complex in soil environments. Appl. Environ. Microbiol. 68, 3750–3758.

Mitchell, P.S., Germer, J.J. and Patel, R. 2004. Molecular Microbiology: Diagnostic Principles and Practice. (Washington, DC: ASM Press).

Newby, D.T., Hadfield, T.L. and Roberto, F.F. 2003. Real-time PCR detection of Brucella abortus: a comparative study of SYBR green I, 5' -exonuclease, and hybridisation probe assays. Appl. Environ. Microbiol. 69, 4753–4759.

Nguyen, L.T., Gillespie, B.E., Nam, H.M., Murinda, S.E. and Oliver, S.P. 2004. Detection of Escherichia coli O157,H7 and Listeria monocytogenes in beef products by real-time polymerase chain reaction. Foodborne Pathog. Dis. 1, 231–240.

Nitsche, A., Ellerbrok, H. and Pauli, G. 2004. Detection of orthopoxvirus DNA by real-time PCR and identification of variola virus DNA by melting analysis. J. Clin. Microbiol. 42, 1207–1213.

Novak, R.T., Glass, M., Gee, J., Gal, D., Mayo, M., Currie, B. and Wilkins, P. 2006. Development and evaluation of a real-time PCR assay targeting the type III secretion system of Burkholderia pseudomallei. J. Clin. Microbiol. 44, 85–90.

Panning, M., Asper, M., Kramme, S., Schmitz, H. and Drosten, C. 2004. Rapid detection and differentiation of human pathogenic orthopox viruses by a fluorescence resonance energy transfer real-time PCR assay. Clin. Chem. 50, 702–708.

Pedley, S. and Pond, K. 2003. Emerging issues in water and infectious disease.: World Health Organisation.

Perdue, M. L. 2003. Molecular diagnostics in an insecure world. Avian Dis. 47, 1063–1068.

Perelle, S., Dilasser, F., Malorny, B., Grout, J., Hoorfar, J. and Fach, P. 2004. Comparison of PCR-ELISA and LightCycler real-time PCR assays for detecting

Salmonella spp. in milk and meat samples. Mol. Cell Probes *18*, 409–420.

Persing, D.H. 1991. Polymerase chain reaction: trenches to benches. J. Clin. Microbiol. *29*, 1281–1285.

Queipo-Ortuno, M.I., Colmenero, J., Reguera, J., Garcia-Ordonez, M., Pachon, M., Gonzales, M. and Morata, P. 2005. Rapid diagnosis of human brucellosis by SYBR Green I-based real-time PCR assay and melting curve analysis in serum samples. Clin. Microbiol. Infect. *11*, 713–718.

Ramisse, V., Patra, G., Garrigue, H., Guesdon, J. and Mock, M. 1996. Identification and characterisation of *Bacillus anthracis* by multiplex PCR analysis of sequences on plasmids pXO1 and pXO2 and chromosomal DNA. FEMS Microbiol. Lett. *145*, 9–16.

Ryu, C., Lee, K., Yoo, C., Seong, W.K. and Oh, H.B. 2003. Sensitive and rapid quantitative detection of anthrax spores isolated from soil samples by real-time PCR. Microbiol. Immunol. *47*, 693–699.

Safi Ibrahim, M., Kulesh, D., Saleh, S., Damon, I., Esposito, J., Schmaljohn, A. and Jahrling, P. 2003. Real-time PCR assay to detect smallpox virus. J. Clin. Microbiol. *41*, 3835–3839.

Salles, J., De Souza, F. and van Elsas, J. 2002. Molecular method to assess the diversity of *Burkholderia* species in environmental samples. Appl. Environ. Microbiol. *68*, 1595–1603.

Schaad, N.W., Frederick, R.D., Shaw, J., Schneider, W.L., Hickson, R., Petrillo, M.D. and Luster, D.G. 2003. Advances in molecular-based diagnostics in meeting crop biosecurity and phytosanitary issues. Annu. Rev. Phytopathol. *41*, 305–324.

Shepard, J.E., Egan, C. and Cirino, N.M. 2006. Water Biosurveillance Systems (San Diego: Elsevier Science/Academic Press).

Sjostedt, A., Eriksson, U., Berglund, L. and Tarnvik, A. 1997. Detection of *Francisella tularensis* in ulcers of patients with tularemia by PCR. J. Clin. Microbiol. *35*, 1045–1048.

Sutlovic, D., Definis, G. M., Andelinovic, S., Gugic, D. and Primorac, D. 2005. Taq polymerase reverses inhibition of quantitative real time polymerase chain reaction by humic acid. Croat. Med. J. *46*, 556–562.

Synder, J., Jungkind, D. and Persing, D. 2004. Molecular Microbiology: Diagnostic Principles and Practice. (Washington, DC: ASM Press).

Szabo, E. A., Pemberton, J., Gibson, A., Eyles, M. and Desmarchelier, P. 1994. Polymerase chain reaction for detection of *Clostridium botulinum* types A, B and E in food, soil and infant faeces. J. Appl. Bacteriol. *76*, 539–545.

Templeton, K. E., Scheltinga, S., Beersma, M., Kroes, A., Claas, E. 2004. Rapid and sensitive method using multiplex real-time PCR for diagnosis of infections by influenza a and influenza B viruses, respiratory syncytial virus, and parainfluenza viruses 1, 2, 3, and 4. J. Clin. Microbiol. *42*, 1564–1569.

Teshale, E. H., Painter, J., Burr, G., Mead, P., Wright, S., Cseh, L., Zabrocki, R., Collins, R., Kelley, K., Hadler, J., Swerdlow, D. 2002. Environmental sampling for spores of *Bacillus anthracis*. Emerg. Infect. Dis. *8*, 1083–1087.

Tomaso, H., Reisinger, E., Al, D., Frangoulidis, D., Rakin, A., Landt, O. and Neubauer, H. 2003. Rapid detection of *Yersinia pestis* with multiplex real-time PCR assays using fluorescent hybridisation probes. FEMS Immunol. Med. Microbiol. *38*, 117–126.

Tsai, Y.L. and Olson, B.H. 1992. Rapid method for separation of bacterial DNA from humic substances in sediments for polymerase chain reaction. Appl. Environ. Microbiol. *58*, 2292–2295.

van De Rijke, F., Zijlmans, H., Li, S., Vail, T., Raap, A., Niedbala, R. and Tanke, H. 2001. Up-converting phosphor reporters for nucleic acid microarrays. Nat. Biotechnol. *19*, 273–276.

Versage, J. L., Severin, D., Chu, M. and Petersen, J. 2003. Development of a multitarget real-time TaqMan PCR assay for enhanced detection of *Francisella tularensis* in complex specimens. J. Clin. Microbiol. *41*, 5492–5499.

Volkmann, H., Schwartz, T., Bischoff, P., Kirchen, S. and Obst, U. 2004. Detection of clinically relevant antibiotic-resistance genes in municipal wastewater using real-time PCR (TaqMan). J. Microbiol. Methods *56*, 277–286.

Wannamaker, B., Denio, L., Dodson, W., Dreifuss, F., Crosby, C., Santilli, N., Duffner, P., Ryan-Dudeck, P., Conboy, C., Ellis, E. *et al.* 1989. Immunochromatographic measurement of phenobarbital in whole blood with a non-instrumented assay. Neurology *39*, 1215–1218.

Whelen, A.C. and Persing, D.H. (1996). The role of nucleic acid amplification and detection in the clinical microbiology laboratory. Annu. Rev. Microbiol. *50*, 349–373.

Wilson, I.G. 1997. Inhibition and facilitation of nucleic acid amplification. Appl. Environ. Microbiol. *63*, 3741–3751.

Yamamoto, Y. 2002. PCR in diagnosis of infection: detection of bacteria in cerebrospinal fluids. Clin. Diagn. Lab. Immunol. *9*, 508–514.

Yoon, S.Y., Chung, G., Kang, D., Ryu, C., Yoo, C. and Seong, W. 2005. Application of real-time PCR for quantitative detection of *Clostridium botulinum* type A toxin gene in food. Microbiol. Immunol. *49*, 505–511.

Real-time PCR: Application to Food Authenticity and Legislation

16

Gordon Wiseman

Abstract

Real-time PCR is now an accepted analytical tool within the food industry. Its principal role has been one of assisting the legislative authorities, major manufacturers and retailers to confirm the authenticity of foods. The most obvious role is the detection and quantification of GMOs, but real-time PCR makes a significant contribution to many other areas of the food industry, including food safety and other speciality analyses such as the detection of common wheat adulteration in pasta and the detection of allergenic species. The role of quantitative real-time PCR in determining the actual amount of these materials, which are subject to considerable regulation, is discussed together with a consideration of the uncertainty of the methods.

Why real-time PCR?

Analysing processed food and ingredients for the presence of illegal materials or simply to confirm its authenticity has never been straight forward and is often considerably more difficult than the analysis of materials of either pharmaceutical or medical origin. This is due to the large number of difficulties that must be overcome before a proper analysis can be affected. These difficulties include the isolation of DNA that is often substantially degraded by thermal or manufacturing processes which consequently limits the size of the amplicon. Occasionally the DNA has been partially depurinated by acidic conditions limiting the amount of target DNA available for amplification which in turn affects the limits of detection and quantification. A major problem area associated with the analysis of food is the purification of the DNA to remove the many potent PCR inhibitors such as polyphenolic compounds, metal ions and complex carbohydrates that are abundant in most foodstuffs. Most matters of food authenticity attempt to address the misdescription of the food and aim to stop unscrupulous manufacturers from defrauding customers by substituting cheaper ingredients and degrading product quality. It is, however, rare for a food safety issue to develop from such a substitution. An exception to this is the detection of the smallest amount of neuronal tissue in processed meats. The regulatory authorities and to a lesser extent the food industry have been interested in developing the most sensitive and accurate methods of analysis to address these issues. The advent of PCR revolutionized the way many of these analyses were performed and gave the industry a means of detecting materials that previously were considered impossible to analyse. However, as is often the case, the matter of a positive result is followed by the inevitable question of how much? Unfortunately, the end-point detection of specific amplicons cannot provide a quantitative result due to the logarithmic nature of the amplification reaction unless significant effort is put into the development of an appropriate competitive PCR method and which at best generates a semi-quantitative result. The development of real-time PCR has offered the food industry a significant advance over these original methods. The advent of fluorescent in-tube detection of specific amplicon formation coupled with the inherent specificity that is generated by incorporating a labelled third oligonucleotide (e.g. TaqMan™ chemistry) has led to the development of a number of convenient methods that are applicable to addressing the issues prevalent within the industry. The major reasons behind this success relate to the inherent processes of real-time PCR and are associated with the kinetics of specific small amplicon production. It is often the case that food processing degrades the DNA by mechanical strand cleavage to fragments that are significantly smaller (100–500 bp) than the molecules of DNA present in native material. Hence, it is essential that the specific amplicons used as diagnostic tools within food analysis are within this size range and it is recommended that these amplicons are generally 80–120bp.

Food authenticity

The substitution of high quality expensive components of food with those of lesser quality and cost has been a constant problem within the food industry and has frequently been as a consequence of organized crime. Consumers insist in exercising preference with regard to their diets and make their decisions on the basis of health, religious and lifestyle grounds. These choices are usually made on the basis of product labelling with a proportion of the decisions being made in favour of premium products and prices. Customer choice should therefore be supported by firm labelling regulations which address any misdescription by the food producer or vendor and lead to the prosecution of intentional fraud. The detection of fraud and the enforcement of labelling regulations have therefore driven the development of methods capable of confirming identity and detecting adulteration. This development, in conjunction with the need to be able to carry out the analysis on processed foods, has utilized real-time PCR. An obvious role is in the confirmation of meat species (Lopez-Andreo et al., 2004), particularly where the meat has been minced or has been subjected to severe processing conditions. While a large number of methods exist for meat speciation, most use the heterogeneity of the mitochondrial cytochrome b gene to effect detection. This gene was used to differentiate between horse and donkey (Chisholm et al., 2005) allowing their presence to be detected in imported commercial products.

A similar real-time method to detect bovine material has been used help reinforce the legislation that prohibits the presence of ruminant materials such as bovine bone meal or meat in cattle feeds (Renson et al., 2005) following the epidemic of bovine spongiform encephalopathy (BSE). The detection system is reliable in detecting contaminating bovine material in such feeds at 0.05%. However, given the seriousness of this topic, it is evident that more development is required to improve the sensitivity of detection. Recently, real-time PCR was applied to the detection and quantification of adulteration within goose foie gras (Rogriguez et al., 2004) a particularly expensive product. Normally, the morphology of the whole liver is used to confirm authenticity but this ability is lost once processing begins making additional testing methods essential to protect the product. Multi-copy mitrochondrial sequences allow an increased sensitivity over single copy genomic DNA sequences, making their use appropriate for highly degraded or processed samples. In general, within real foods these methods will reliably detect 1% contamination, which is well below the level that anyone wishing to deliberately adulterate for financial gain would employ. Recent changes in the EU Food Labelling Regulations mean that food products now require a quantitative ingredient declaration (QUID) (European Directive., 2001) for any ingredient that is mentioned in the name of the food. Hence, foods which have a specific kind of meat in the name of the product must indicate the amount of that meat present in the product and this contrasts to the former regulations which required only the total minimum meat content to be declared on the label. This change in the law applying to product labelling places significant further requirements on the analytical techniques used to support the legislation. Real-time PCR is often the method of choice for analysts determining the qualitative presence of adulterating meats in food products; however, while there have been several attempts to quantify meat in processed products, all truly quantitative investigations have lacked the required accuracy required for legislative uses. This lack of success has in part been due to a large number of factors that influence the total amount of DNA available for analysis within the tissues, including the age of the animal, the natural variation between tissue types within any individual animal but also between animals. Currently, no suitable DNA sequence has been identified that is suitable for use as a means of normalizing the data obtained.

The authentication of fish species, now for the first time, includes a quantitative real-time PCR method for the white fish (haddock) in commercial products (Hird *et al.*, 2005). The analytical errors involved in this particular analysis are quite large but take into account the natural variations in DNA and protein content of the fish as a result of seasonal changes.

Real-time PCR (Hernandez *et al.*, 2005) has augmented traditional PCR methods for the identification of small cereal grains (Terzi *et al.*, 2004, 2005) which was instrumental in assisting the identification of cereals of industrial and commercial importance. These methods too are aimed at reducing fraud and guaranteeing the quality of products through due diligence testing of raw materials and ingredients and thereby contributing to the traceability of these commodities.

Organic foods have presented a particular problem with regard to verifying their authenticity and reliable methods have yet to be devised, however, the presence of genetically modified species within such foods is unacceptable to both retailers and consumers alike. The real-time PCR screening methods for detecting the presence of GMOs have made a significant contribution to safeguarding these assumptions.

Food safety applications

Detection of food pathogens and food spoilage organisms

The conventional methods that are employed in assessing microbiological food safety whilst being generally cost effective are limited by a number of factors. They can be time consuming as most methods take several days to isolate presumptive micro-organisms, with further time required for definitive identification. Conventional methods also employ a great deal of individual interpretation, involving inclusion and exclusion of both typical and atypical organisms on agar plates which can lead to variation in results.

Current food manufacturing processes involving 'At Risk' foods rely on accurate and timely results in terms of food safety and quality. Positive release is rarely used in modern manufacturing, except where a particularly high risk has been identified, with Hazard Analysis and Critical Control Points (HACCP) forming the bulk of control measures in the majority of cases. This means that the product is in the supply chain before the results of most conventional methods have been received. Any assessment of food safety therefore needs to be fast, reliable and accurate, as the implications of recalling unsafe product from the retailer, resulting in the loss of brand image, reputation and potential fines to the supplier must be clearly justified.

As a result of these considerable pressures, faster and more reliable testing methods have been sought to detect pathogenic bacteria resulting in a number of approaches including real-time PCR. While some advantage can be gained by employing real-time methods to detect pathogens, their sensitivity and usefulness can be limited by the need for pre-enrichment. Pre-enrichment does, however, negate two of the most obvious problems with any DNA method: firstly, 'false-positive' results originating from the amplification of DNA from dead cells (Wolffs *et al.*, 2005) and, secondly, the possibility of the more serious 'false negative' is diminished by the removal of potent inhibitors of PCR that can be present in complex food products. There are not many real-time PCR assays for food pathogens in general use (Table 16.1). Likewise similar methods have been applied to the detection of food spoilage organisms (Table 16.2). Due to their economic importance, several real-time assays including those for *Salmonella*, *Escherichia coli* O157:H7, *Listeria* spp. and *Listeria monocytogenes* have been subjected to commercialization and can be purchased in kit format for a variety of real-time instruments, some of these methods have been subjected to extensive validation i.e. AOAC approved, while others have not.

Allergens

One of the major issues facing the food industry at present is the potential for induced health problems in sensitized consumers due to the presence of undeclared allergenic materials in

Table 16.1 Real-time PCR detection of food-borne pathogens.

	Food type	Reference(s)
Bacterial pathogens		
Salmonella	General	Malorny *et al.* (2004), Liming *et al.* (2004), Seo *et al.* (2004)
Escherichia coli O157:H7	Meat	Fu *et al.* (2005), Spano *et al.* (2005), O'Hanlon *et al.* (2004)
Various	General	Fukushima *et al.* (2003)
Staphylococcus	Milk	Hein *et al.* (2005)
Clostridium botulinum	General	Yoon *et al.* (2005)
Enterobacter sakazakii	Infant formula	Seo *et al.* (2005)
Listeria monocytogenes	Meat products	Rodriguez-Lazaro *et al.* (2004a,b)
Thermophilic *Campylobacter*	Meat and dairy products	Perelle *et al.* (2004)
Campylobacter jejuni	Poultry, milk and water	Yang *et al.* (2003)
Yersinia enterocolitica	Pork	Wolffs *et al.* (2004)
Vibrio parahaemolyticus	Shellfish and sea water	Takahashi *et al.* (2005), Kaufman *et al.* (2004)
Viral pathogens		
Norovirus	Shellfish	Loisy *et al.* (2005)
Enteric viruses	General	Beuret (2004)

food as a result of inadvertent contamination during the manufacturing process. This situation has promoted reliable and sensitive testing methods to ensure compliance with food labelling regulations and improve consumer protection. These methods are directed at either the allergens themselves which are specific proteins or other markers such as specific DNA sequences that can be used to indicate the presence of the allergenic species. Due to their adaptability and applicability to routine screening, enzyme-linked immunosorbent assays (ELISAs) are the method of choice to detect and quantify the presence of allergenic or marker proteins. However, as a result of the demand for more sensitive testing methods a number of real-time PCR methods have been devised to detect allergenic species. These methods include the detection of peanut (Stephan *et al.*, 2004 a) which is the principal allergen in both the United States and Western Europe, celery (Stephan *et al.*, 2004 b), hazelnut and almond. In an attempt to improve the sensitivity of detection, an extremely sensitive assay for wheat gliadin, one of the principal causes of celiac disease, has been devised using the hybrid technique of real-time immuno-PCR (Henterich

Table 16.2 Real-time PCR detection of food spoilage organisms

Organism	Food type	Reference(s)
Alicylobacillus	Apple juice	Connor *et al.* (2005), Luo *et al.* (2004)
Lactic acid bacteria	Fermented milk products	Grattenpanche *et al.* (2005), Furet *et al.* (2004)
Potyviruses	Garlic	Lunello *et al.* (2004)
Yeast	Fruit juice	Casey *et al.* (2004)
Oenococcus oeni	Wine	Pinzani *et al.* (2004
Dekkera bruxellensis	Wine	Phister *et al.* (2003)
Yeasts and moulds	Yoghurts and pasteurized foods	Bleve *et al.* (2003)

et al., 2003). This method generated a 30-fold increase in sensitivity above the comparative ELISA method.

Recently, as in the case with pathogen testing some of these allergen methods have been commercialized and it is now possible to purchase assays in kit format. It should be noted that while the detection of species-specific DNA can be used to detect contaminating material and hence infer the presence of allergen it does not equate to detecting the allergen itself, hence, extreme care must be used in interpreting the data generated. It is generally acknowledged that some allergens can be effective in pre-sensitized individuals at minute concentrations, well below that detectable by ELISA (typically 1–5 ppm). Real-time PCR on the other hand is dependant upon the presence of amplifiable DNA, and would fail to detect depurinated and other degraded DNA target sequences that could occur as a result of manufacturing conditions.

Quantitative real-time PCR and the food industry

The application of quantitative real-time PCR to food uses has been limited in terms of the analyses performed; the most widely used application has been to the analysis and control of genetically modified organisms (GMOs). The other application has been speciality analyses such as the quantification of soft wheat adulteration of pasta where each of these materials is subject to strict legislation. Due to the inherent complexity in terms of both appropriate knowledge and suitable instrumentation, these analyses have usually been carried out in specialist laboratories.

Real-time PCR and the detection and quantification of GMOs

GMOs and EU Legislation

The current EU GMO legislation is unambiguous regarding the labelling of food products that contain ingredients that have been obtained from genetically modified (GM) sources. Product labelling is solely based upon the origin of the material and is not dependant upon the presence of either transgenic protein or DNA and there is a requirement for traceability throughout the food chain for all ingredients of GM origin. The need for traceability with regard to non-GM soya beans and derived products has led to a complex system of identity preservation (IP) where the non-GM origin of the material is assured. This system however, does not guarantee 'GM-free' status as a low level of adventitious contamination can be expected. All foods made from ingredients that are known to be of GM origin and have not been the subject of an identity-preserved scheme must be labelled. All materials of non-GM origin and which are produced under an identity-preserved scheme are subject to a *de minimis* threshold of 0.9% for adventitious contamination with EU approved GM varieties, any products containing approved GM varieties below this limit do not need to be labelled. For non-EU approved varieties the limit is now 0% as the 3-year period which allowed a threshold of 0.5% for non-EU authorized GMOs that had received a favourable scientific review elsewhere in the world has now elapsed (European Commission Regulation 1829/2003: 2003). Ingredients without traceability data to indicate their GM status and which have not been produced under an identity-preserved system activate labelling if the smallest amount of GM DNA is detected.

GMO analysis

The EU Regulations (European Commission Regulations 1997 and 2003) mean that all food manufacturers, retailers and enforcement agencies require access to the most accurate means of determining the levels of GM materials in foods (Gachet *et al.*, 1998). The method of choice for determining the levels of GM material present in food and its ingredients is real-time PCR (Wiseman, 2002) which is due to the sensitivity, precision and throughput of the method (Wurtz *et al.*, 1999). In general, the amount of a GM tissue present in a sample is determined by comparison of the fluorescence signals with those obtained from calibrants of known concentration. The most common method of constructing a calibration curve uses DNA extracted from mixtures of ground GMO with known levels of the GMO in question. The Certified Reference Materials (CRMs), available from the Institute of Reference and Measurement (IRRM) generally contain 0–5% GM material and are the only fully traceable materials suitable for this purpose, generating a linear calibration curve. The production of CRMs is costly and has been fraught with difficulty particularly with regard to stability, leading to the withdrawal of the Bt11 and Bt176 maize CRMs (Trapmann et al., 2002). Adding to the difficulty of the situation is the slowness of new CRM production and the small number of CRMs available compared to the number of approved GM varieties and those under consideration for approval. A further major restriction is that the largest GM standard is generally 5% which limits the working range of the calibration curve (Van den Eede et al., 2002; Taverniers et al., 2004). Consequently, plasmids have recently been investigated as an alternative calibration source to the CRMs (Taverniers et al., 2001, 2004; Kuribara et al., 2002; Shindo et al., 2002, Weighardt, 2002; Block et al., 2003) The plasmids used in these studies contained specific GM sequences and endogenous gene sequences allowing the potential detection of Roundup-Ready[TM] soya and five GM maize lines. The authors came to the conclusion that the plasmid method was reliable for the detection of these GM lines. A recent comparison of genomic and plasmid-based calibrants (Burns et al., 2006) concluded that plasmid calibration gave a closer mean estimate of the expected %GM content of the sample and exhibited less variation. Plasmid calibrants also gave more accurate results in terms of trueness and precision when assessed using an inter-laboratory study. These findings are timely as the EU Commission (European Commission Recommendation, 2004) has provided guidance for the sampling and detection of GMOs and material produced from GMOs in context of Regulation 1830/2003 (European Commission Regulation, 2003). The recommendation is that the amount GM material present should be expressed as the percentage of the number of GM target DNA sequence per target taxon specific sequence calculated in terms of haploid genomes. Given that this is impossible with the current CRMs, plasmid calibrants with a certified copy number will need to be made or the current CRMS recalibrated to be labelled in terms of haploid genome equivalents.

While a large number of real-time methods exist for the detection of GMOs in foodstuffs, at the practical level these analyses are performed in three stages to avoid the unnecessary expense of inappropriate PCR analysis. While it is impossible to generalize, it is usual for qualitative real-time PCR to act as the first stage of a general screen determining which commonly modified species (soya, maize, oils seed rape etc) are present in the test material. Unified qualitative nucleic acid based methods for the detection of genetically modified organisms and derived products have been developed (EN ISO 21569: 2005). At the same time analysis is carried out to detect a number of the most commonly used promoter and terminator sequences used in genetic modification such as the cauliflower mosaic virus (35S CaMv) promoter and terminator sequences and the nopaline synthetase (NOS) terminator sequence. It should be noted that due to the multiple use of some of these sequences by the biotechnology companies releasing GMOs, it is not possible to use their presence in a quantitative manner in complex mixtures. Following a positive result, a second stage of analysis should be performed using event-specific qualitative real-time PCR to determine which events are present in the sample (EN ISO 21569: 2005). This stage is particularly arduous with respect to maize, due to the large number of approved and non-approved events that may be present. Once the identities of the GM cultivars that are present are known, a third stage of quantitative event-specific qualitative real-time PCR can be carried out to determine the amounts present. (EN ISO 21570: 2005).

The list of GM cultivars approved by the EU is constantly updated as is the list of cultivars that have been put forward for approval. An up to date list can be found at the EU Community Reference Laboratory (CRL) website (http://gmo-crl.jrc.it/statusofdoss.htm) together with links to the validation reports performed on the GMOs and several of the real-time methods used to detect them. This source of information is welcome as one of the major difficulties encountered when detecting GMOs is the unavailability of DNA sequence information which is generally withheld on propriety grounds. Exceptions are the DNA sequences of the junction regions of the Roundup Ready[TM] insert present in the Monsanto's soya bean line GTS40-3-2 (Windels et al., 2001), the junction sequences of Monsanto's maize MON 810 (Hernández et al., 2003; Holck et al., 2002), Aventis maize CBH-351 (Starlink) (Windels et al., 2003) and Syngenta Bt11 maize (Zimmermann et al., 2000; Rønning et al., 2003).

Measurement uncertainty

The usefulness of quantitative real-time methods has now been recognized by both the regulatory bodies and by manufacturers. Consequently, as a result of increased regulatory compliance, considerable attention is now paid to the reliability and quality of analytical results with those commissioning analysis demanding to know the level of confidence that can be applied to the results that are reported. Hence, results should now be reported with a statement of the uncertainty of the method (U) which allows the results from different laboratories to be compared. It is essential that recipients of analytical data from a range of laboratories have confidence in the results as they will inevitably be used as the basis of decision making. An incorrect decision may lead to the failure to detect an illegal practice, expose consumers to a health risk or simply incur higher costs.

Central to the concept of quality is that a method must be validated, i.e. it must be proved that it measures that which is intended and consequently is fit for specific use, the European and International standards such as ISO/IEC 17025 have gone some way to ensuring this situation. The expanded uncertainty (U) should therefore be seen as indication of fitness-for-purpose and reliability. However, it should not be confused with analytical error which is the difference between a measured result and a true value; it consequently is a single value. The expanded uncertainty is expressed as a range and applies only to the specific type of sample being analysed and the specific method used.

The processes that are used to determine U from first principals are complex and are too lengthy to be discussed here, for a full explanation of the expanded uncertainty and how it can be determined see http://www.measurementuncertainty.org/. Alternatively, U can be determined for any analysis that is carried out 'in-house' where the uncertainty can be derived from intermediate precision, repeatability and run errors. A more robust measure of U is derived from the reproducibility, repeatability, run and laboratory effects that are established during complex inter-laboratory trials.

Real-time PCR detection and quantification of common wheat adulteration of pasta

Legislation and pasta

The majority of pasta products that are manufactured and offered for sale within the European Community (EC) are made solely from durum wheat (*Triticum durum* Desf) semolina and are considered to be superior products to those manufactured from common wheat (*Triticum aestivum* L), or mixtures of the two species of wheat. The Italian law (Gazzetta Ufficiale, 1967) prohibits the manufacture of pasta containing *T. aestivum* for sale in Italy but surprisingly, not for export. Many European countries, France, Italy and Spain included assume that the inclusion of common wheat in pasta products constitutes adulteration. However, pasta is frequently manufactured for sale in Germany and Holland containing both wheat species. Currently the United Kingdom does not have specific regulations regarding the composition of pasta. However, the UK legislation (Food Safety Act, 1990: UK Food Labelling Regulations, 1996) makes it clear that it is an offence to mis-describe a product or to offer it for sale in a misleading manner. Therefore, any common wheat present in a pasta product must be declared clearly on the product label. Pasta that is manufactured from durum wheat and which is destined for export outside the EC is allowed to contain a maximum of 3% common wheat (EC Commission Regulation, 1994: International Standard, 1994) to allow for 'adventitious' cross-contamination during the agricultural process. It has also been suggested by the EC that all pasta products should be labelled to show the extent of the inclusion of each wheat species in the product. The requirement for correct labelling assumes a further importance within the EC as the accurate descriptions of products form the basis for their classification in the Common Customs and Tariff system and hence may be used to apply for an export refund under the terms of the Common Agricultural

Policy. Common wheat (or bread wheat) while varying considerably in cost depending upon global commodity prices, usually trades at a substantially lower price than durum wheat and this has consequently acted as an incentive to those wishing to adulterate pasta for financial gain.

Wheat genetics

Before discussing the adulteration of pasta made from *T. durum* with *T. aestivum*, it is worth considering the evolution of wheat to understand the origins of the various genomes that are present (Wiseman, 2001). *T. durum* is tetraploid containing two copies each of the seven chromosomes that make up both the A and B genomes. *T. durum* has its historical origins in a spontaneous hybridization between wild einkorn wheat, the A genome donor and a species of goat grass (*Aegilops* sp.) that donated the B genome (Fig. 16.1). *T. aestivum*, however, is hexaploid containing two copies of each of the seven chromosomes that make up the A, B and D genomes. While the exact origins are unknown it is believed that *T. aestivum* is the product of a second spontaneous hybridization between the tetraploid wheat *T. dicoccum* and a second goat grass (*Aegilops squarrosa*) which donated the D-genome. The D-genome encodes the many proteins that are responsible for the viscoelastic properties of bread dough as well as a variety of soluble proteins that are responsible for the sticky properties of pasta that has been made using *T. durum* contaminated with *T. aestivum*.

Quantification of T. aestivum adulteration in pasta

The authenticity of *T. durum* pastas and the detection and quantification of deliberate adulteration of *T. durum* pastas by common wheat (*T. aestivum*) continues to be an important issue throughout Europe. As the geographical origin of the grain diversifies, resulting from economic trading agreements and occasional adverse harvest conditions, the strict laws and agreements regarding the adventitious contamination of *T. durum* by *T. aestivum* require enforcement to protect both the retailers and the consumer. This resulted in the application of a number of protein based techniques to address the situation, most of which have substantial shortcomings resulting in difficulties in quantification (Wiseman, 2001).

It has now been demonstrated that it is possible to address the problems associated with the detection and quantification of common wheat (*T. aestivum*) adulteration of *T. durum* pastas by the techniques of molecular biology. Initial research showed that a repetitive but specific D-genome DNA sequence (Dgas44) (Bryan *et al.*, 1995; Bryan *et al.*, 1998; Wiseman *et al.*, 1999) could be used to detect the presence of adulterating common wheat in pastas that had been subjected to a range of drying processes. Following difficulties associated with a multiple copy sequence further investigation was performed using real-time PCR and the DNA sequence derived from PSR 128 (Larson *et al.*, 1996).

PSR 128 is a homologous single copy cDNA probe which has been mapped to the 5A, 5B and 5D chromosomes of wheat and which shows a significant level of sequence polymorphism between each of the three genomes when compared. The D-genome sequence contains a 53bp intron sequence that is absent from the other two genomes (Fig. 16.2). The A, B genome sequences of PSR128 present in *T. durum* and the A, B and D genome sequences present in *T. aestivum* show little or no polymorphism between the wheat varieties examined. This lack of polymorphism makes the sequences ideal for PCR amplification and hence detection of D-genome sequence where the contaminating *T. aestivum* cultivar is unknown. The variation in the intron sequences between the genomes was exploited to construct PCR primers to amplify a 117-bp D-genome-specific amplicon (Fig. 16.2) to effect *T. aestivum* detection and quantification in analytical samples. A second set of PCR primers were designed to anneal to the coding region of PSR 128 and which amplify all three genomes equally to generate a 121bp amplicon that can be used to normalize for the amount of total amplifiable wheat DNA present in the sample and subsequently effect quantification. The analysis of a large number of commercial cultivars of both *T. durum* and *T. aestivum* from around the world together with a number of related species and amphiploid lines of wheat indicated that the detection method was highly unlikely to be compromised by cross-reaction with other unrelated genomes.

Real-time PCR of other targets including the purindoline b gene (Alary *et al.*, 2002) and gliadin and glutinin genes (Terzi *et al.*, 2003) have

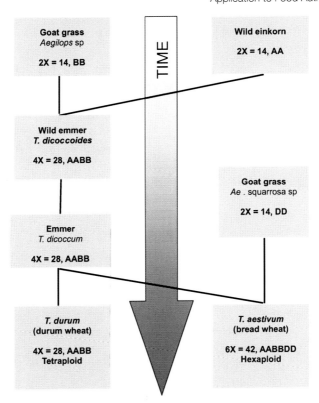

Figure 16.1 The evolution of *T. durum* and *T. aestivum* wheats.

been used in the detection and quantification of *T. aestivum* contamination in pasta and semolina. It was generally noted that decreased yield of DNA was obtained from heat-treated pasta and that it was necessary to dilute the DNA prior to amplification to remove the adverse effects of inhibitors that co-purified with the wheat DNA. While it is clear that real-time PCR is an applicable tool for the analysis of dried pasta, it is also evident that the method requires validation as a prerequisite of becoming a robust method of analysis. To date, there are very few statistical data that underpin the characteristics of these new analytical methods and which can be used to appraise their fitness for purpose. However, it is well known that the coefficients of variation (CV) in similar assays that employ PCR are typically in the region of 17–42% (CEN/TC 275.

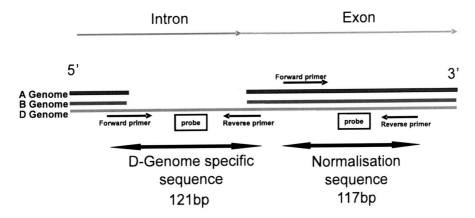

Figure 16.2 Schematic of the single copy DNA sequence PSR128.

WG., 2002). It is also known that the cvs are not linear throughout the analytical range, increasing dramatically with increasing measured analyte concentration.

Standard curve

Since the emergence of quantitative real-time PCR as an analytical tool there has been considerable discussion relating to the method by which standard curves are generated. There are two general methods by which this can be achieved, firstly the delta C_t method. Here, DNA extracts from standards of known composition are used as the target DNA in appropriate PCRs. Two PCR reactions are performed; one to amplify a sequence that defines the analyte and another which amplifies an endogenous sequence and which in effect acts as a normalizing reaction. The numerical values for the individual threshold cycles (Cts) that are obtained are subtracted and plotted against the known composition of the standards to generate a calibration curve. It is usual for the average values of several PCR replicates to be used for this purpose. A disadvantage is that any test sample result must fall with the values of the standards used to generate the standard curve.

Secondly there is the 'relative' copy number method. Here, serial dilutions of a single standard are made over a very wide dynamic range and the 'relative' copy number of both the specific sequence used to detect the analyte together with the 'relative' copy number of the endogenous gene sequence are calculated in each dilution. As with the delta C_t method two individual PCRs are performed and the C_t values for each of the dilutions noted for both DNA sequences. Two calibration curves are then generated, one for the analyte-specific sequence and another for the indigenous gene sequence (Fig. 16.3). Unknown samples are treated similarly and the 'relative' copy number of both sequences determined (at an appropriate dilution) using the two calibration curves. The percentage of the analyte can then be calculated. The advantage of this method is that any analyte concentration can be determined and is not limited to the range and availability of reference materials.

The relative copy number method of analysis is now used as a reliable method for the analysis

of pasta in which DNA extracted from reference pasta or blended flour standard is serially diluted to generate calibration curves. DNA extracted from the adulterated flour test samples is usable at a dilution of 1 in 500 enabling the relative copy number of both normalizing sequence and D-genome sequence to be established from their respective calibration curves. An example of the 'relative' copy number calibration curve generated from the pasta standard containing 10% *T. aestivum* is shown (Fig. 16.3). In each case, PCR inhibition can be detected by divergence from a linear response.

Calculation of the percentage *T. aestivum* present in *T. durum* pasta using the copy number method

When calculating the percentage of a component present in a mixture using the 'relative' copy number approach it is imperative that the number of copies of each of the sequences present in the genome is known. In many cases such as the quantification of Roundup Ready™ soya the ratio is normally accepted to be 1:1, making the ensuing calculation straight forward. However, in the case of the quantification of *T. aestivum* present as a contaminant in *T. durum*, the case is a more complicated. Considering the D-genome-specific sequence:- *T. aestivum* has two copies and *T. durum* has zero copies. The situation for the normalization sequence is somewhat different where *T. aestivum* has six copies and *T. durum* four copies.

Following construction of the two calibration curves the amount of *T. aestivum* present in a test sample may be calculated using the following formula:

$$R = \frac{[(100) - A) \times 4] + [6A]}{2A}$$

Hence,

$$A = \frac{200}{(R-1)}$$

where R = 'relative' copy number ratio and A = % *T. aestivum*

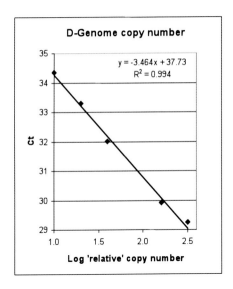
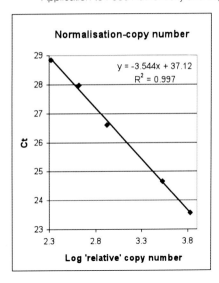

Figure 16.3 Calibration curves for the D-genome-specific and normalizing sequences.

Accuracy

As with all measurement methods accuracy is important. In this instance it was determined by analysing a sample of known composition (2% *T. aestivum*) ten times, giving a mean value of 2.35% ± 0.24% (95% confidence limit) with a standard deviation of 0.34%. The 95% confidence limit of the mean result (2.35% ± 0.24%) does not include the known value of the test material (2%). Hence on the basis of this sample, the analysis has a small positive bias in the range of 0.1–0.6%.

Intermediate precision

When reporting the result of a measurement of a physical quantity, it is essential that some quantitative indication of the quality of the result is given so that the person responsible for utilizing and acting upon this information can assess its reliability. Without such an indication, measurement results cannot be compared, either among themselves or with reference values given in a specification or standard (ISO 1995, section 0.1, ISBN 92-67-10188-9). In the case of real-time PCR as the calculations are based upon two amplifications, this is not an easy matter and the precision of the reported result should be presented rather than the precision of the inter-mediate values (Cts). In this case, the uncertainty associated with the measurement of *T. aestivum* content of *T. durum* pasta was determined using multiple analyses, performed on different oc-

casions by several people using two ABI 7700 instruments. The analysis was not expanded to include other laboratories and hence is described as 'in-house', or intermediate, precision. The value of this uncertainty is likely to vary with the magnitude of the measurement and hence was determined at two analytical levels. An es-sential component of any quantitative method is the inclusion of a performance standard that is co-analysed with the test samples. It is usual for these performance standards to have an assigned value that has been derived by multiple analyses over an extended period. The data obtained from our two pasta performance standards (0.2% and 5.89%) has been used to estimate the uncer-tainty of the method. The lower performance standard (mean value of 0.19%, $n = 36$, standard deviation = 0.04%) gave a coefficient of varia-tion of 21% corresponding to an uncertainty at an approximate 95% confidence limit of 0.11 to 0.26% (expansion factor 2). Hence, for a single analytical determination of a material known to contain 0.19% contamination, the result could be expected to be in the range 0.11–0.26% 19 times in every 20 analyses. The higher performance standard (mean value of 5.89%, $n = 12$, standard deviation = 1.9%) had a coefficient of variation of 33% corresponding to an uncertainty at an approximate 95% confidence limit of 2.02 to 9.75% (expansion factor 2). Hence, for a single analytical determination of a material known to contain 5.89% contamination, the result could be expected to be in the range 2.02–9.75% 19 times in every 20 analyses. Assuming that the coefficient of variation varies between the two standard values (21–33%), at 3% contamination it would be approximately 27% (standard deviation = 0.81). Hence, a pasta containing the legal limit of 3% *T. aestivum* would have an uncertainty of 3% ± 1.6 at 95% confidence limit, i.e. 1.4–4.6%.

Conclusion and future direction

Real-time PCR has now become an accepted analytical tool by the food industry where the number of applications for the technology is rapidly increasing, despite the perception that it is costly. This success is due mainly to the specificity and speed of analysis, together with the ability to amplify DNA sequences from highly fragmented DNA, that are often present in processed foods, enabling analysis to be carried out that were hitherto thought impossible. It has been long suspected that those wishing to deliberately adulterate food for financial gain or incorporate illegal materials or species, have appreciated the deleterious effect that processing has upon proteins and other biomarkers used to detect adulteration, Undoubtedly, this has led to deliberately adulterated materials going undetected. As more DNA sequence information becomes available through internet databases this situation should improve significantly.

An aspect of DNA-based analysis that requires immediate attention is the development and availability of suitable reference materials. In the particular case of emerging GMOs, while the difficulties of both the biotech companies and the legislative bodies are realized, it is essential that the process must become more streamlined to enable effective analysis and for consumers to have confidence in the procedures. A potential solution, although not ideal, is the development of plasmids carrying a range of target sequences appropriate to a suite of analyses.

The quantitative detection of target sequences by real-time PCR raises the question of comparison of results from DNA-based analyses with other analytical methods which have been reported in a traditional manner (e.g. mg/kg,% or ppm). Quantitative real-time PCR results are typically presented as either a relative percentage or as a relative copy number on a haploid genome basis. It is usual for these calculations to include a comparison of a specific target sequence with an endogenous DNA sequence extracted from both the sample and a certified reference material. To do this a number of assumptions are made concerning the lack of variation in the target DNA content, extractability and degradative state between the reference material and the sample being tested. Either new standards must be agreed or a means of comparison found.

The increasing popularity of real-time methods in the detection of microorganisms has its own difficulties to which robust solutions must be sought. These include rapid detection without pre-enrichment and differentiation between live and dead organisms.

The application of real-time PCR methods based upon RNA using reverse transcriptase has yet to impact upon the food industry with the exception of the assay for avian flu (H5), which is available commercially. Given current trends in food safety and customer protection, it would seem prudent to investigate similar methods to detect food-borne viruses which are responsible for a number of food poisoning out-breaks each year.

Acknowledgement

The author gratefully acknowledges the generous support of the UK Food Standards Agency in funding research into the quantification of *T. aestivum* adulteration of *T. durum* pasta (projects 2A 015, AN0667 and Q01085).

References

Alary, R., Serin, A., Duviau, M.-P., Joudrier, P and Gautier, M.-F. 2002. Quantification of common wheat adulteration of durum wheat pasta using real-time quantitative polymerase chain reaction (PCR). Cereal Chem. 79, 553–558.

Beuret, C. 2004. Simultaneous detection of enteric viruses by multiplex real-time RT-PCR. J. Virol. Methods *115*, 1–8.

Bleve, G., Rizzotti, L., Dellaglio, F. and Torriani, S. 2003. Development of reverse transcription (RT)-PCR and real-time RT- PCR assays for rapid detection and quantification of viable yeasts and moulds contaminating yogurts and pasteurized food producers. Appl. Environ. Microbiol. 69, 4116–4122.

Block, A. and Schwarz, G. 2003. Validation of different genomic and cloned DNA calibration standards for construct-specific quantification of LibertyLink in rapeseed by real-time PCR. Eur. Food Res. Technol. *216*, 421–427.

Bryan, G.J., Gale, M.D., Dixon, A. and Wiseman, G. 1995. Report on project 2A 015 – The detection and quantification of common wheat (*T. aestivum*) in durum (*T. durum*) pastas and semolinas. MAFF, UK.

Bryan, G. J., Dixon, A., Gale., M.D. and Wiseman, G. 1998. A PCR-based Method for the detection of hexaploid bread wheat adulteration of durum wheat and pasta. J. Cereal Sci. *28*, 135–145.

Burns, M., Corbisier, P., Wiseman, G., Valdivia, H., McDonald, P., Bowler, P., Ohara, K., Schimmel, H., Charels, D. and Harris, N. 2006. Comparison of plasmid and genomic DNA calibrants for the assessment of genetically modified ingredients. European Food Res. Technol. *224*, 249–258.

Casey, G.D. and Dobson, A.D.W. 2004. Potential of using real-time PCR-based detection of spoilage yeast in fruit juice-a preliminary study. Int. J. Food Microbiol. *91*, 327–335.

CEN/TC 275.WG in Vienna (4th–6th February 2002. Foodstuffs-Methods of analysis for the detection of genetically modified organisms and derived products-Quantitative nucleic acid based methods CEN/TC275/WG 11N 186.

Chisholm, J., Conyers, C., Booth, C., Lawley, W. and Hird, H. 2005. The detection of horse and donkey using real-time PCR. Meat Sci. *70*, 727–732.

Connor, C.J., Luo, H., Gardener, B.B. and Wang, H.H. 2005. Development of a real-time PCR-based system targeting the 16S Rrna gene sequence for rapid detection of *Alicyclbacillus* spp. in juice products. Int. J. Food Microbiol. 99 (3), 229–235.

EC Commission Regulation 1222/94 (Annex C) 199). Official Journal of the European Communities. L*136*, 31 May 1994, p5.

EN ISO 21569,2005. Foodstuffs- Methods of analysis for the detection of genetically modified organisms and derived products-Qualitative nucleic acid based methods.

EN ISO 21570,2005. Foodstuffs – Methods of analysis for the detection of genetically modified organisms and derived products-Quantitative nucleic acid based methods.

European Commission Recommendation 2004/787/EC 2004. Official Journal of the European Communities, L 348/18.

European Commission Regulation 258/97 (1997) Official Journal of the European Communities. L 43;14.2.

European Commission Regulation 1829/2003. 2003. Official Journal of the European Communities L 268/1;18.10.

European Commission Regulation 1830/2003. 2003. Official Journal of the European Communities L 268/1;24–28.

European Directive 2001/101 Food Safety Act. (1990). HMSO.

Furet, J.P., Quenee, P. and Tailliez, P. 2004. Molecular quantification of lactic acid bacteria in fermented milk products using real-time quantitative PCR. Int. J. Food Microbiol. 97, 197–207.

Fu, Z., Rogelj, S. and Kieft, T.L. 2005. Rapid detection of *Escherichia coli* O157,H7 by immunomagnetic separation and real-time PCR. Int. J. Food Microbiol. 99, 47–57.

Fukushima, H., Tsunomori, Y. and Seki, R. 2003. Duplex real-time SYBR green PCR assays for the detection of 17 species of food or water borne pathogens in stools. J. Clin. Microbiol. *41*, 5134–5146.

Gachet, E., Martin, G., Vigneau, F. and Meyer, G. (1998). Detection of genetically modified organisms (GMOs) by PCR: a brief review of methodologies available. Trends Food Sci. Technol. 9, 380–388.

Gazzetta Ufficiale. (1967). Disciplina per la lavorazione e commercio dei cereali, degli sfarinati, del pane e delle paste alimentary. Law No *580*, 4 July.

Grattenpanche, F., Lacroix, C., Audet, P. and Lapointe, G. 2005. Quantification by real-time PCR of *Lactococcus lactis* subsp. *cremortis* in milk fermented by a mixed culture. Appl. Microbiol. Biotechnol. 66, 414–421.

Hein, I., Jorgensen, H.J., Loncarvic, S. and Wagner, M. 2005. Quantification of *Staphylococcus aureus* in unpasteurised bovine and caprine milk by real-time PCR. Res. Microbiol. 156, 554–573.

Hird, H.J., Hold, G.L., Chisholm, J., Reece, P., Russell, V.J., Brown, J., Goodier, R. and MacArthur, R. 2005. Eur. Food Res. Technol. 220, 633–637.

Henterich, N., Osman, A.A., Mendez, E. and Mothes, T. 2003. Assay of gliadin by immunopolymerase chain reaction. Nahrung *47*, 345–348.

Hernández, M., Pla, M., Esteve, T, Prat, S., Puigdomènech, P. and Ferrando, A. 2003. A specific real-time quantitative PCR detection system for event MON810 in maize YieldGard® based on the 3-transgene integration sequence. Transgenic Res. *12*, 179–189.

Hernandez, M., Esteve, T. and Pla, M. 2005. Real-time polymerase chain reaction based assays for quantitative detection of barley, rice, sunflower and wheat. J. Agric. Food. Chem. 53, 7003–7009.

Holck, A., Vaïtilingom, M., Didierjean, L. and Rudi, K. 2002. 5'-Nuclease PCR for quantitative event-specific detection of the genetically modified Mon810 MaisGard maize. Eur. Food Res. Technol. *214*, 449–453.

http://gmo-crl.jrc.it/statusofdoss.htm (accessed 19 June 2008).

http://www.measurementuncertainty.org/ (accessed 19 June 2008).

International Standard (ISO 11051,1994 (E)). ISO 1995, section 0.1, ISBN 92-67-10188-9.

Kaufman, G.E., Blackstone, G.M., Vickery, M.C., Bej, A.K., Bowers, J., Bowen, M.D., Meyer, R.F. and DePaola. 2004. Real-time PCR quantification of *Vibrio parahaemolyticus* in oysters using an alternative matrix. J. Food Prot. 67, 2424–2429.

Kuribara, H., Shindo, Y., Matsuoka, T., Takubo, K., Futo, S., Aoki, N., Hirao, T., Akiyama, H., Goda, Y., Toyoda, M. and Hino, A. 2002. Novel reference molecules for quantitation of genetically modified maize and soybean. J. AOAC Int. 85, 1077–1089.

Larson, S., Bryan, G., Dyer, W and Blake, T. 1996. Evaluating gene effects of a major barley seed dormancy QTL in reciprocal backcross populations. J. Quantitative Trait Loci, 2.

Liming, S.H. and Bhagwat, A.A. 2004. Application of a molecular beacon-real-time PCR technology to detect Salmonella species contaminating fruits and vegetables. Int. J. Food Microbiol. 95, 177–187.

Loisy, F., Atmar, R.L.,Guillon, P., Le Cann, P., Pommepuy, M. and Le Guyader, F.S. 2005. Real-time RT-PCR for norovirus screening in shellfish. J. Virol. Methods 123, 1–7.

Lopez-Andreo, M., Lugo, L., Garrido-Pertierra, A., Prieto, M.I. and Puyet, A. 2004. Identification and quantitation of species in complex DNA mixtures by real-time polymerase chain reaction. Anal. Biochem. 339, 73–82.

Lunello, P., Mansilla, C., Conci, V and Ponz, F. 2004. Ultra-sensitive detection of two garlic potyviruses using a real-time fluorescent (TaqMan) RT-PCR assay. J. Virol. Methods 118, 15–21.

Luo, H., Yousef, A.E. and Wang, H.H. 2004. A real-time polymerase chain reaction –based method for the rapid and specific detection of spoilage Alicyclobacillus spp. In apple juice. Lett. Appl. Microbiol. 39, 376–382.

Malorny, B., Paccassoni, E., Fach, P., Bunge, C., Martin, A. and Helmuth, R. 2004. Diagnostic real-time PCR for detection of Salmonella in food. Appl. Environ. Microbiol. 70, 7046–7052.

O'Hanlon, K.A., Catarame, T.M., Duffy, G., Blair, I.S. and McDowell, D.A 2004. RAPID detection and quantification of E. coli O157/O26/O111 in minced beef by real-time PCR. J. Appl. Microbiol. 96, 1013–1023.

Perelle, S., Josefsen, M., Hoofar, J., Dilasser, F., Grout, J. and Fach, P. 2004. A Light-Cycler real-time PCR hybridization probe assay for detecting food-borne thermophilic Campylobacter. Mol Cell Probes. 18, 321–327.

Phister, T.G. and Mills, D.A. 2003. Real-time PCR assay for detection and enumeration of Dekkera bruxellensis in wine. Appl. Environ. Microbiol. 69, 7430–7434.

Pinzani, P., Bonciani, L., Pazzagli, M., Orlando, C., Guerrini, S. and Granchi, L. 2004. Rapid detection of Oenococcuc oeni in wine by real-time quantitative PCR. Lett. Appl. Microbiol. 38, 118–124.

Renson, G., Smith, W., Ruzante, J., Sayer, M., Osburn, B. and Cullor, J. 2005. Development and evaluation of a real-time fluorescent polymerase chain reaction assay for the detection of bovine contaminates in cattle feed. Foodborne Pathog. Dis. 2, 152–159.

Rodriguez, M.A., Garcia, T., Gonzalez, I., Asensio, L., Hernandez, P.E. and Martin, R. 2004. Quantitation of Mule duck in goose foie gras using Taqman real-time polymerase chain reaction. J. Agric. Food Chem. 52, 1478–1483.

Rodriguez-Lazaro, D., Jofre, A., Aymerich, T., Hugas, M. and Pla, M. 2004a. Rapid quantitative detection of Listeria monocytogenes in meat products by real-time PCR. Appl. Environ. Microbiol. 70, 6299–6301.

Rodriguez-Lazaro, D., Hernandez, M. and Pla, M. 2004b. Simultaneous quantitative detection of Listeria spp. and Listeria monocytogenes using a duplex real-time PCR-based method. FEMS Microbiol. Lett. 233, 257–267.

Rønning, S.B., Vaïtilingom, M., Berdal, K.G. and Holst-Jensen, A. 2003. Event specific real-time quantitative PCR for genetically modified Bt11 maize (Zea mays). Eur. Food Res. Technol. 216, 347–354.

Seo, K.H. Valentin,-Bon, I.E., Brackett, R.E. and Holt, P.S. 2004. Rapid, specific detection of Salmonella enteritidis in pooled eggs by real-time PCR. J. Food Prot. 67, 864–869.

Seo, K.H. and Brackett, R.E. 2005. Rapid specific detection of Enterobacter sakazakii in infant formula using a real-time PCR assay. J. Food Prot. 68, 59–63.

Shindo, Y., Kuribara, H., Matsuoka, T., Futo, S., Sawada, C., Shono, J., Akiyama, H., Goda, Y., Toyoda, M. and Hino, A. 2002. Validation of real-time PCR analyses for line-specific quantitation of genetically modified maize and soybean using new reference molecules. J. AOAC Int. 85, 1119–1126.

Spano, G., Beneduce, L., Terzi, V., Stanca, A.M. and Massa, S. 2005. Real-time PCR for the detection of Eschericia coli O157,H7 in dairy and cattle wastewater. Lett. Appl. Microbiol. 40, 164–171.

Stephan, O. and Vieths, S 2004a. Development of a real-time PCR and a sandwich ELISA for the detection of potentially allergenic trace amounts of peanut (Arachis hypogaea) in processed foods. J. Agric. Food Chem. 52, 3754–3760.

Stephan, O., Weisz, N., Vieths, S., Weiser, T., Rabe, B. and Vatterott, W. 2004b. Protein quantification, sandwich ELISA, and real-time PCR used to monitor industrial cleaning processes for contamination with peanut and celery allergens. J. AOAC Int. 87, 1448–1457.

Takahashi, H., Iwade, Y., Konuma, H. And Hara-Kudo, Y. 2005. Development of a quantitative real-time PCR method for estimation of the total number of Vibrio parahaemolyticus in contaminated shellfish and seawater. J. Food Prot. 68, 1083–1088.

Taverniers, I., Windels, P., Van Bockstaele, E. and De Loose, M. 2001. Use of cloned DNA fragments for event-specific quantification of genetically modified organisms in pure and mixed food products. Eur. Food Res. Technol. 213, 417–424.

Taverniers, I., Van Bockstaele, E., De Loose, M. 2004. Cloned DNA fragments as calibrators for controlling GMOs: different real-time duplex quantitative PCR methods. Anal. Bional. Chem. 378, 1198–1207.

Terzi, V., Malnati, M., Barbanera, M., Stanca, A.M. and Faccioli, P. 2003. Development of analytical systems based on real-time PCR for Triticum species-specific detection and quantification of bread wheat contamination in semolina and pasta. J. Cereal Sci. 38, 87–94.

Terzi, V., Infascelli, F., Tudisco, R., Russo, G., Stanca, A.M. and Faccioli, P. 2004. Quantitative detection of Secale cereale by real-time PCR. Lebensm.-Wiss.u.-Technol. 37, 239–246.

Terzi, V., Morcia, C., Gorrini, A., Stanca, A.M., Shewry, P.R. and Faccioli, P. 2005. DNA-based methods for the identification and quantification of small grain cereal mixtures and fingerprinting of varieties. J. Cereal Sci. 41, 213–220.

Trapmann, S., Catalani, P., Corbisier, P., Van Iwaarden, P., Schimmel, H., Zelany, R. 2002. Withdrawal of Bt-11 and Bt-176 GMO CRMs. Presentation at

ENGL 4th Plenary Meeting, JRC, Ispra 29th–30th April.

UK Food Labelling Regulations. (1996) (SI No. 1499).

Van den Eede, G., Kay, S., Anklam, E. and Schimmel H. 2002. Analytical challenges: bridging the gap from regulation to enforcement. J. AOAC Int. 85, 757–761.

Weighardt, F.2002. development of a plasmid standard methodology. Presentation at ENGL 4th Plenary Meeting, JRC. Ispra, 29th–30th April.

Windels, P., Taverniers, I., Depicker, A., Van Bockstaele, E. and De Loose, M. 2001, Characterisation of the Roundup Ready soyabean insert. Eur. Food Res. Technol. *213*, 107–112.

Windels, P., Bertrand, S., Depicker, A., Moens, W., Van Bockstaele, E. and De Loose, M. 2003. Qualitative and event-specific PCR real-time detection methods for StarLink maize. Eur. Food Res. Technol. *216*, 259–263.

Wiseman, G., Stephenson, P., Kirby, J and Gale, M. D. 1999. Report AN0667-Quantitative PCR detection of T. aestivum adulteration in commercial T. durum pasta using PSR 128 primers: optimisation. MAFF, UK.

Wiseman, G. 2001. Durum eheat. In: Pasta and Semolina Technology (Roy Kill and Keith Turnbull, eds), Blackwell Science Ltd, Oxford, pp. 11–42.

Wiseman, G. 2002. State of the art and limitations of the quantitative polymerase chain reaction. J. AOAC Int. 85, 792–796.

Wolffs, P., Knutsson, R., Norling, B. and Radstrom, P. 2004. Rapid quantification of Yersinia enterocolitica in pork samples by a novel sample preparation method, flotation, prior to real-time PCR. J Clin Microbiol. *42*, 1042–1047.

Wolffs, P., Norling, B. and Radstrom, P. 2005. Risk assessment of false positive quantitative PCR results in food, due to detection of DNA originating from dead cells. J. Microbiol. Methods 60, 315–323.

Wurtz, A., Bluth, A., Zeltz, P., Pfeifer, C. and Willmund, R. (1999). Quantitative analysis of genetically modified organisms (GMO) in processed food by PCR-based methods. Food Control *10*, 385–389.

Yang, C., Jiang, Y., Huang, K., Zhu, C. and Yin, Y. 2003. Application of real-time PCR for quantitative detection of Campylobacter jejuni in poultry, milk and environmental water. FEMS Immuno.l Med. Microbiol. *38*, 265–271.

Yoon, S.Y., Chung, G.T., Kang, D.H., Ryu, C., Yoo, C.K. and Seong, W.K. 2005. Application of real-time PCR for quantitative detection of Clostridium botulinum type A toxin gene in food. Microbiol. Immunol. *49*, 505–511.

Zimmermann, A., Luthy, J. and Pauli, U. 2000. Event specific transgene detection of Bt11 corn by quantitative PCR at the integration site. Lebensm-Wiss.u.-Technol. *33*, 210–216.

Molecular Haplotyping by Real-time PCR

17

Genevieve Pont-Kingdon, Alison Millson and Elaine Lyon

Abstract
Molecular real-time PCR methods can determine whether two or more mutations are on the same or different chromosomes. This ability to haplotype without family studies is useful for research and clinical purposes and can give an advantage over genotyping. Haplotyping by real-time PCR with hybridization probes has been demonstrated for adjacent repeats and single base alterations, with a probe that covers both sites. However, base alterations may be separated by distances greater than a traditional hybridization probe will cover. We described a probe design that covers both (or all) sites, but does not include the entire sequence between the sites. When hybridized with the template, the template is forced to form a loop. This 'loop-out' probe will dissociate from the template as a unit, therefore allowing haplotyping of base alterations separated by over 80 bp. Examples of haplotyping by traditional probes for adjacent sequence variants, as well as examples of 'loop-out' probes are presented.

Introduction

Genotyping assays interrogate nucleotide bases at one or more positions. However, genotyping assays alone are incapable of predicting if two or more nucleotide changes occur on the same (*in cis*) or on different (*in trans*) chromosomes. Two or more variants together in linkage disequilibrium are often referred to as a haplotype or chromosome phase. Recently, efforts have been made to determine haplotype blocks in the human genome (Zhang *et al.*, 2004). These haplotypes can be used in case/control studies as well as in family studies to show association with disease (Zhang *et al.*, 2004; Zhao *et al.*, 2003). Their power is in their ability to identify genetic variants that contribute to complex genetic diseases. Haplotypes are typically assigned by first identifying common combinations in populations. Computational models and algorithms have then been developed to assign haplotypes statistically. These computational approaches can accurately determine common variants, but often mis-assign rare haplotypes (Sabbagh and Darlu, 2005). For diagnostic purposes, a statistical approach is often inadequate, since rare haplotypes may also be clinically relevant. Traditionally laboratories try to obtain parental samples to be able to assign chromosome phase. A molecular haplotyping approach is desirable to test directly the chromosome phase in one individual. This approach eliminates the need to obtain parental samples and can distinguish common and rare haplotypes. During this chapter we will discuss clinical applications and techniques available for molecular haplotyping and present design and applications for real-time haplotyping methods.

Clinical haplotyping applications

The combination of variants can show a stronger association with disease than a single variant alone (Frank *et al.*, 2005; Zhao *et al.*, 2003). For example, haplotype or genotype variations in the beta2 adrenergic receptor gene (ADRB2) have been associated with asthma, cardiovascular diseases and obesity (Kirstein and Insel, 2004). Haplotype analysis also indicates different risks from beta-blockers treatment of asthma (Israel et al., 2004) and acute coronary syndrome (Lanfear et al., 2005). Alternately, one variant may modify the effect of another, contributing to its pathogenesis or providing a protective effect (Lee et al., 2003). For example, the R117H mutation in the cystic fibrosis transregulator (CFTR) gene is considered a mild mutation, unless it is on the same chromosome as a variant consisting of five thymidines (5T) (Kiesewetter et al., 1993). The 5T variant (as opposed to seven or nine thymidines) reduces the splicing efficiency of exon 9. The combined effect when the R117H and 5T are together in cis is a severe cystic fibrosis (CF) chromosome. When 5T and R117H are found together, family studies are recommended to determine phase (Grody et al., 2001). A third polymorphic site, a poly(TG) tract adjacent to the 5T, further modifies the splicing of exon 9 (Cuppens et al., 1998). Haplotyping of the TG/T tracts will be discussed in greater detail later. In addition, when two CFTR mutations are found in a person suspected of having CF, the two mutations should be confirmed as being in trans. This is particularly important if at least one of the mutations is rare. If the two mutations are in cis, only one causative chromosome has been identified.

Pharmacogenetics offers another application for molecular haplotyping. The goal for these assays is to predict an individual's response to drugs by identifying genetic variants in genes coding for drug-metabolizing enzymes. In the cytochrome 2D6 gene, the allelic variants have been defined according to several SNPs typically inherited together (Sim, 2006, accessed 02/23). Available assays determine the genotype of each nucleotide change and then assign the combination as an allele. However, in some cases, the haplotype cannot be inferred from the genotype. The *4 allele is identified by the presence of the 1846G>A mutation. A *4K subtype will also have a common 2850C>T variant, that is present also in a number of other alleles. When a 1846G>A and 2850C>T are detected together, the patient could be compound heterozygous

*4 with another allele containing 2850C>T, which could be associated with either functional, decreased function or non-functional alleles. Conversely, if found in cis, the patient would be heterozygous for the *4K allele and unlikely to have an adverse reaction. A similar case is found in the thiopurine S-methyl transferase gene which metabolizes thiopurine drugs used for the treatment of acute leukaemia. A haplotype assay is needed to distinguish between a simple heterozygous *3A (c.460G>A and c.719A>G in cis) and a compound heterozygous *3B (c. 460G>A) and *3C (719A>G) (Von Ahsen et al., 2004).

Review of available molecular haplotyping technologies

Molecular haplotyping describes direct approaches to establishing haplotypes on single samples. It excludes techniques that need pedigree testing or infer haplotypes using statistical analysis. Three main approaches are used to ensure molecular haplotyping: physical separation of homologues prior to analysis, specific analysis of only one haplotype (allele specific) or direct distinction of haplotypes by an analytical method (haplo-specific).

Physical separation of homologues prior to analysis

Haplotypes of any loci can be readily determined if the two homologues from diploid cells are spatially separated. This approach has been championed by the conversion technology (Yan et al., 2000) in which whole human chromosomes are transferred into mouse hybridoma cells. Cloning of DNA fragments in microorganisms has also been reported (Kim et al., 2003). Genomic DNA samples have also been diluted to single DNA molecules in aqueous phase prior to analysis (Ding and Cantor, 2003; Paul and Apgar, 2005). An interesting approach is the physical separation of genomic DNA in a gel, making the two homologues identifiable by fluorescent PCR (Mitra et al., 2003). Recently, single templates isolated in emulsion droplets have been successfully haplotyped using 'linking-PCR' to create mini-chromosomes (Wetmur et al., 2005). Finally, individual DNA templates, visualized by atomic electron microscopy have been haplotyped using a series of tagged oligonucleotides (Woolley *et al.*, 2000).

Allele-specific PCR approach

Many variations of the basic haplotyping method by allele-specific amplification (Ruano and Kidd, 1989) have been published. Multiplex reactions (Eitan and Kashi, 2002) and recent applications to long range PCR (Pont-Kingdon *et al.*, 2004; Wu *et al.*, 2005; Yu *et al.*, 2004) have increased the power of allele-specific amplification for haplotyping. Several polymorphisms located kilobases apart have been haplotyped by an elegant allele-specific amplification using a long-range ligated PCR product as template (McDonald *et al.*, 2002). Allele-specific amplification approaches have the flexibility to allow analysis of polymorphisms linked with the selected allele by any 'genotyping' technology, making them easily implemented in virtually any laboratory setting. Detection systems such as real time PCR (Proudnikov *et al.*, 2004), mass spectrometry (Tost *et al.*, 2002), fragment length analysis and melting curves of hybridization probes (Pont-Kingdon *et al.*, 2004), pyrosequencing (Pettersson *et al.*, 2003) and restriction enzyme digestion (McDonald *et al.*, 2002) have been described.

Haplo-specific approaches

We define haplo-specific as techniques that directly differentiate different haplotypes from a mixture present in heterozygous samples. For example, the simultaneous capture of several polymorphisms by allele-specific capture beads allows haplotyping of up to 11 loci in the NAT2 gene (Hurley *et al.*, 2004). Properties intrinsic to DNA such as single strand conformation and double strand stability have also been used for haplotype determination. Denaturing gel based assays that analyse the different conformations of single strand DNA fragments carrying different polymorphisms (SSCP) or the different stability of heteroduplexes have both been described reviewed recently in Szantai *et al.* (2005). The real-time approach we will review in this chapter relies on the melting properties of fluorescent probes hybridized on templates with different haplotypes.

Haplotype of two adjacent sequences

Principle

Real-time fluorescent monitoring allows mutation detection via sequence-specific hybridization probes labelled with fluorophores. Two probes are used, one as an anchor, the other as a reference probe. Both are labelled with a single fluorophore. Binding of both probes to the amplicon brings the fluorophores into close proximity, allowing energy transmission from one to the other, resulting in direct fluorescent detection of the amplification. After amplification is complete, the probes are allowed to bind to the amplicon and fluorescence is continually monitored while slowly raising the temperature. Depending on the homology of the template to the reference probe, the probe 'melts' off at a characteristic temperature (T_m). Genotyping is possible because the stability of a probe with its template partially depends on the presence or absence of a mismatch between them. The presence of multiple mutations between a probe and a template will further decrease the stability, with each combination having a signature T_m and thereby allowing haplotyping.

Probe design

As mentioned earlier, a poly(TG) tract followed immediately 3′ by a polyT tract influences the splicing of exon 9 in the CFTR gene (Cuppens *et al.*, 1998). Due to the clinical importance of this region, it was targeted as a prime example of two adjacent sequences requiring haplotyping (Millson *et al.*, 2005). The poly(TG) tract is a run of 10, 11, 12 or 13 TG repeats followed by a tract of 5, 7 or 9 Ts. Various combinations exist, with TG11–7T being the most common and TG13–5T being the least. A traditional anchor/reference probe pair design was used to cover this region. We designed our anchor probe to bind to the sequence 5′ of the tract and into the TG tract, while our reference probe interrogated the 3′ end of the TG tract and extended through and beyond the T tract. The probe binding to various templates can produce specific base mismatches, loop-out bases or create a dangling end on the reference probe due to length incompatibility

(Fig. 17.1). In our experience with the TG-T polymorphisms we see examples of all of these destabilizing events, resulting in the lowering, to a greater or lesser extent, of the melting temperature. The melting temperature for each unique template is very specific (Fig. 17.2). Within- and between-run precision studies all had standard deviations within ± 0.3°C with T_m values ranging from 42.4°C to 62.9°C.

The TG-T haplotype becomes even more clinically informative when haplotyped to a third mutation, R117H, residing 17.7 kb upstream. Long-range allele-specific PCR targeting the R117H site is used to amplify a 17.7 kb product from each chromosome (Pont-Kingdon *et al.*, 2004), which are then analysed with the TG-T haplotyping probes (Millson *et al.*, 2005). This assay determines the phase of all three loci, alleviating the need for traditional family studies. A chromosome containing the R117H mutation, 5T and 12 or 13 TGs can lead to classical CF due to the combined effects of the three variants.

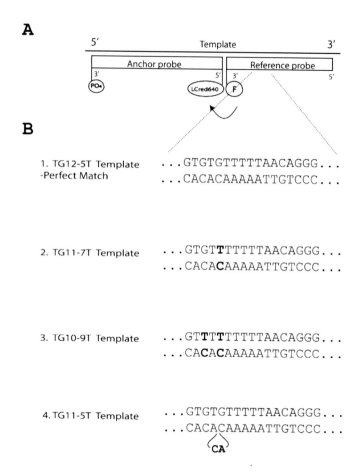

Figure 17.1 Probe design and destabilizing effects of different haplotypes. A. Anchor and reference probe binding to template illustrating fluorescence resonance energy transfer (FRET). The probes are brought into close proximity upon binding to the template. The FITC-labelled probe is excited via a light source, emits energy of a slightly higher wavelength, exciting the LCred640-labelled probe, which then emits light at an even higher wavelength that is detected by the LightCycler. B. Four partial illustrations of the reference probe binding to the TG-T tract in the IVS-8 region of the CFTR gene. The reference probe is a perfect match to a TG12–5T template. 1. A perfect match with a TG12–5T template. 2. TG11–7T template binding with the probe, resulting in one T:C mismatch due to the 'extra' Ts in the template T-tract. With one fewer TG the probe and template stay aligned. 3. TG10–9T template, resulting in two T:C mismatches due to the 'extra' Ts, but having two fewer TGs the alignment again is maintained. 4. TG11–5T template. This template is two bases shorter than the reference probe, the ends of which are firmly matched to the template. We theorize that the probe 'loops out' to accommodate the template.

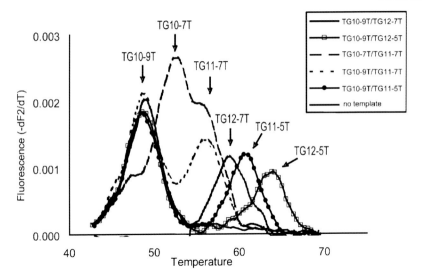

Figure 17.2 Derivative melting curves displaying combinations of TG and T repeats. Reproduced by permission from Millson, *et al.* (2005) Direct molecular haplotyping of the IVS-8 poly(TG) and polyT repeat tracts in the cystic fibrosis gene by melting curve analysis of hybridization probes.

Haplotype of multiple loci using loop-out haplotyping probes

Principle

As reported in the above example, in the HFE gene (Phillips *et al.*, 2000) and the TNFα promoter region (Song *et al.*, 2005) two polymorphisms in close proximity (approximately 20 nt) can be haplotyped by a reporter probe that covers both.

We have demonstrated that probes hybridizing on discontinuous sequences of a template allow haplotyping of loci too far apart to hybridize under a continuous probe (Pont-Kingdon and Lyon, 2005). In these probes, called loop-out (lpo) haplotyping probes, sequences found in the template between the different loci are omitted. Upon probe hybridization, the template sequence loops-out allowing continuous binding of the lpo probe. This is demonstrated in Fig. 17.3 for lpo hybridization probes using FRET. We have demonstrated that lpo probes dissociate as units from the template with a stability distinctive for each haplotype.

This section addresses considerations in lpo haplotyping probe design, real-time detection methods of different haplotypes and maximum loop length.

Probe design

As observed with hybridization probes used in genotyping (Lyon, 2001), discrimination of haplotypes depends on probe design and choice of matches and mismatches.

After successfully designing several lpo probes, we have seen some key elements of probe design emerge. Probes were from 23 to 29 nucleotides (nt) long with predicted T_m values (Integrated DNA Technologies; www.idtdna. com) varying between 56 °C to 69°C. Note that these predictions do not take into account the presence of the template loop. The effect of the loop on T_m is difficult to predict and we have observed T_m differences from 2°C to 13°C. Each half of the probe (from the position of the loop to the ends) was between 10 and 13 nt with T_m values varying from 32°C to 48°C. On each side of the polymorphic site the probe extends from 4 to 8 nt.

Haplotype discrimination with an lpo probe depends also on the probe sequence. This is illustrated by an example of haplotyping two polymorphisms of the beta-2-adrenergic receptor gene (ADRB2) (Fig. 17.4). The polymorphism at position 46 of ADRB2 is a G or an A and the polymorphism at position 79 is a G or a C. The polymorphisms are separated by 33 nucleotides. Four haplotypes are possible; GG, GC, AC and

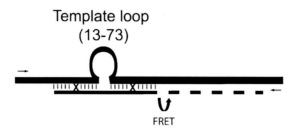

Figure 17.3 Schematic representation of a loop out probe design. Thin arrows represent PCR primers. The PCR template is represented by the thick black line interrupted by the loop. Tested loop sizes are indicated. The probes are represented underneath; the loop out haplotyping probe is the black line and the broken line represents the anchor probe. Matches between the lpo probe and the template are shown by short vertical lines and positions of possible mismatches by the crosses.

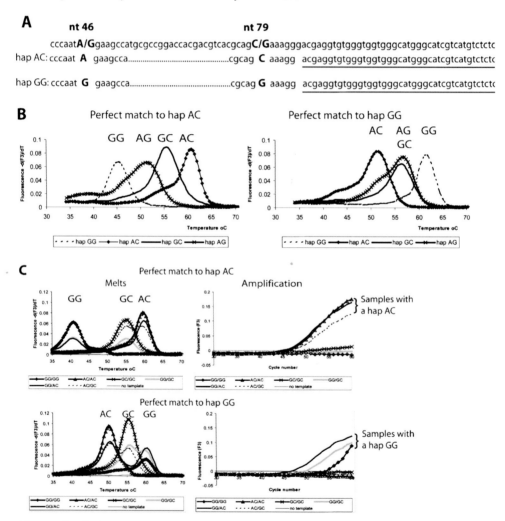

Figure 17.4 Probe design and detection method using ADRB2 haplotypes. (A) Top line: Partial sequence of ADRB2 covering the region with SNP 46 and SNP 79. Polymorphisms are indicated in bold. Hap AC and Hap GG lines have the probes sequences. The Hap AC probe matches perfectly to the polymorphism of the AC haplotype and the Hap GG probe matches perfectly to the polymorphisms of the GG haplotype. The sequence omitted in the probe and looped out upon binding is indicated by a dotted line. The underlined sequences are anchor probes. (B) Derivative melting curves of the Hap AC (left) and Hap GG probe (right) annealed on complementary oligonucleotides identical to the four possible haplotypes listed in the figures legends. (C) Derivative melting curves (left) and amplification curves (right) from samples with different haplotypes analysed with the Hap AC probe (top) or the Hap GG probe.

AG. Lpo probes were designed to haplotype these two positions (Pont-Kingdon and Lyon, 2005) (Fig. 17.4A). The following experiment compares the ability of the lpo probe matching the AC haplotype and the lpo probe matching the GG haplotype to discriminate the CG and AG haplotypes, both with one mismatch with the lpo-probes. The three haplotypes GG, GC and AC are common in human populations (Drysdale et al., 2000) while the AG haplotype is rare. For this reason, the templates used in Fig. 17.4B are simple oligonucleotides corresponding to the four haplotypes and complementary to the lpo probes and the anchor probes. Probes were mixed with the four templates in the presence of PCR buffer and melted in capillaries in the LightCycler. The conditions were as follows: 95°C for 30 s, 30°C for 120 s followed by an increase to 75°C at a rate of 0.1°C/s. Fluorescence was continuously monitored during the melt. Both lpo probes clearly distinguish the three major haplotypes but only the AC-matched probe distinguishes the GC and AG haplotypes (Fig. 17.4B). The different abilities of the probes are possibly due to the different mismatches with the GC and AG haplotypes. Mismatches with the AC-matched probe are A:C and C:C, both known for their strong destabilizing effect on duplexes, while the mismatches with the GG probe are G:T and G:G, known for their weaker destabilizing effect (Lyon, 2001).

Detection methods

Detection of polymorphisms (genotyping) by real-time PCR can be accomplished by monitoring probe hybridization (or quenching), either during PCR or after PCR by melting curve analysis. The following example shows that this is also partially true for haplotyping using probes. Again, probes matched to the AC or the GG haplotypes of ADRB 2 are used. Experiments are performed in a LightCycler using de-identified human DNA samples. Fluorescence is monitored during PCR at the annealing temperature of 60° C and continuously during the melt.

1. Haplotyping using melting curve analysis
Single-step molecular haplotyping is possible in individual samples by melting curve analysis. Fig. 17.4C (left panels) shows examples of ho-

mozygous and heterozygous samples haplotyped for the two polymorphisms in ADRB2 using the AC haplotype matched probe or the GG haplotype matched probe.

2. Haplotyping by amplification curve analysis
The annealing temperature used during PCR is 60°C. At this temperature, with the AC matched probe, only the AC haplotype is stable, and probe binding is recorded during PCR only in samples with an AC haplotype (homozygous and heterozygous; Fig. 17.4C, right). If the lpo probe is designed to match perfectly the GG haplotype (bottom of Fig. 17.4C) only this haplotype is detected during PCR. Analysis of amplification curves can be used to pre-screen or confirm haplotypes determined by the melting analysis. A drawback of this approach is that only the haplotype most stable with the probe is detectable making distinction between homozygotes and heterozygotes impossible. A combination of several probes matching with the different haplotypes and emitting at different wavelengths could render this approach feasible.

3. Utilization of lpo probes with other detection systems
The principle of lpo probes relies on the dissociation of the probe from its template as a unit. The examples above use hybridization probes but it should be possible to design lpo probes using fluorescein-labelled probes, MGB-Eclipse probes, molecular beacons and unlabelled probes in presence of a double-stranded specific dye (data not shown).

Maximum length loop
To determine the maximum length between polymorphisms allowing an lpo-haplotyping-probe to hybridize on both sides of the sequence deleted in the probe and dissociate as a unit, we used a series of artificial templates. Two polymorphisms, WIAF 1537 (C/T) and WIAF 1538 (A/G), separated by 27 nt on chromosome 21 were used (Fig. 17.5A). Four haplotypes are possible: CA, CG, TA and TG. The lpo probe perfectly matches the TG haplotype and a loop of 13 nt of the template is forced upon binding (Fig. 17.5A). Double-stranded artificial templates representing all possible haplotypes were

A

WIAF 1537   13-33-53-73 nt   WIAF 1538

5' gcca **T/C** tttcttctctttta-20-40-60-caatgcagtttc **G/A** acataacat

3' cggt **A** aaagaaga.....................tcaaag **C** tgtattgta

B

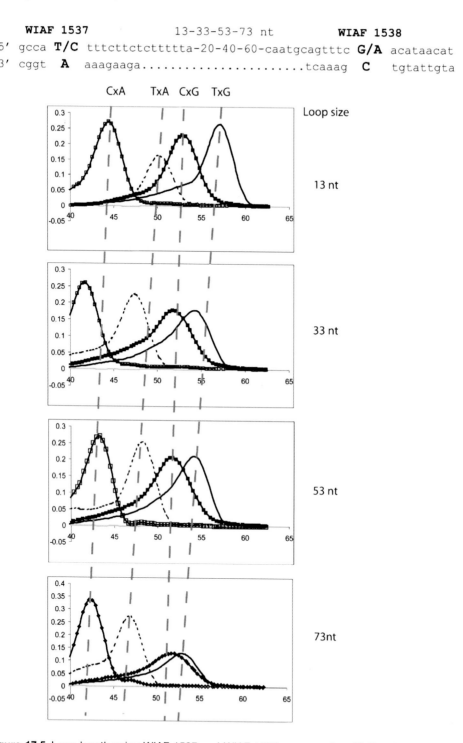

Figure 17.5 Loop length using WIAF 1537 and WIAF 1538 as examples. (A) Top line: Partial sequence of a region of chromosome 21 with both SNPs. The numbers above the sequence indicate the length of the template's loop and the numbers in the sequence the length of nucleotide inserted in the series of artificial templates. In the lpo probe, shown on the complementary strand, the omitted template sequences are shown by the dotted line. (B) Derivative melting curves of the lpo probe hybridized on artificial templates with increasing (top to bottom) length of loop out sequences. Dashed lines indicate the approximate effect of the length of the loop on probe stability.

constructed using variations of the megaprimer approach (Barik, 1996). For each haplotype, a series of templates with loops of 13, 33, 53 and 73 nucleotides was made by introducing random nucleotide sequences between WIAF 1537 and 1538 (Fig. 17.5A). The melting curve analysis of the 16 artificial templates with the lpo probes is shown in Fig. 17.5B. We were able to distinguish the four haplotypes even when a loop of 73 nucleotides must occur in the sample. In general, presence of a larger loop diminished lpo probe stability as indicated by the grey dashed line on Fig. 17.5B. Notice that the X33X templates appear less stable than predicted by the loop length. For example, C33A is less stable than C53A and even C73A. Secondary structures in the loop could affect the stability of the lpo probe.

To mimic samples heterozygous at both loci, we mixed equimolar amounts of artificial templates with the TG and the CA haplotypes or the TA and CG haplotypes. PCR was performed and products were analysed by recording melting of the 1537–1538 lpo haplotyping probe. Two derivative melting curves, corresponding to the premixed haplotypes, were observed in all cases. These are the results from the melting of the haplotyping probe hybridizing on single PCR products (intramolecular reaction). Melting of a probe bridging two different PCR products (intermolecular reaction) would appear as derivative melting curves for the absent haplotypes. We have occasionally observed these events with the X73X templates. In these cases, four curves corresponding to the four haplotypes were observed (data not shown) but dilution of PCR products prior to melting should reduce these occurrences.

Conclusion
Molecular haplotyping is an important technology for clinical laboratories, since statistical approaches developed for population studies cannot be transitioned into a diagnostic setting. Real-time PCR amplification using Temperature-dependent probe hybridization and dynamic melting analysis are useful tools for haplotyping. As shown, both adjacent sequence variants and variants separated by up to 87 nt are successfully haplotyped. In addition, these methods allow haplotyped to be performed as easily as genotyping.

References
Barik, S. (1996). Site-directed mutagenesis in vitro by megaprimer PCR. Methods Mol. Biol. 57, 203–215.

Cuppens, H., Lin, W., Jaspers, M., Costes, B., Teng, H., Vankeerberghen, A., Jorissen, M., Droogmans, G., Reynaert, I., Goossens, M., et al. (1998). Polyvariant mutant cystic fibrosis transmembrane conductance regulator genes. The polymorphic (Tg)m locus explains the partial penetrance of the T5 polymorphism as a disease mutation. J. Clin. Invest. 101, 487–496.

Ding, C. and Cantor, C.R. 2003. Direct molecular haplotyping of long-range genomic DNA with M1-PCR. Proc. Natl. Acad. Sci. USA 100, 7449–7453.

Drysdale, C.M., McGraw, D.W., Stack, C. B., Stephens, J.C., Judson, R.S., Nandabalan, K., Arnold, K., Ruano, G. and Liggett, S.B. 2000. Complex promoter and coding region beta 2-adrenergic receptor haplotypes alter receptor expression and predict in vivo responsiveness. Proc. Natl. Acad. Sci. USA 97, 10483–10488.

Eitan, Y. and Kashi, Y. 2002. Direct micro-haplotyping by multiple double PCR amplifications of specific alleles (MD-PASA). Nucleic Acids Res 30, e62.

Frank, B., Hemminki, K., Shanmugam, K. S., Meindl, A., Klaes, R., Schmutzler, R.K., Wappenschmidt, B., Untch, M., Bugert, P., Bartram, C.R. and Burwinkel, B. 2005. Association of death receptor 4 haplotype 626C-683C with an increased breast cancer risk. Carcinogenesis 26, 1975–1977.

Grody, W.W., Cutting, G. R., Klinger, K.W., Richards, C.S., Watson, M.S. and Desnick, R.J. 2001. Laboratory standards and guidelines for population-based cystic fibrosis carrier screening. Genet Med. 3, 149–154.

Hurley, J.D., Engle, L.J., Davis, J.T., Welsh, A.M. and Landers, J.E. 2004. A simple, bead-based approach for multi-SNP molecular haplotyping. Nucleic Acids Res. 32, e186.

Israel, E., Chinchilli, V.M., Ford, J. G., Boushey, H.A., Cherniack, R., Craig, T.J., Deykin, A., Fagan, J.K., Fahy, J.V., Fish, J., et al. 2004. Use of regularly scheduled albuterol treatment in asthma: genotype-stratified, randomised, placebo-controlled cross-over trial. Lancet 364, 1505–1512.

Kiesewetter, S., Macek, M., Jr., Davis, C., Curristin, S.M., Chu, C.S., Graham, C., Shrimpton, A.E., Cashman, S.M., Tsui, L.C., Mickle, J. et al. (1993). A mutation in CFTR produces different phenotypes depending on chromosomal background. Nat. Genet. 5, 274–278.

Kim, J.H., Leem, S.H., Sunwoo, Y. and Kouprina, N. 2003. Separation of long-range human TERT gene haplotypes by transformation-associated recombination cloning in yeast. Oncogene 22, 2452–2456.

Kirstein, S.L. and Insel, P.A. 2004. Autonomic nervous system pharmacogenomics: a progress report. Pharmacol. Rev. 56, 31–52.

Lanfear, D.E., Jones, P.G., Marsh, S., Cresci, S., McLeod, H.L. and Spertus, J.A. 2005. Beta2-adrenergic receptor genotype and survival among patients receiving beta-blocker therapy after an acute coronary syndrome. JAMA 294, 1526–1533.

Lee, J.H., Choi, J. H., Namkung, W., Hanrahan, J. W., Chang, J., Song, S.Y., Park, S.W., Kim, D.S.,

Yoon, J.H., Suh, Y., *et al.* 2003. A haplotype-based molecular analysis of CFTR mutations associated with respiratory and pancreatic diseases. Hum. Mol. Genet. *12*, 2321–2332.

Lyon, E. 2001. Mutation detection using fluorescent hybridization probes and melting curve analysis. Expert Rev. Mol. Diagn *1*, 92–101.

McDonald, O.G., Krynetski, E.Y. and Evans, W.E. 2002. Molecular haplotyping of genomic DNA for multiple single-nucleotide polymorphisms located kilobases apart using long-range polymerase chain reaction and intramolecular ligation. Pharmacogenetics *12*, 93–99.

Millson, A., Pont-Kingdon, G., Page, S. and Lyon, E. 2005. Direct molecular haplotyping of the IVS-8 poly(TG) and polyT repeat tracts in the cystic fibrosis gene by melting curve analysis of hybridization probes. Clin. Chem. *51*, 1619–1623.

Mitra, R.D., Butty, V.L., Shendure, J., Williams, B.R., Housman, D.E. and Church, G.M. 2003. Digital genotyping and haplotyping with polymerase colonies. Proc. Natl. Acad. Sci. USA *100*, 5926–5931.

Paul, P. and Apgar, J. 2005. Single-molecule dilution and multiple displacement amplification for molecular haplotyping. Biotechniques *38*, 553–554, 556, 558–559.

Pettersson, M., Bylund, M. and Alderborn, A. 2003. Molecular haplotype determination using allele-specific PCR and pyrosequencing technology. Genomics *82*, 390–396.

Phillips, M., Meadows, C.A., Huang, M.Y., Millson, A. and Lyon, E. 2000. Simultaneous detection of C282Y and H63D hemochromatosis mutations by dual-colour probes. Mol. Diagn. *5*, 107–116.

Pont-Kingdon, G., Jama, M., Miller, C., Millson, A. and Lyon, E. 2004. Long-range (17.7 kb) allele-specific polymerase chain reaction method for direct haplotyping of R117H and IVS-8 mutations of the cystic fibrosis transmembrane regulator gene. J. Mol. Diagn. *6*, 264–270.

Pont-Kingdon, G. and Lyon, E. 2005. Direct molecular haplotyping by melting curve analysis of hybridization probes: beta 2-adrenergic receptor haplotypes as an example. Nucleic Acids Res, *33*, e89.

Proudnikov, D., LaForge, K.S. and Kreek, M.J. 2004. High-throughput molecular haplotype analysis (allelic assignment) of single-nucleotide polymorphisms by fluorescent polymerase chain reaction. Anal. Biochem. *335*, 165–167.

Ruano, G. and Kidd, K.K. (1989). Direct haplotyping of chromosomal segments from multiple heterozygotes via allele-specific PCR amplification. Nucleic Acids Res. *17*, 8392.

Sabbagh, A. and Darlu, P. 2005. Inferring haplotypes at the NAT2 locus: the computational approach. BMC Genet. *6*, 30.

Sim, S.C. 2006, accessed 02/23). www.imm.ki.se/cypalleles. Human Cytochrome P450 (CYP) Allele Nomenclature Committee.

Song, Y., Araki, J., Zhang, L., Froehlich, T., Sawabe, M., Arai, T., Shirasawa, T. and Muramatsu, M. 2005. Haplotyping of TNFalpha gene promoter using melting temperature analysis: detection of a novel -856(G/A) mutation. Tissue Antigens *66*, 284–290.

Szantai, E., Ronai, Z., Szilagyi, A., Sasvari-Szekely, M. and Guttman, A. 2005. Haplotyping by capillary electrophoresis. J Chromatogr A *1079*, 41–49.

Tost, J., Brandt, O., Boussicault, F., Derbala, D., Caloustian, C., Lechner, D. and Gut, I.G. 2002. Molecular haplotyping at high throughput. Nucleic Acids Res. *30*, e96.

Von Ahsen, N., Armstrong, V. W. and Oellerich, M. 2004. Rapid, Long-Range Molecular Haplotyping of Thiopurine S-Methyltransferase (TPMT*) *3A, *3B, and *3C. Clin. Chem. *50*, 1528–1534.

Wetmur, J.G., Kumar, M., Zhang, L., Palomeque, C., Wallenstein, S. and Chen, J. 2005. Molecular haplotyping by linking emulsion PCR: analysis of paraoxonase 1 haplotypes and phenotypes. Nucleic Acids Res. *33*, 2615–2619.

Woolley, A. T., Guillemette, C., Li Cheung, C., Housman, D.E. and Lieber, C.M. 2000. Direct haplotyping of kilobase-size DNA using carbon nanotube probes. Nat. Biotechnol. *18*, 760–763.

Wu, W.M., Tsai, H.J., Pang, J.H., Wang, T.H., Wang, H.S., Hong, H.S. and Lee, Y.S. 2005. Linear allele-specific long-range amplification: a novel method of long-range molecular haplotyping. Hum. Mutat. *26*, 393–394.

Yan, H., Papadopoulos, N., Marra, G., Perrera, C., Jiricny, J., Boland, C. R., Lynch, H.T., Chadwick, R.B., de la Chapelle, A., Berg, K., *et al.* 2000. Conversion of diploidy to haploidy. Nature *403*, 723–724.

Yu, C.E., Devlin, B., Galloway, N., Loomis, E. and Schellenberg, G. D. 2004. ADLAPH: A molecular haplotyping method based on allele-discriminating long-range PCR. Genomics *84*, 600–612.

Zhang, K., Qin, Z.S., Liu, J. S., Chen, T., Waterman, M.S. and Sun, F. 2004. Haplotype block partitioning and tag SNP selection using genotype data and their applications to association studies. Genome Res. *14*, 908–916.

Zhao, L.P., Li, S.S. and Khalid, N. 2003. A method for the assessment of disease associations with single-nucleotide polymorphism haplotypes and environmental variables in case-control studies. Am. J. Hum. Genet. *72*, 1231–1250.

Current Books of Interest

Caister Academic Press   www.caister.com

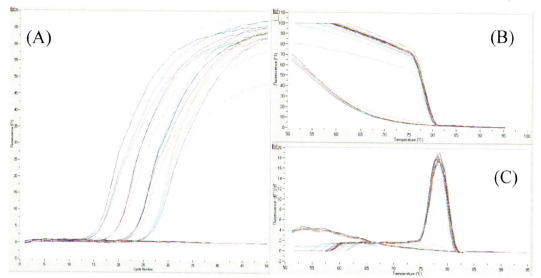

Figure 3.1 Amplification using generic binding dyes (minor-groove binding dye SYBR®Gold) on the LightCycler®. (A) The amplification plot shows 50 cycles (to completion) of amplification of four dilutions (four replicates) of a 10-fold dilution series and four no-template controls, one of which produces a low-amplification signal. The reaction utilized anti-Taq antibody hot start and UNG carry-over protection and the data utilized the background subtraction algorithm. (B) Melting point analysis of the amplified products and (C) the first negative differential of the fluorescence with respect to temperature, plotted against temperature. From this plot it can be observed that the signal in the positive no-template control has the same melting point as the specific amplification in positive samples, and therefore amplification is most likely a result of cross-contamination.

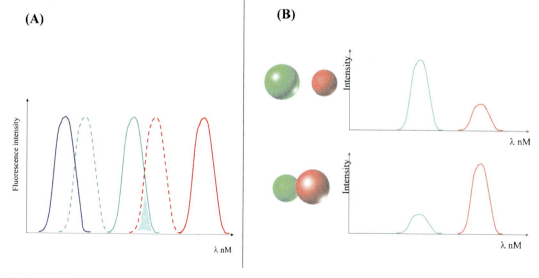

Figure 3.2 Fluorescence energy resonance transfer. (A) The emissions of a light source (blue line) falls within the excitation spectrum (green line dashed) of a fluorophore and cause this dye to fluoresce. The emissions (green solid) of this fluorophore fall within the excitation spectrum (red dashed line) of a second fluorophore that emits fluorescence at longer wavelengths (red solid). When the second fluorophore is close to the first on the same or neighbouring molecule, the energy that would be emitted by the first fluorophore ((B) top) can be transferred to the second through a number of energy transfer mechanisms such that the second fluorophore emits this energy at its own emission wavelengths ((B) bottom).

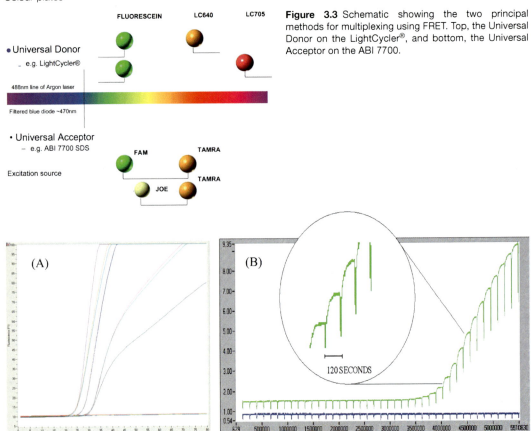

FLUORESCEIN   LC640   LC705

• Universal Donor
  – e.g. LightCycler®

488nm line of Argon laser

Filtered blue diode ~470nm

• Universal Acceptor
  – e.g. ABI 7700 SDS

FAM   TAMRA

Excitation source

JOE   TAMRA

Figure 3.3 Schematic showing the two principal methods for multiplexing using FRET. Top, the Universal Donor on the LightCycler®, and bottom, the Universal Acceptor on the ABI 7700.

(A)

(B)

120 SECONDS

Figure 3.4 Amplification using the 5′ nuclease assay on the LightCycler®. (A) The amplification plot shows 80 cycles (in excess of completion) of thermal cycling of three dilutions (three replicates) of a 10-fold dilution series and three no-template controls on the Roche LightCycler®. The reaction utilized anti-Taq antibody hot start and UNG carry-over protection. The reaction used the primers and probes from the Applied Biosystems Human β-Actin kit which utilizes a FAM-TAMRA internally quenched TaqMan® probe. The amplification plot shows raw data (no analysis/background subtraction) and illustrates the excellent signal: noise ratio achievable with this chemistry. (B) The same reaction amplified on the Idaho LightCycler® carried out by continuous monitoring of a positive and negative sample throughout all stages of amplification. The data shows that ~120 seconds are required each cycle to take the hydrolysis of bound probe to completion (plateau) each cycle.

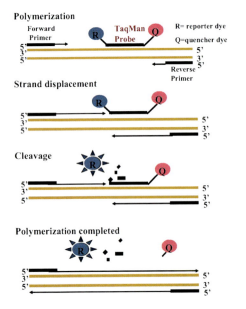

Polymerization

Forward Primer   TaqMan Probe   R = reporter dye
                                  Q = quencher dye

Reverse Primer

Strand displacement

Cleavage

Polymerization completed

Figure 11.1 Principle of SNP detection with double–dye oligonucleotide TaqMan® probe. The fluorophore linked to inact TaqMan® probe is efficiently quenched by the quencher moiety. Polymerization: TaqMan® probe hybridizes to DNA target during PCR extension. Allelic discrimination is achieved by selective annealing of match probe and template sequence. Strand displacement and Cleavage: The 5′–3′ exonuclease activity of Taq polymerase promotes cleavage of the probe between fluorophore and quencher components, separating the two moieties. Once separated from the quencher, the fluorophore emits fluorescent signal. Polymerization completed: Once the TaqMan® probe is cleaved from the targeted DNA, PCR extension continues.

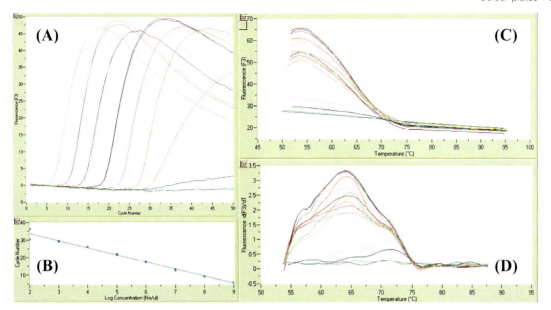

Figure 3.5 Amplification using dual hybridisation probes on the LightCycler®. (A) The amplification plot shows 50 cycles of thermal cycling of replicates of eight dilutions of a 10-fold dilution series of a reference for the target, and no-template controls on the Roche LightCycler®. The amplification shows the data analysed using the background subtraction algorithm. The reaction utilised anti-*Taq* antibody hot-start and UNG carry-over protection. The reaction used the primers and probes for the Human β-globin gene using a 3′ labelled fluorescein donor probe and a 5′ Cy5.5 acceptor probe. The acceptor probe was blocked to prevent 3′ extension using an octendiol moiety. A 5′–3′ exonuclease minus *Taq* polymerase was used to illustrate that the "hook" effect that is observed in linear hybridisation probes as a result of probe displacement by product in later cycles. (B) The derived standard curve from the amplification plot. (C) The melt curve showing probe disassociation from product. (D) The first negative differential of fluorescence with respect to temperature showing the peak at which the maximum rate of probe dissociation occurs.

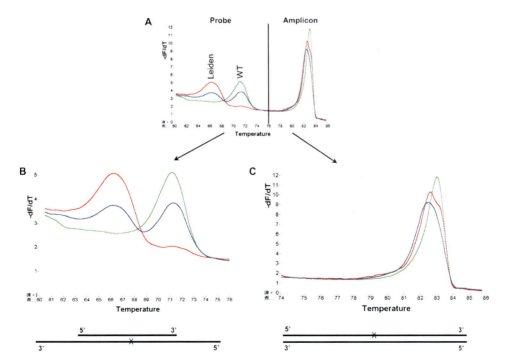

Figure 11.5 High-resolution melting analysis of an unlabelled probe and amplicon using an intercalating dye. (A) The melting range from 60°C to 90°C showing both probe and amplicon melts. (B) An enlargement of the probe melt used for genotyping. (C) An enlargement of the amplicon melt. Courtesy of Dr L.-S. Chou. The diagram below shows probe and amplicon hybridization with 'x' indicating potential mismatches.

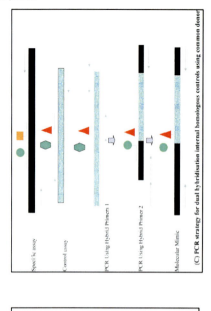

(C) **PCR strategy for dual hybridisation internal homologous controls using common donor**

Specific assay

Control assay

PCR Using Hybrid Primers 1

PCR Using Hybrid Primer 2

Molecular Mimic

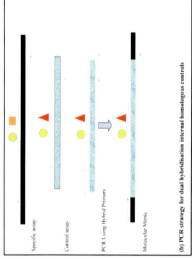

(b) **PCR strategy for dual hybridisation internal homologous controls**

Specific assay

Control assay

PCR Using Hybrid Primers

Molecular Mimic

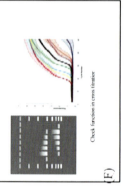

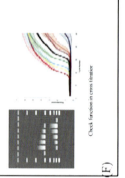

(G)

Clone into vector

Transform

Culture and purify extract

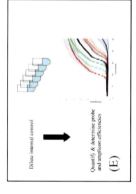

(F)

Check functions in cross titration

Diluae internal control

Quantify & determine probe and amplicon efficiencies

(E)

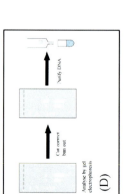

Analyse by gel electrophoresis

Cut correct ban out

Purify DNA

(D)

PCR strategy for generation of 5' nuclease internal homologous controls

Specific assay

Control assay

PCR Using Hybrid Primers

Molecular Mimic

(a)

Figure 7.1 Schematic showing strategies for rapidly generating and evaluating competitive internal homologous controls by PCR. (A) For assays such as the 5' nuclease where there is only one probe, (B) For dual hybridization assays using two additional probes, (C) Dual hybridization probes using one common probe for both the control and specific assay, (D) Amplify the IC using PCR and extract the correct band from the gel, (E) Serial dilute the product and determine the amplification and probe efficiencies (Fig. 7.2), (F) Cross-titre the IC with specific target to assess the efficacy of the IC in multiplex PCR before cloning into a suitable vector and (G) Transformation, culture, extraction and use.